Nonextensive Entro
Interdisciplinary Ap

Santa Fe Institute
Studies in the Sciences of Complexity

Nonextensive Entropy— Interdisciplinary Applications

Editors

Murray Gell-Mann
Santa Fe Institute

Constantino Tsallis
Brazilian Center for Physics Research

Santa Fe Institute
Studies in the Sciences of Complexity

UNIVERSITY PRESS

2004

OXFORD
UNIVERSITY PRESS

Oxford   New York
Auckland   Bangkok   Buenos Aires   Cape Town   Chennai
Dar es Salaam   Delhi   Hong Kong   Istanbul   Karachi   Kolkata
Kuala Lumpur   Madrid   Melbourne   Mexico City   Mumbai   Nairobi
São Paulo   Shanghai   Taipei   Tokyo   Toronto

Copyright © 2004 by Oxford University Press, Inc.

Published by Oxford University Press, Inc.
198 Madison Avenue, New York, New York 10016

www.oup.com

Oxford is a registered trademark of Oxford University Press

Library of Congress Cataloging-in-Publication Data
Nonextensive entropy : interdisciplinary applications / editors,
Murray Gell-Mann, Constantino Tsallis.
   p. cm. — (Santa Fe Institute studies in the sciences
of complexity. Proceedings volumes)
Includes bibliographical references and index.
ISBN 0-19-515976-4; ISBN 0-19-515977-2 (pbk.)
1. Statistical mechanics. I. Gell-Mann, Murray. II. Tsallis, Constantino.
III. Proceedings volume in the Santa Fe Institute studies in the sciences of complexity
QC174.8 .N675 2003
530.13—dc22   2003053684

9 8 7 6 5 4 3 2 1

Printed in the United States of America
on acid-free paper

About the Santa Fe Institute

The *Santa Fe Institute* (SFI) is a private, independent, multidisciplinary research and education center, founded in 1984. Since its founding, SFI has devoted itself to creating a new kind of scientific research community, pursuing emerging science. Operating as a small, visiting institution, SFI seeks to catalyze new collaborative, multidisciplinary projects that break down the barriers between the traditional disciplines, to spread its ideas and methodologies to other individuals, and to encourage the practical applications of its results.

All titles from the *Santa Fe Institute Studies in the Sciences of Complexity* series carry this imprint which is based on a Mimbres pottery design (circa A.D. 950–1150), drawn by Betsy Jones. The design was selected because the radiating feathers are evocative of the outreach of the Santa Fe Institute Program to many disciplines and institutions.

Contributors List

Claudia Acquisti, *Department of Animal Biology and Genetics, Via Romana 17, 50125, Firenze Italy*

Sumiyoshi Abe, *University of Tsukuba, Institute of Physics, Ibaraki, 305-8571, Japan; e-mail: suabe@sf6.so-net.ne.jp*

Arthur B. Adib, *Box 1843, Brown University, Providence, RI 02912; e-mail: arthur_adib@brown.edu*

Murilo P. Almeida, *Departamento de Física, Universidade Federal do Ceará, Caixa Postal 6030, CEP 60451-970, Fortaleza, Ceará, Brazil; e-mail: murilo@fisica.ufc.br*

José Soares de Andrade, Jr., *Departament de Física, Universidade Federal de Ceará, Caixa Postal 6030, CEP 60451-970, Fortaleza, Ceará, Brazil; e-mail: soares@fisica.ufr.br*

Fulvio Baldovin, *Ctr. Brasileiro de Pesquisas Fisicas - CMF, Rua Dr. Xavier Sagaud, 150, Urca - Rio de Janeiro -RJ, CEP: 22290-180, Rio de Janeiro, Brazil; e-mail: baldovin@cbpf.br*

Patrizia Bogani, *Department of Animal Biology and Genetics, Via Romana 17, 50125, Firenze, Italy; e-mail: patrizia.bogani@unifi.it*

Lisa Borland, *Evnine-Vaughan Associates, Inc., 456 Montgomery Street, Suite 800, San Francisco, CA 94109; e-mail: lisa@sphinx.com*

Marcello Buiatti, *Department of Animal Biology and Genetics, Via Romana 17, 50125, Firenze Italy; e-mail: mbuiatti@dbag.unifi.it*

Armin Bunde, *Universitaet Giessen, Institut fuer Theoretische Physik, Heinrich-Buff-Ring 16, Giessen, D-35392, Germany; e-mail: bunde@uni-giessen.de*

Sergio A. Cannas, *FaMAF, Universidad Nacional de Cordoba, Cuidad Universidad, Cordoba 5000, Argentina; e-mail: cannas@famaf.unc.edu.ar*

Lukasz Debowski, *Instytut Podstaw Informatyki PAN, ul. Ordona 21, Warszawa 01-237, Poland; e-mail: ldebowsk@ipipan.waw.pl*

Jan Eichner, *Universitaet Giessen, Institut fuer Theoretische Physik, Heinrich-Buff-Ring 16, Giessen, D-35392, Germany; e-mail: Jan.F.Eichmer@physik.uni-giessen.de*

Gil A. Farias, *Departament de Física, Universidade Federal de Ceará, Caixa Postal 6030, CEP 60451-970, Fortaleza, Ceará, Brazil; e-mail: gil@fisica.ufr.br*

Leone Fronzoni, *Physics Department, Pisa University, Via Buonarroti 2, Pisa, Italy; e-mail: leone.franzoni@unifi.it*

Murray Gell-Mann, *Santa Fe Institute, 1399 Hyde Park Road, Santa Fe, NM 87501; e-mail: mgm@santafe.edu*

W. M. Gonçalves, *Laboratório de Feñomenos Näo-Lineares, Departamento de Física, Universidade de São Paulo, Caiza Postal 66318, 05315-970, São Paulo, SP, Brazil; e-mail: whilk@yahoo.com*

Rathinaswamy Govindan, *Universitaet Giessen, Institut fuer Theoretische Physik, Heinrich-Buff-Ring 16, Giessen, D-35392, Germany; e-mail: Govindan@uni-giessen.de*

P. Grigolini, *Physics Department, University of North Texas, Denton, TX 76204, USA; Dipartimento di Fisica dell'Universita di Pisa and INFM, Pisa 56127, Italy; e-mail: grigo@unt.edu*

P. Hamilton, *Center for Nonlinear Science of Nursing Department, Texas Woman University, Denton, TX 76204, USA; e-mail: patti.hamilton@exchange.twu.edu*

Shlomo Havlin, *Minerva Center and Department of Physics, Bar-Ilan University, Ramat Gan, Israel; e-mail: havlin@ophir.ph.biu.ac.il*

Eva Koscielny-Bunde, *Universitaet Giessen, Institut fuer Theoretische Physik, Heinrich-Buff-Ring 16, Giessen, D-35392, Germany; e-mail: Eva.Koscielny-Bunde@physik.uni-giessen.de*

Vito Latora, *Dipartimento di Fisica e Astronomia, Università di Catania, and INFN sezione di Catania, Catania, Italy; e-mail: latora@ct.infn.it*

Seth Lloyd, *Massachusetts Institute of Technology, Department of Mechanical Engineering, Cambridge, MA: e-mail: slloyd@mit.edu*

Marcelo L. Lyra, *Universidade Federal de Alagoas, Departamento de Física, Maceió - AL 57072-970, Brazil; e-mail: marcelo@fis.ufal.br*

Massimo Marchiori, *W3C and Laboratory for Computer Science, Massachusetts Institute of Technology, Cambridge, USA; e-mail: massimo@w3.org*

Diana E. Marco, *IIB-INTECH, Camino Circunv. Laguna Km 6. CC 164 (7130) Chascomús, Pcia. de Buenos Aires, Argentina; e-mail: dmarco@intech.gov.ar*

M. T. Martin, *La Plata Physics Institute, National University La Plata and Argentina National Research Council (CONICET), C. C. 727, 1900 La Plata, Argentina*

Giuseppe Mersi, *Department of Animal Biology and Genetics, Via Romana 17, 50125, Firenze, Italy*

Marcelo A. Montemurro, *Condensed Matter Group, ICTP, 34014 Trieste, Italy; e-mail: mmontemu@ictp.trieste.it*

André A. Moreira, *Departament de Física, Universidade Federal de Ceará, Caixa Postal 6030, CEP 60451-970, Fortaleza, Ceará, Brazil; e-mail: auto@fisica.ufr.br*

Roberto Osorio, *Evnine-Vaughan Associates, Inc., 456 Montgomery Street, Suite 800, San Francisco, CA 94109; e-mail: roberto@sphinx.com*

Sergio A. Páez, *IIB-INTECH, Camino Circunv. Laguna Km 6. CC 164 (7130) Chascomús, Pcia. de Buenos Aires, Argentina; e-mail: spaez@intech.gov.ar*

Thadeu J. P. Penna, *Universidad Federal Fluminense, Instituto de Fisica, Niteroi, RJ, Brazil; e-mail: tjpp@if.uff.br*

Juan Pérez-Mercader, *Centro de Astrobiologia, Carretera de Ajalvir, s/n, km. 4, 28850 Torrejon de Ardoz, 28020 Madrid, Spain; e-mail: mercader@laeff.esa.es*

R. D. Pinto, *Laboratório de Feñomenos Näo-Lineares, Departamento de Física, Universidade de São Paulo, Caiza Postal 66318, 05315-970, São Paulo, SP, Brazil; e-mail: reynaldo@if.usp.br*

Angel Plastino, *National University La Plata, C. C. 727, La Plata, 1900, Argentina; e-mail: plastino@venus.fisica.unlp.edu.ar*

Andrea Rapisarda, *Universita di Catania, Dipartimento de Fisica e Astronomia, and INFN via S. Sofia 64 95123 Catania, Italy; e-mail: andrea.rapisarda@ct.infn.it*

Alberto Robledo, *Instituto de Fisica, UNAM, Apartado, Postal 20-364- 01000 Mexico DF, Mexico; e-mail: robledo@fisica.unam.mx*

O. Rosso, *Instituto de Cálculo, Facultad de Ciencias Exactas y Naturales (UBA), Cuidad universitaria, Pabellón II, 1428 Buenos Aires, Argentina*

Diego Rybski, *Universitaet Giessen, Institut fuer Theoretische Physik, Heinrich-Buff-Ring 16, Giessen, D-35392, Germany; e-mail: diego.rybski@physik.uni-giessen.de*

J. C. Sartorelli, *Instituto de Física, Universidade de São Paulo, Caiza Postal 66318, 05315-970, São Paulo, SP, Brazil; e-mail: sartorelli@if.usp.br*

N. Scafetta, *Pratt School of EE Department and Physics Department, Duke University, Durham, NC 27708, USA; e-mail: ns2002@duke.edu*

Robin Stinchcombe, *Oxford University, Theoretical Physics, 1 Keble Road, Oxford OX1 3NP, United Kingdom; e-mail: stinch@thphys.ox.ac.uk*

Hugo Touchette, *McGill University, School of Computer Science, 3480 University Street, McConnell Engineering Bldg, Rm 232, Montreal, Quebec H3A 2A7, Canada; e-mail: htouc@cs.mcgill.ca*

Constantino Tsallis, *Centro Brasileiro de Pesquisas Fisicas—Rua Xavier Sigaud 150 22290-180 Rio de Janeiro, Brazil; e-mail: tsallis@cbpf.br*

Dmitry Vjushin, *Minerva Center and Department of Physics, Bar-Ilan University, Ramat Gan, Israel; e-mail: vjushin@ory.ph.biu.ac.il*

B. J. West, *Pratt School of EE Department and Physics Department, Duke University, Durham, NC 27708, USA; e-mail: WestB@aro.arl.army.mil*

Horacio Wio, *Centro Atomico Bariloche, 8400 San Carlos de Bariloche, Argentina; e-mail: wio@cab.cnea.gov.ar*

Contents

Preface

The present book constitutes a pedagogical effort that reflects the presentations and discussion at the International Workshop on "Interdisciplinary Applications of Ideas from Nonextensive Statistical Mechanics and Thermodynamics," held at the Santa Fe Institute in New Mexico from April 8–12, 2002. The participants, close to 60 in number, were scientists at both junior and senior levels from Argentina, Brazil, Canada, Germany, Great Britain, Italy, Mexico, Poland, and the U.S.A. The subjects of the chapters relate to dynamical, physical, geophysical, biological, economic, financial, and social systems, and to networks, linguistics, and plectics.

Some of the contributions focus directly on a specific nonextensive entropic form proposed by one of us as a basis for generalizing Boltzmann-Gibbs (BG) statistical mechanics to a formalism sometimes called nonextensive statistical mechanics. Such is the case for the chapters by Abe; Andrade, Almeida, Moreira, Adib, and Farias; Baldovin; Borland; Lyra; Montemurro; Osorio, Borland, and Tsallis; Penna, Sartorelli, Pinto, and Gonçalves; Rapisarda and Latora; Robledo; Plastino, Martin, and Rosso; Scafetta, Grigolini, Hamilton, and West;

Touchette; Wio; and Tsallis. Other contributions analyze problems that point toward possible fruitful uses of nonextensive concepts. Such is the case for the manuscripts by Cannas, Marco, Páez, and Montemurro; Gell-Mann and Lloyd; Latora and Marchiori; and Stinchcombe. Finally, other contributions, such as those of Buiatti, Bogani, Acquisti, Mersi, and Fronzoni; Bunde, Eichner, Govindan, Havlin, Koscielny-Bunde, Rybski, and Vjushin; Debowski, and Pérez-Mercader, focus on rich complex systems that might or might not be related to nonextensive entropy.

It should be clear that nonextensive statistical mechanics is by no means intended to replace BG statistical mechanics for systems such as those in stationary states characterized by thermal equilibrium consistent with ergodicity. The nonextensive alternative is proposed, instead, as a way of dealing, through mathematical methods that are quite similar to the usual ones, with anomalous systems. These include a wide class of nonergodic systems, with stationary (or rather quasistationary) states that are metastable and long-lived, for example in many-body Hamiltonian systems with long-range forces. Some strong analogies with the BG theory emerge, as well as some important physical differences. Nonextensive statistical mechanics exhibits apparent success (in a sense that we discuss below) for certain closed systems as well as a variety of open systems— in biology, economics, linguistics, the physics of turbulence, and other fields. An intriguing question that remains unanswered is: exactly what do all these systems have in common? One suspects, of course, that the deep explanation must arise from microscopic dynamics. The various cases could all be associated with something like a scale-free dynamical occupancy of phase space, but this certainly deserves further investigation.

Are other, somewhat similar generalizations of statistical mechanics also possible? What would be the physical entropy to be used, and what would be its relation to the symmetry of occupation of phase space? Answering such questions is by no means trivial. However, it is interesting to remark that for q positive the present nonextensive entropy shares with the BG entropy three important properties, namely concavity (related to thermodynamic stability, or robustness with respect to fluctuations of energy and other quantities), stability or continuity (related to experimental robustness, in the sense that similar experiments should provide quantitatively similar results), and finiteness of the entropy production per unit time (conveniently characterizing the gradual exploration of the available phase space). In order to appreciate the difficulty of satisfying all of these conditions, it is worth mentioning that Rényi entropy violates all three.

Since the nonextensive entropy S_q and the Rényi entropy are related through a monotonic function, the distinction between them is irrelevant to the determination of the probability distribution, but for other physical quantities, such as those connected with the flow of entropy, the distinction can be crucial. In general, one can vary the form of the entropy and also the choice of average quantities to be kept fixed as the entropy is maximized, while keeping the probabilities unchanged. Again, quantities that explicitly involve the entropy will come out different. It is important to remember that the probabilities alone do not distinguish in a rigorous manner between conventional and unconventional entropy. However, the probabilities can suffice to distinguish different formulae for entropy if the average quantities kept fixed are restricted to particular ones that are regarded as natural, for example linear or quadratic expressions in the variable under study. It is in this specific sense that nonextensive entropy has scored much of its success.

We are extremely pleased to acknowledge the very valuable support of Ellen Goldberg, the editorial assistance of Ronda Butler-Villa, Della Ulibarri, and Laura Ware, the practical help on so many occasions of Olivia Posner, Kevin Drennan, and Andi Sutherland, and—last but not least—the generous financial support of the Santa Fe Institute, International Program grant, without which the meeting could not have occurred.

Constantino Tsallis
Centro Brasileiro de Pesquisas Fisicas

Murray Gell-Mann
Santa Fe Institute

Nonextensive Statistical Mechanics: Construction and Physical Interpretation

Constantino Tsallis

1 INTRODUCTION

Statistical mechanics is clearly mechanics (classical, quantum, special or general relativistic, or any other) *plus* the theory of probabilities, as is well known. It is our understanding, however, that it is more than that. It is *also* the adoption of a specific entropic functional, which will, in some sense, adequately shortcut the vast, and for most practical purposes useless, detailed microscopic mechanical information on the system. It is, in particular, through this functional that the connection with thermodynamics and its macroscopic laws will be established. This particular functional is determined by the specific type (or geometry) of occupation of the phase space (or Hilbert space or analogous space). This geometrical structure depends in turn not only on the microscopic dynamics that the system obeys, but also on the initial conditions at which the system is placed at $t = 0$. In colloquial terms, we could say that the microscopic dynamics determine where the system is *allowed to live*, whereas the initial conditions determine where it *likes to live* within the allowed region. This viewpoint is consistent with

Nonextensive Entropy—Interdisciplinary Applications
edited by Murray Gell-Mann and Constantino Tsallis, Oxford University Press

Einstein's perspective on classical statistical mechanics, and especially with his criticism [82, 92][1] of the celebrated *Boltzmann principle*

$$S = k \ln W \ . \tag{1}$$

However, the problem is that, up to now, no systematic manner exists for univocally determining the entropic functional to be used, given the dynamics and the initial conditions. The optimization of this entropy under the physically appropriate constraints is expected to provide the correct probability distribution for the microscopic states of the macroscopic stationary state of the system. Boltzmann, then complemented by Gibbs, proposed the celebrated form which is the foundation of standard statistical mechanics. This form is (in its discrete version) fruitfully explored by Shannon:

$$S_{BG} = -k \sum_{i=1}^{W} p_i \ln p_i \tag{2}$$

with

$$\sum_{i=1}^{W} p_i = 1 \ . \tag{3}$$

At equiprobability (i.e., $p_i = 1/W, \forall i$), this (nonnegative) entropy achieves its maximum, given by eq. (1). For simplicity, and without loss of generality, from now on we shall use units such that $k = 1$. The form (2) has been at the origin of uncountable successes concerning the statistical mechanics of ubiquitous systems. It is known to be the correct entropic form for ergodic systems, that is, systems which, whenever isolated, dynamically visit with equal probability all the allowed microscopic states. In other words, these systems like equally to live everywhere that they are allowed. This happens each time the system is chaotic enough, as first envisioned by Boltzmann himself [56, 57] (the *Stosszahlansatz* or *molecular chaos hypothesis*).

The question follows naturally: *is it not possible to describe on similar grounds dynamical systems which visit phase space in a manner more complex than prescribed by ergodicity?* (In fact, one expects *nonergodicity* to be the generic case for many types of complex systems [102]). We believe the answer to be *yes*. If so, the next question is *what is the entropic form to be used instead of eq. (2)?* As already anticipated, no systematic way has yet been found which uniquely determines the answer. In spite of this inexistence, the generalization

---

[1] "Usually W is put equal to the number of complexions. . . . In order to calculate W, one needs a *complete* (molecular-mechanical) theory of the system under consideration. Therefore, it is dubious whether the Boltzmann principle has any meaning without a complete molecular-mechanical theory or some other theory which describes the elementary processes. $S = R/\mathcal{N} \log W +$ const. seems without content, from a phenomenological point of view, without giving, in addition, such an *Elementartheorie*."

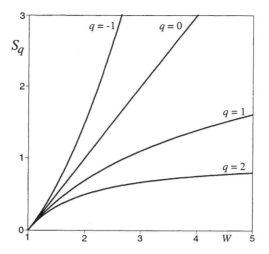

FIGURE 1 The equiprobability entropy S_q as a function of the number of states W, for typical values of q. For $q > 1$, S_q saturates at the value $1/(q-1)$ if $W \to \infty$; for $q \leq 1$, it diverges. For $q \to \infty$ ($q \to -\infty$), it coincides with the abscissa (ordinate).

of the Boltzmann-Gibbs (BG) statistical mechanics, inspired by multifractals, has been heuristically proposed [210] by postulating the following entropic form:

$$S_q = \frac{1 - \sum_{i=1}^{W} p_i^q}{q - 1} \quad (q \in \mathcal{R}) . \tag{4}$$

It can be straightforwardly checked that the $q \to 1$ limit precisely reproduces S_{BG} as given by eq. (2) (use $p_i^q = p_i e^{(q-1)\ln p_i} \sim p_i[1 + (q-1)\ln p_i]$ for $q \to 1$). We are, therefore, not talking of an alternative to, but of a generalization of the BG entropy. The (nonnegative) form (4) achieves its extreme value at equiprobability ($p_i = 1/W, \forall i$), and this value is

$$S_q = \frac{W^{1-q} - 1}{1 - q} . \tag{5}$$

This value (see fig. 1) corresponds to a maximum for $q > 0$ and to a minimum for $q < 0$ (for $q < 0$, only microscopic states with nonvanishing probability are to be considered in the sum appearing in eq. (4)). Also, if A and B are two systems which are independent in the sense of the theory of probabilities, such that $p_{ij}^{A+B} = p_i^A p_j^B$, $\forall(ij)$, then we easily verify the following property

$$S_q(A + B) = S_q(A) + S_q(B) + (1 - q)S_q(A)S_q(B) . \tag{6}$$

Therefore, S_q is, *in this specific sense*, generically nonextensive (subextensive for $q > 1$ and superextensive for $q < 1$) unless $q = 1$, which precisely corresponds to the BG entropy.[2] From this property stands the denomination *nonextensive statistical mechanics* sometimes used to refer to the present formalism. It must, however, be clear that, if A and B are not independent, there is no reason for property (6) to hold. More than that, it is conceivable that, for a specific given correlation between A and B, a special value of $q = q^*$ might exist for which $S_{q^*}(A + B) = S_{q^*}(A) + S_{q^*}(B)$! To make this possibility transparent, let us consider the following examples. If we throw N independent coins, then $W = 2^N$; hence, from eq. (1), we obtain (under the assumption of equiprobability, i.e., that all the coins are *fair*) $S_{BG} = N \ln 2$. If we have N dices, we will get $S_{BG} = N \ln 6$. For an isolated system with N particles which only have short-range correlations (i.e., each particle is correlated to a few other particles inside its immediate neigborhood, defined through some simple metric), we expect, for $N \to \infty$, $W \sim \mu^N$ with $\mu > 1$; hence $S_{BG} \sim N \ln \mu$. Summarizing, in all these cases, S_q is (at least asymptotically) proportional to N only for $q = q^* = 1$. Indeed, all other possible values of q destroy the proportionality with N. Let us now consider a different situation, in which all N particles are correlated among them in such a way that, for $N \to \infty$, $W \sim N^\gamma$ with $\gamma > 0$. In this case, a special value of q does exist such that $S_q \propto N$. This value is $q^* = 1 - 1/\gamma$, as immediately obtained by using eq. (5). In other words, when our problem is such that $W(N + 1) \simeq W(N) \times W(1)$ (more precisely, whenever $1 < \lim_{N \to \infty}[W(N + 1)/W(N)] < \infty$), we expect $q^* = 1$. There is clearly no reason for expecting such a simple result for more complex combinatorics (e.g., if $\lim_{N \to \infty}[W(N + 1)/W(N)] = 1$).

This remark provides a good hint for the physical motivation for eq. (4), which, together with more than 20 different entropic forms available in the literature (see Tsallis [211, 213, 214, 215, 216] for details), has a long history of temptative generalizations of S_{BG}. We believe, however, that further mathematical characterizations of S_q can only improve our intuition about the present proposal for generalization of BG statistical mechanics. We shall proceed to that in the following section.

---

[2]Let be a physical quantity $\mathcal{O}$ associated with a given system, and a composition law for physical systems A and B (the composition law might or might not be probabilistic independence between A and B). Many authors use the term *additive* if and only if $\mathcal{O}(A+B) = \mathcal{O}(A)+\mathcal{O}(B)$. If so, it is clear that if we have N equal systems, then $\mathcal{O}(N) = N\mathcal{O}(1)$. A weaker condition is $\mathcal{O}(N) \sim N\Omega$ $(0 < |\Omega| < \infty)$, i.e., $\lim_{N \to \infty} \mathcal{O}(N)/N$ is finite (generically $\Omega \neq \mathcal{O}(1)$). In this case, the term *extensive* is usually preferred. In other words, any observable which is additive with regard to a given composition law, is extensive (with $\Omega = \mathcal{O}(1)$), but the opposite is not necessarily true. Unless otherwise specified, we shall use the word *extensive* to indistinguishably mean both the weak and the strong meanings.

2 THE EQUIPROBABILITY ENTROPIC FORM: CHARACTERIZATIONS

2.1 FIRST CHARACTERIZATION: ORDINARY DIFFERENTIAL EQUATIONS

Let us consider the simplest differential equation, namely

$$\frac{dy}{dx} = 0 \quad (y(0) = 1) \,. \tag{7}$$

Its solution is

$$y = 1 \quad (\forall x) \,, \tag{8}$$

and its inverse function (in the sense of being symmetrical with respect to the $y = x$ axis) is given by

$$x = 1 \quad (\forall y) \,. \tag{9}$$

The next simplest differential equation can be considered to be

$$\frac{dy}{dx} = 1 \quad (y(0) = 1) \,. \tag{10}$$

Its solution is

$$y = 1 + x \,, \tag{11}$$

and its inverse function is

$$y = x - 1 \,. \tag{12}$$

One more natural step in increasing difficulty is to consider the differential equation

$$\frac{dy}{dx} = y \quad (y(0) = 1) \,. \tag{13}$$

Its solution is

$$y = e^x \,, \tag{14}$$

and its inverse function is

$$y = \ln x \,. \tag{15}$$

It satisfies the well-known additive property

$$\ln(x_A \, x_B) = \ln x_A + \ln x_B \,. \tag{16}$$

Let us assume we want to unify all three previous cases (eqs. (7), (10), and (13)) in one more general form. One trivial way is to preserve the *linearity* of the right-hand member of the differential equations, thus writing $dy/dx = a + by$, which involves two parameters, namely a and b. Another, more compact, unification is to precisely violate this linearity, thus writing

$$\frac{dy}{dx} = y^q \quad (y(0) = 1; \, q \in \mathcal{R}) \,. \tag{17}$$

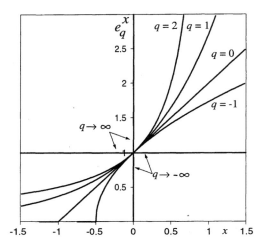

FIGURE 2   The q-exponential function e_q^x for typical values of q. For $q > 1$, it is defined in the interval $(-\infty, (q-1)^{-1})$; it diverges if $x \to (q-1)^{-1} - 0$. For $q < 1$, it is defined for all x, and vanishes for all $x < -(1-q)^{-1}$. In the limit $x \to 0$, it is $e_q^x \sim 1 + x$ ($\forall q$).

Its solution is

$$y = [1 + (1-q)\, x]^{\frac{1}{1-q}} \equiv e_q^x \quad (e_1^x = e^x) , \tag{18}$$

(see figs. 2, 3, 4, and 5) and its inverse function is

$$y = \frac{x^{1-q} - 1}{1 - q} \equiv \ln_q x \quad (\ln_1 x = \ln x) \tag{19}$$

(see fig. 1).

It satisfies the *pseudo-additive* property

$$\ln_q(x_A\, x_B) = \ln_q x_A + \ln_q x_B + (1-q)(\ln_q x_A)(\ln_q x_B) . \tag{20}$$

Within this unification, the three initial cases are reobtained by taking the *single* parameter $q \to -\infty$, $q = 0$, and $q = 1$, respectively. From this viewpoint, we have thus obtained eq. (5) as a possible natural generalization of eq. (1). In other words

$$S_q = \ln_q W . \tag{21}$$

As an aside, let us mention a property similar to that of eq. (20), which can be of some utility. It can be easily verified that $\ln_q(x_A^{1-q} + x_B^{1-q} - 1)^{1/(1-q)} = \ln_q x_A + \ln_q x_B$ and that $\lim_{q\to 1}(x_A^{1-q} + x_B^{1-q} - 1)^{1/(1-q)} = x_A x_B$.

As a second aside, let us analyze the *cumulative* function frequently appearing in applications. We first introduce the probability distribution $p(x) \equiv$

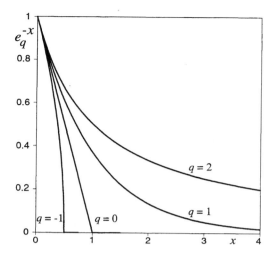

FIGURE 3   The function e_q^{-x} for typical values of q: linear-linear scale. For $q > 1$, it vanishes like $[(q-1)x]^{-1/(q-1)}$ for $x \to \infty$. For $q < 1$, it vanishes for $x > (1-q)^{-1}$ (*cutoff*).

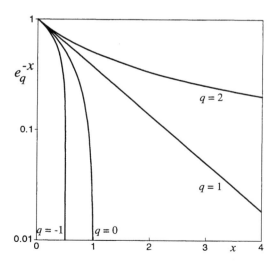

FIGURE 4   The function e_q^{-x} for typical values of q: log-linear scale. It is convex (concave) if $q > 1$ ($q < 1$). For $q < 1$ it has a vertical asymptote at $x = (1-q)^{-1}$.

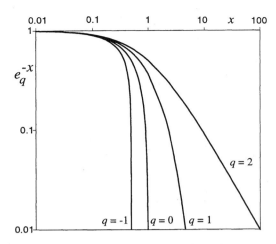

FIGURE 5   The function e_q^{-x} for typical values of q: log-log scale. For $q > 1$, it has an asymptotic slope equal to $-1/(q-1)$.

$(2-q)e_q^{-x}$ $(x \geq 0)$ with $\int_0^\infty dx\, p(x) = 1$. We define now the *cumulative probability* $P(X) \equiv \int_X^\infty dx p(x)$. It can be trivially shown that this also is a q-exponential function, namely $P(X) = e_{q_M}^{-X/q_M}$, with

$$q_M \equiv 1/(2-q) \quad (q < 2) \; ; \tag{22}$$

for $q \geq 2$, $p(x)$ is not normalizable.

2.2   SECOND CHARACTERIZATION: MEAN VALUES

Equation (2) can be written as

$$S_{BG} = \left\langle \ln \frac{1}{p_i} \right\rangle , \tag{23}$$

with $\langle \ldots \rangle \equiv \sum_{i=1}^W (\ldots)p_i$. The quantity $\ln(1/p_i)$ can be called *surprise* [239] or *unexpectedness* [48]. Indeed, $p_i = 1$ corresponds to certainty; hence there is no surprise if that event does occur. On the other hand, $p_i \to 0$ corresponds to almost impossibility; hence the surprise is enormous if that event does occur. We can now introduce the q-surprise or q-unexpectedness as $\ln_q(1/p_i)$, and hence the q-entropy as

$$S_q = \left\langle \ln_q \frac{1}{p_i} \right\rangle , \tag{24}$$

which can easily be verified to coincide with eq. (4). The use of property (20) immediately yields eq. (6).

2.3 THIRD CHARACTERIZATION: BIAS

This procedure was originally used in the 1988 paper [210]. Let us consider $0 < p_i < 1$ and $q > 0$. We have that $p_i^q > p_i$ if $q < 1$, $p_i^q < p_i$ if $q > 1$, and of course $p_i^q = p_i$ if $q = 1$ (BG). We want to introduce a *bias*, so instead of using p_i, we shall use p_i^q. The entropy $S_q(\{p_i\})$ should be invariant under permutations of the events. The simplest manner is to be $S_q(\{p_i\}) = f(\sum_{i=1}^{W} p_i^q)$, where $f(x)$ is some smooth function. The simplest function is the linear one, hence we can expect

$$S_q(\{p_i\}) = a + b \sum_{i=1}^{W} p_i^q . \tag{25}$$

Certainty corresponds to $p_{i_0} = 1$ and $p_i = 0, \forall i \neq i_0$; hence $\sum_{i=1}^{W} p_i^q = 1$. For this case, we desire $S_q = 0$, so $a + b = 0$, and

$$S_q(\{p_i\}) = a \left(1 - \sum_{i=1}^{W} p_i^q \right) . \tag{26}$$

Finally, we desire S_1 to coincide with S_{BG}. By taking into account that, in the $q \to 1$ limit, it is $p_i^q = p_i e^{(q-1) \ln p_i} \sim p_i[1 + (q-1) \ln p_i]$, we obtain $S_1 \sim -a(q-1) \sum_{i=1}^{W} p_i \ln p_i$. The simplest choice is $a(q-1) = 1$, which replaced into eq. (26), yields once again the form (4).

2.4 FOURTH CHARACTERIZATION: RESPONSE TO BIAS

As Abe [1] showed, we can easily verify that

$$S_{BG} = -\left[\frac{d}{dx} \sum_{i=1}^{W} p_i^x \right]_{x=1} , \tag{27}$$

which can be understood as a *reaction* to *translation* of the bias x, since derivation is the "reaction" of a function under translation of its abscissa. Jackson introduced in 1909 a generalization of the standard derivation. The Jackson derivative is defined as

$$D_q h(x) \equiv \frac{h(qx) - h(x)}{qx - x} \quad \left(D_1 h(x) = \frac{dh(x)}{dx} \right) , \tag{28}$$

which is the reaction of the function under *dilatation* of the abscissa (or, alternatively, under finiteness of the modification of the abcissa). It can be easily verified that

$$S_q \equiv -\left[D_q \sum_{i=1}^{W} p_i^x \right]_{x=1} \tag{29}$$

precisely yields eq. (4).

2.5 FIFTH CHARACTERIZATION: GROUPING

We may arbitrarily divide the W possibilities into $W_L + W_M$ possibilities. Consistently, $p_L \equiv \sum_{i=1}^{W_L} p_i$ (W_L terms) and $p_M \equiv \sum_{i=W_L+1}^{W} p_i$ (W_M terms); hence $p_L + p_M = 1$. We then consider the *conditional* probabilities $\{p_i/p_L\}$ ($\sum_{i=1}^{W_L}(p_i/p_L) = 1$) and $\{p_i/p_M\}$ ($\sum_{i=W_L+1}^{W}(p_i/p_M) = 1$). It can be straightforwardly proven [85] that

$$S_q(\{p_i\}) = S_q(p_L, p_M) + p_L^q S_q(\{p_i/p_L\}) + p_M^q S_q(\{p_i/p_M\}), \qquad (30)$$

which recovers, for $q = 1$, the celebrated Shannon property.

The fifth characterization of S_q refers to the fact that it can be proven [185] that eq. (4) is *necessary and sufficient* for S_q satisfying the following properties:

1. S_q is a continuous function of the $\{p_i\}$,
2. $S_q(p_i = 1/W, \forall i)$ is a monotonic function of W,
3. eq. (6), and
4. eq. (30).

2.6 SIXTH CHARACTERIZATION: CORRELATIONS

Let us assume that we have a system composed of subsystems A (with W_A possibilities) and B (with W_B possibilities), whose joint probabilities are $\{p_{ij}^{A+B}\}$ ($i = 1, 2, \ldots, W_A$; $j = 1, 2, \ldots, W_B$). We define the *marginal* probabilities $p_i^A \equiv \sum_{j=1}^{W_B} p_{ij}^{A+B}$ (hence $\sum_{i=1}^{W_A} p_i^A = 1$) and $p_j^B \equiv \sum_{i=1}^{W_A} p_{ij}^{A+B}$ (hence $\sum_{j=1}^{W_B} p_j^B = 1$). In general, $p_i^A p_j^B \neq p_{ij}^{A+B}$; if they happen to be equal, then A and B are said to be probabilistically *independent*. Otherwise they are *dependent* or *correlated*.

We can then define $S_q(A) \equiv S_q(\{p_i^A\})$ and $S_q(B) \equiv S_q(\{p_i^B\})$. We can also define *conditional* entropies [8, 14, 223] $S_q(A|B)$ and $S_q(B|A)$. If, and only if, A and B are independent, we have that $S_q(A|B) = S_q(A)$ and $S_q(B|A) = S_q(B)$. It can be proven [8, 14] that $S_q(A + B) \equiv S_q(\{p_{ij}^{A+B}\})$ in general satisfies

$$\begin{aligned} S_q(A + B) &= S_q(A) + S_q(B|A) + (1 - q)S_q(A)S_q(B|A) \\ &= S_q(B) + S_q(A|B) + (1 - q)S_q(B)S_q(A|B), \end{aligned} \qquad (31)$$

which reproduces eq. (6) when A and B are independent.

The sixth characterization of S_q refers to the fact that it can be proven [3] that eq. (4) is *necessary and sufficient* for S_q satisfying the following properties:

1. S_q is a continuous function of the $\{p_i\}$,
2. $S_q(p_i = 1/W, \forall i)$ is a monotonic function of W,
3. $S_q(p_1, p_2, \ldots, p_w, 0) = S_q(p_1, p_2, \ldots, p_w)$ and
4. eq. (31).

The various characterizations of S_q that we have presented should provide a reasonably good intuition of its physical and mathematical contents. To make this description more complete, let us mention two more remarkable properties: *concavity* and *stability*.

2.7 CONCAVITY

S_q is *concave* (*convex*) for *all* probability distributions and *all* $q > 0$ ($q < 0$). To be more precise, let us consider two arbitrary probability distributions $\{p_i\}$ and $\{p_i'\}$ ($i = 1, 2, \ldots, W$), and let us construct a so-called *intermediate* distribution as follows:

$$p_i'' \equiv \lambda p_i + (1 - \lambda)p_i' \quad (0 < \lambda < 1) . \tag{32}$$

By concavity we mean that it can be proven that, for all λ,

$$S_q(\{p_i''\}) \geq \lambda S_q(\{p_i\}) + (1 - \lambda)S_q(\{p_i'\}) . \tag{33}$$

By convexity we mean the property where $\geq$ is replaced by $\leq$. What these inequalities essentially guarantee is a most important property, namely *thermodynamic stability*. The entropic functional is supposed to be defined such that stationarity (e.g., thermodynamic equilibrium) makes it extreme (in our case, a maximum for $q > 0$ and a minimum for $q < 0$) in the presence of the generic constraints imposed on the system. Any occasional perturbation of the $\{p_i\}$ which makes the entropy extreme is necessarily followed by a tendency toward $\{p_i\}$ once again, since the entropy is concave (convex). Things and people have names because, within some time scale, they are thermodynamically stable! Furthermore, concavity also is relevant to the fact that when we put two systems at different temperatures in thermal contact, they naturally tend to approach a common temperature; in other words, they tend to *equilibrate*.

A remark is worthy at this point. Renyi entropy is another generalization of the BG entropy. It is defined by $S_q^R \equiv [\ln \sum_{i=1}^W p_i^q]/[1 - q]$, and for a long time has been fruitfully used in the geometrical characterization of multifractals (generated for example by nonlinear dynamical systems). This (extensive) entropy is a monotonically increasing function of S_q, namely $S_q^R = \ln[1 + (1 - q)S_q]/[1 - q]$. Another generalization of the BG form is the *normalized* nonextensive form $S_q^N \equiv S_q / \sum_{i=1}^W p_i^q$, independently introduced by Landsberg and Vedral [119] and by Rajagopal and Abe [173]. Also, this entropy is a monotonically increasing function of S_q, namely $S_q^N = S_q/[1 + (1 - q)S_q]$. Monotonicity makes S_q, S_q^R, and S_q^N extreme for the *same* probability distribution. Then, *why not base the generalization of thermodynamics on S_q^R or on S_q^N?*. Well, it happens that they are *not* concave for all positive values of q, but only for $0 < q \leq 1$. It is clear that this is a considerable disadvantage, especially since several physically meaningful phenomena are known which demand $q > 1$. In general, monotonicity does not preserve concavity. But even if some specific choices do, we would not be allowed to arbitrarily consider such monotonic functions: statistical mechanics is much more than an optimizing probability distribution. Nobody would even think of basing BG statistical mechanics on $(-\sum_{i=1}^W p_i \ln p_i)^3$ just because it is both concave and monotonically increasing with S_{BG}!

2.8 STABILITY

S_q is *stable* for *all* probability distributions, *all* $q > 0$ and (hopefully) *all* metric exponents μ. Let us be precise about what we mean by this. Two probability distributions $\{p_i\}$ and $\{p_i'\}$ ($i = 1, 2, \ldots, W$) will be said to be μ-*close* if their μ-*distance* (see Hardy et al. [108]) $d_\mu(p_i, p_i')$ is smaller than a small δ, in other words, if, for fixed $\mu > 0$,

$$d_\mu(p_i, p_i') \equiv \left[\sum_{i=1}^{W} |p_i - p_i'|^\mu \right]^{1/\mu} < \delta \quad (0 < \delta \ll 1) \,. \tag{34}$$

Strictly speaking, this quantity defines a metric only for $\mu \geq 1$; indeed, for $0 < \mu < 1$, the triangle inequality is violated.

We can define also the *entropy relative discrepancy*

$$\Delta(\mu, W, \delta) \equiv \frac{S(\{p_i\}) - S(\{p_i'\})}{\sup[S(\{p_i\})]} \tag{35}$$

where $\sup[S(\{p_i\})]$ is the maximal value that the entropy can achieve while varying $\{p_i\}$. For positive q, this value equals $\ln_q W$ for S_q, $\ln W$ for S_q^R, and $\ln_{2-q} W$ for S_q^N. One normally expects for any entropy that $\lim_{W \to \infty} \lim_{\delta \to 0} \Delta(\mu, W, \delta) = 0$ for all $\mu > 0$. If the entropy also satisfies, for a given $\mu > 0$,

$$\lim_{\delta \to 0} \lim_{W \to \infty} \Delta(\mu, W, \delta) = 0 \,, \tag{36}$$

it is said to be $\mu - stable$ (strictly speaking, the commutability of these two limits does not guarantee uniform convergence, whereas the opposite is true). Lesche showed in 1982 [137] that S_{BG} is 1-stable, whereas S_q^R ($q \neq 1$) is not. It must be clearly understood that this property is independent from concavity: for $q < 1$, S_q^R and S_q^N are concave but not stable, whereas for $q > 1$ they are neither concave nor stable. Abe has recently proved [7] that S_q is 1-stable for all $q > 0$, whereas S_q^R and S_q^N are not. We have indications [91], but not yet a proof, that Abe's remark is valid for all $\mu > 0$, for example, that S_q is μ-stable for all $\mu > 0$ and all $q > 0$, whereas S_q^R and S_q^N are not. This point is in progress.

The present stability (not to be confused with thermodynamical stability, which is, in turn, related to concavity, as was already mentioned) is a very deep (and quite unknown) property that, as far as we know, had never been characterized before Lesche [137]. It has to do with the necessary robustness of experimental results. An experimentally observable quantity depending on probability distributions should exhibit a small relative error with regard to small changes of the probability distributions. This should hold for the entropy functional itself. More than that, S_{BG}, as well as S_q, S_q^R, and S_q^N (for $q > 0$), are *expansible* in the sense that the entropy does not vary if we add new possibilities (thus increasing W) whose probabilities are zero. Consistently, the relative variation of the entropy should be small if, during the experimental measure, we do not take into account remote possibilities with almost vanishing probability, or do not take into account the possibilities of another system (e.g., a star far away from a crystal sample that we are measuring on a table) whose correlation with the actual system is practically zero. In this sense, taking first the limit $W \to \infty$ and then the limit $\delta \to 0$ should not alter the result. Lesche [137] has argued that this restrictive property should be demanded only for $\mu = 1$. We really do not see a strong reason why we should not expect this property to hold for all $\mu > 0$.

2.9 REMARKS

S_q, S_q^R, and S_q^N satisfy one more nontrivial property, namely *composability* [4, 5], which we will not describe here. But independently of this, the demands of concavity and stability are strongly restrictive. In fact, we do not know of any entropy other than S_q which simultaneously satisfies both ($\forall q > 0$). *A priori* it might well exist, but up to now none other than S_q is known.

Before concluding the present section, let us present table 1 where S_{BG}, S_q, S_q^R, and S_q^N are compared. Although several other entropic forms do exist in the literature which generalize S_{BG}, these are the principal ones whenever the focus is on thermodynamics. This table includes a variety of properties that we believe to be quite relevant. Let us briefly address each line of the table for $q \neq 1$, since for $q = 1$, all three coincide.

The first line concerns extensivity with regard to independent subsystems. S_q^R composes like S_{BG} ($\forall q$), and is, therefore, extensive. S_q satisfies eq. (6), and is, therefore, nonextensive as already discussed. S_q^N satisfies eq. (6) but with the prefactor $(q - 1)$ instead of $(1 - q)$; it is, therefore, nonextensive.

The second line concerns concavity. S_q is concave for all values of $q > 0$, whereas S_q^R and S_q^N only in the interval $0 < q \leq 1$. All three generalized entropies are convex for $q < 0$.

The third line concerns stability (surely for $\mu = 1$, apparently $\forall \mu > 0$) and $q > 0$. Since none of these generalized entropies is expansible for $q < 0$, this property is to be discussed only for $q > 0$. Only S_q shares (for all $q > 0$) with S_{BG} this basic property.

The next line concerns eq. (29) and is self-explanatory.

The fifth line concerns, as we shall illustrate in detail, the time evolution of the entropy associated with systems such as dissipative maps and other nonlinear dynamical systems. We start at $t = 0$ with N initial conditions all being inside a little cell among the W cells into which the phase space has been partitioned. We then calculate the time evolution of the entropy, and average (noted as $\langle \ldots \rangle$) over all W cells. For S_q we verify the existence of a special value of q, noted as q_{sen} "sen" stands for "sensitivity," such that $\lim_{t \to \infty} \lim_{W \to \infty} \lim_{N \to \infty} \langle S_q(t) \rangle / t$ is finite for both strong (exponential type) and weak (power-law type) chaos. No such property exists for S_q^R and S_q^N.

TABLE 1   Comparison of S_q, S_q^R (R stands for *Renyi*), and $S_q^{LVRA} \equiv S_q^N$ ($LVRA$ and N stand for *Landsberg-Vedral-Rajagopal-Abe* [119, 173] and *normalized*, respectively) with regard to important properties. NO and <u>NO</u> correspond, respectively, to what, in our opinion, are thermodynamically allowed and forbidden violations of the BG properties. This, of course, does not exclude possible utility of the entropic forms S_q^R and S_q^N for other purposes. For instance, the utility of the Renyi entropy S_q^R in the characterization of multifractal structures has long been known. The first seven properties depend exclusively on the entropic form; the last three also depend on the constraints under which the entropy is optimized. In the last line, by power law we mean precisely the q-exponential.

Property	Entropy			
	S_{BG}	S_q	S_q^R	S_q^{LVRA}
	(see [a])	(see [b])	(see [c])	(see [d])
Extensive $(\forall q)(p_{ij}^{A+B} = p_i^A P_j^B)$	YES	NO	YES	NO
Concave $(\forall q > 0)$	YES	YES	<u>NO</u>	<u>NO</u>
Stable $(\forall q > 0)$	YES	YES	<u>NO</u>	<u>NO</u>
Jackson q-derivative application on $-\sum_i p_i^x (\forall q)$	YES	YES	NO	NO
Finite entropy production per unit time $(\forall q = q_{\text{sen}} \leq 1)$	YES	YES	NO	NO
$\ni \hat{S}/\hat{S}$ and $S \equiv \langle \hat{S} \rangle$ obey same composition law $(\hat{\rho}^{A+B} = \hat{\rho}^A \otimes \hat{\rho}^B)(\forall q)$	YES	YES	NO	NO
$\ni \hat{S}/\hat{S}(\hat{\rho}^{-1})$ has same functional form as $S(p_i = 1/W)(\forall q)$	YES	YES	NO	NO
Same functional form for $Z_q(\beta F_q)$ and $Z_q p(\beta E_i)(\forall q)$	YES	YES	NO	NO
Energy eigenvalues scaling temperature coincides with inverse Lagrange parameter	YES	NO	YES	NO
Optimizing distribution $(\forall q)$	Exponential	Power Law	Power Law	Power Law

[a]Where $S_{BG} \equiv -\sum_i p_i \ln p_i = S_1 = S_1^R = S_1^{LVRA}$.
[b]Where $S_q \equiv (1 - \sum_i p_i^q)/(q-1)$.
[c]Where $S_q^R \equiv (\ln \sum_i p_i^q)/(1-q) = \ln[1 + (1-q)S_q]/(1-q)$
[d]Where $S_q^{LVRA} \equiv (S_q)/(\sum_i p_i^q) = (1 - [\sum_i p_i^q]^{-1})/(1-q) = S_q/[1 + (1-q)S_q]$.

The next line concerns the comparison of the entropy operator $\hat{S}$ and the entropy itself $S = Tr\hat{\rho}\hat{S}$, where $\hat{\rho}$ is the density operator (we are using here the quantum notation for convenience, but of course this property also holds for a classical system; remember that, in its diagonal form, $\hat{\rho}$ is a $W \times W$ matrix whose eigenvalues are the $\{p_i\}$). For S_q, there is $\hat{S}_q \equiv \ln_q \hat{\rho}^{-1}$ such that $S_q = Tr\hat{\rho}\hat{S}_q$. For independent systems, this operator composes as follows:

$$\hat{S}_q(A + B) = \hat{S}_q(A) + \hat{S}_q(B) + (1 - q)\hat{S}_q(A)\hat{S}_q(B) \quad (\forall q) . \tag{37}$$

In other words, the entropy operator composes precisely as S_q itself (see eq. (6)). No such property exists for S_q^R and S_q^N.

The seventh line points to the fact that the definition $\hat{S}_q \equiv \ln_q \hat{\rho}^{-1}$ reproduces, for all values of q, the same function as S_q for equiprobability, i.e., eq. (5) ($S_q(p_i = 1/W) = \ln_q W$). No such property is verified for S_q^R and S_q^N.

Up to this line, the properties were related exclusively to the functional form of the entropy, not to the constraints to be imposed at the optimization step which eventually constructs a (meta)equilibrium statistical mechanics connected to thermodynamics. The eighth line uses, in addition to the entropic form, these constraints. By anticipating a little bit, from eq. (39) (to come in the next section), we see that the connection between $Z_q p_i$ and the eigenvalues E_i is provided by the q-exponential function ($\forall q$). This same function connects the partition function Z_q with the free energy F_q, as trivially implied by eq. (47). For S_q^R, the first of these connections is still given by the q-exponential function, but the second one is given by the exponential function, as shown in Lenzi et al. [136]. A similar lack of symmetry occurs for S_q^N.

The ninth line refers to the following. For all these entropies it is obtained $1/T = \partial S/\partial U$ with $T \equiv 1/(k\beta)$, β being the Lagrange parameter associated with the energy constraint, as will be illustrated in the next section. The temperature which, in the (meta)equilibrium distribution, scales with the eigenvalues E_i is this same T if we use S_q^R, whereas it is T_q if we use S_q (see Lenzi et al. [136] and Toral [207]). A similar fact occurs for S_q^N. At first sight, this looks like a possible countraindication for using S_q. This point, however, still needs clarification since, as we shall see for many-body long-range interacting Hamiltonians, the temperature associated with the metastable state is *not* the same as that associated with the standard equilibrium state.

The tenth line constitutes a good stage for clarifying *why* we have very specifically compared, in this table, S_q, S_q^R, and S_q^N. All three entropies depend on $\sum_i p_i^q$; hence, we can write any of them as a function of any of the other two (in the table itself we have indicated S_q^R and S_q^N as functions of S_q). It can easily be checked that these functions are monotonically increasing ones. Therefore, through any optimization procedure using the same constraints, the three entropic functionals under analysis are *simultaneously* extremized by the *same* probability distribution. In other words, all three lead to the same (meta)equilibrium or stationary distribution. Since it is the ubiquity of this q-

exponential distribution which is, in fact, verified empirically (see the many examples to come), the question remains open at this level about *which* entropic form should be used to construct a statistical mechanics which would naturally connect to thermodynamics. It is very clear that we cannot argue that any of them will do the same job. Indeed, as already mentioned, it does not occur to anyone to found BG statistical mechanics on S_{BG}^3 instead of on S_{BG}, just because it is a monotonic function. Statistical mechanics and thermodynamics is much more than just an equilibrium distribution. So, which entropy should we use in connection with thermodynamics? We observe in the table that S_q is *minimalist* in the sense that it modifies very little with regard to the BG entropy. Moreover, it is the only one of the three entropies analyzed here which preserves the very basic properties of *concavity* and *stability*. These facts, in our opinion, make S_q a very strong candidate, if not unique in the present context. But until the entire theory is fully and definitively clarified, the choice belongs to the reader.

3   NONEXTENSIVE STATISTICAL MECHANICS

3.1   INTRODUCTION

In contrast to cybernetics, control theory, and related areas where the entropy is the unique central mathematical object, statistical mechanics is more than just the adoption of an entropy (in our present case, S_q). We must also figure out how the constraints that commonly appear in physics must be written, and in particular, the constraint concerning energy (in the expression "statistical mechanics," we may say that the entropy represents "statistical" and the energy represents "mechanics"). Let us consider the most familiar situation, namely the *canonical ensemble*; in other words, the system is assumed to be in contact with a thermostat (by definition infinitely large). Following Gibbs, we intend to optimize S_q with the constraint (3) and some constraint related to energy. In BG statistics we just use

$$\sum_{i=1}^{W} p_i E_i = U \, , \tag{38}$$

where $\{E_i\}$ are the eigenvalues of the Hamiltonian $\mathcal{H}$ with the appropriate boundary conditions, and the *internal energy* U is a *finite* fixed value. We must decide now what form is to be used for this constraint. The first reasonable attempt, of course, is just to leave it as it stands. It was done this way in 1988 [210], and the stationary (*metaequilibrium*) distribution came out to be the q-exponential function (e_q^x). However, it soon became relatively obvious that this constraint is not adequate for anomalous situations such as symmetric Lévy-like superdiffusion. Indeed, the natural constraint for such a problem, in principle, would be to fix $\int_{-\infty}^{\infty} dx \, x^2 p_q(x)$ (remember that symmetry imposes $\int_{-\infty}^{\infty} dx \, x \, p_q(x) = 0$), but this quantity *diverges* for the optimizing p_q (which turns out essentially to be $\propto e_q^{-\beta x^2}$) whenever we allow for the possibility of long-range jumps, which is precisely the kind of situation we want to focus on! Even worse, for such a choice for the constraint form, it can be seen [85] that no natural connection emerges from thermodynamics (essentially because the Lagrange parameter α representing eq. (3) cannot be factorized out, and, therefore, no partition function can be defined which replaces the α parameter). It has already been noticed [210] that another natural choice would be to use p_i^q in the constraint, in other words, to fix $\sum_{i=1}^{W} p_i^q E_i$. This choice simultaneously solved both difficulties just mentioned: $\int_{-\infty}^{\infty} dx\, x^2 [p_q(x)]^q$ is a sensible *finite* constraint for symmetric Lévy-like superdiffusion [25, 71, 170, 224, 247, 248], and a partition function can be naturally defined [85], thus establishing the desired connection with thermodynamics. However, new mathematical difficulties emerged together with this choice for the constraint, mainly:

1. the metaequilibrium distribution was *not* invariant under the change of the zero level for the energies $\{E_i\}$;
2. since $\sum_{i=1}^{W} p_i^q$ generically differs from unity, the constraint applied to a constant did not yield that same constant; and
3. the assumption set $p_{ij}^{A+B} = p_i^A p_j^B$ and $E_{ij}^{A+B} = E_i^A + E_j^B$ does not yield $U_q^{A+B} = U_q^A + U_q^B$, which constitutes a treatment of the energy conservation principle which *differs* if done microscopically or macroscopically (objection raised to me by F.C. Alcaraz during a private conversation in Caxambu).

When a new theoretical formalism is being constructed, it is not necessarily trivial to know which of the usual beliefs is violated and which is to be retained. The theory was used with this choice for the constraint, until the investigation of a variety of nonlinear dynamical systems made it quite clear that the geometrical interpretation of $q \neq 1$ had to do with the nontrivial nature of the *support* in phase space. Under these conditions, it became obviously unacceptable that the constraint, applied to a constant, does not reproduce this very same constant. This naturally led to what we believe to be a final formulation of the theory, introduced in 1998 [227], that is currently used. We will focus on this one from now on. Constraint (38) is generalized as follows:

$$\langle \mathcal{H} \rangle_q \equiv \frac{\sum_{i=1}^{W} p_i^q E_i}{\sum_{j=1}^{W} p_j^q} = U_q \,, \tag{39}$$

with $\langle \ldots \rangle_1 = \langle \ldots \rangle$; $p_i^q / \sum_{j=1}^{W} p_j^q$ is referred to as the *escort distribution* [52]. This form for the energy constraint turned out to solve all three remaining difficulties simultaneously (1, 2, and 3 above)! Furthermore, Abe and Rajagopal have recently produced a variety of alternative ways to arrive at the same (meta)equilibrium distribution [10, 11, 12, 13, 15] that we derive in the next subsection. Along with this effort, the necessity of using p_i^q, and not p_i, for expressing the constraints naturally re-emerged. It is basically related to the simple property $de_q^x/dx = (e_q^x)^q$. Summarizing, it seems fairly well established by now that constraints should be expressed, for the, as they appear in eq. (39).

The physical interpretation of this type of constraint can easily be illustrated in the case of the Lévy-like superdiffusion already mentioned. If we show a practically minded scientist a Gaussian $(p(x) \propto e^{-ax^2})$ and a Lorentzian $(p(x) \propto 1/(1+bx^2))$ distribution, and ask him (her) roughly what are the widths, he (she) will promptly check the width at about half the value of the maximum for *both* cases, quite independently from the fact that the second moment of the first one is finite, whereas it diverges for the second one. This is a *robust* manner of characterizing this particular constraint. Fixing, for the present Lévy example,

$$\frac{\int_{-\infty}^{\infty} dx\ x^2 [p_q(x)]^q}{\int_{-\infty}^{\infty} dx\ [p_q(x)]^q}$$

does precisely that [30, 67, 218]!

3.2 THEORY

We follow here the theory in Tsallis et al. [227] and derive the (meta)equilibrium distribution associated with the canonical ensemble. The optimization of the entropic form (4) with the constraints (3) and (39) straightforwardly yields

$$p_i = \frac{e_q^{-\beta_q(E_i - U_q)}}{\bar{Z}_q}, \tag{40}$$

with

$$\bar{Z}_q \equiv \sum_{j=1}^{W} e_q^{-\beta_q(E_j - U_q)}, \tag{41}$$

and

$$\beta_q \equiv \frac{\beta}{\sum_{j=1}^{W} p_j^q}, \tag{42}$$

$\beta \equiv 1/kT$ being the Lagrange parameter associated with constraint (39) (to avoid confusion here we have restored k). We easily verify that $q = 1$ recovers the standard BG weight, $q > 1$ implies a power-law tail at high values of E_i, and $q < 1$ implies a cutoff at high values of E_i. We can also verify the invariance of this distribution if we add an arbitrary E_0 to all the $\{E_i\}$ (which implies U_q becoming $U_q + E_0$; hence $E_i - U_q$ does not change). The (meta)equilibrium distribution (40) can be rewritten as follows:

$$p_i = \frac{e_q^{-\beta'_q E_i}}{Z'_q}, \tag{43}$$

with

$$Z_q' \equiv \sum_{j=1}^{W} e_q^{-\beta_q' E_j} \, , \tag{44}$$

and

$$\beta_q' \equiv \frac{\beta_q}{1 + (1-q)\beta_q U_q} \, . \tag{45}$$

This form is particularly convenient for many applications where comparisons with experimental or computational data are involved.

3.3   REMARKS

3.3.1   Connection to thermodynamics.   It can be proved that

$$\frac{1}{T} = \frac{\partial S_q}{\partial U_q} \, , \tag{46}$$

as well as

$$F_q \equiv U_q - TS_q = -\frac{1}{\beta} \ln_q Z_q \, , \tag{47}$$

where

$$\ln_q Z_q = \ln_q \bar{Z}_q - \beta U_q \, . \tag{48}$$

This relation takes into account the trivial fact that, in contrast with what is usually done in BG statistics, the energies $\{E_i\}$ are here referred to U_q in eq. (40). It can also be proved

$$U_q = -\frac{\partial}{\partial \beta} \ln_q Z_q \, , \tag{49}$$

as well as relations such as

$$C_q \equiv T\frac{\partial S_q}{\partial T} = \frac{\partial U_q}{\partial T} = -T\frac{\partial^2 F_q}{\partial T^2} \, . \tag{50}$$

In fact, the entire Legendre transformation structure of thermodynamics is q-invariant, which no doubt is remarkable and welcome.

3.3.2 q-invariant relations.   In addition to the Legendre structure of thermodynamics, many other important theorems are q-invariant, including the six presented here:

1. The H-theorem (macroscopic time irreversibility). Under a variety of irreversible equations such as the master equation, Fokker-Planck equation, and others, it has been proven (see, for instance, Mariz [146] and Ramshaw [177]) that

$$q\frac{dS_q}{dt} \geq 0 \quad (\forall q) \, , \tag{51}$$

the equality corresponding to (meta)equilibrium. In other words, the second principle of thermodynamics holds in the usual way. It is appropriate to remember at this point that, for $q > 0$ (for $q < 0$), the entropy tends to attain its maximum (minimum) since it is a concave (convex) functional, as has already been shown.

2. The Ehrenfest theorem (correspondence principle between quantum and classical mechanics). It can be shown [166] that

$$\frac{d\langle \hat{O} \rangle_q}{dt} = \frac{i}{\hbar} \langle [\hat{\mathcal{H}}, \hat{O}] \rangle_q \quad (\forall q), \tag{52}$$

where $\hat{O}$ is any observable of the system.

3. Factorization of the likelihood function (thermodynamically independent systems). This property generalizes [72, 79, 212] the celebrated one introduced by Einstein in 1910 [92] (reversal of the Boltzmann formula). The likelihood function satisfies

$$W_q(\{p_i\}) \propto e_q^{S_q(\{p_i\})} . \tag{53}$$

If A and B are two probabilistically independent systems, it can be immediately verified that

$$W_q(A + B) = W_q(A)W_q(B) \quad (\forall q) . \tag{54}$$

4. Onsager reciprocity theorem (microscopic time reversibility). It has been shown [70, 80, 172] that the reciprocal linear coefficients satisfy

$$L_{jk} = L_{kj} \quad (\forall q) . \tag{55}$$

5. Kramers and Kronig relation (causality). Its validity has been proved [172] for all values of q.

6. Pesin equality (relation between sensitivity to the initial conditions and the entropy production per unit time). It has been conjectured [229] that the q-generalized Kolmogorov-Sinai entropy K_q and the q-generalized Lyapunov coefficient λ_q are related through

$$K_q = \begin{cases} \lambda_q & \text{if } \lambda_q > 0 ; \\ 0 & \text{otherwise} . \end{cases} \tag{56}$$

Details will be given later.

Properties (1)–(5) essentially reflect something quite basic. In the formalism we are presenting here, we have generalized *nothing* into mechanics, either classical, quantum, or any other type. What we have generalized is the concept of information upon mechanics. Consistently, the properties whose origins lie in mechanics should be expected to be q-invariant, and we verify that they are.

3.3.3 Upper bound for q.

The (meta)equilibrium distribution (40) implies that, for high energy and $q > 1$, the probability p_i vanishes as $1/E^{1/(q-1)}$; hence, $\sum_i p_i$ is asymptotically proportional to $\int_{\text{constant}}^{\infty} dE\, g(E)/E^{1/(q-1)}$, where $g(E)$ is the density of states. Typically, at high energy, $g(E) \propto E^{\delta}$. Consequently, for p_i to be normalizable, it must be $q < (2 + \delta)/(1 + \delta)$, which constitutes an upper bound for the physically acceptable values of q for Hamiltonian systems. The constraint (39) must also be finite. Since $\sum_i p_i^q E_i / \sum_j p_j^q$ is asymptotically proportional to $\int_{\text{constant}}^{\infty} dE\, g(E)\, E/E^{q/(q-1)} = \int_{\text{constant}}^{\infty} dE\, g(E)/E^{1/(q-1)}$, *the admissible upper bound precisely coincides with the one that we have just established.* This feature constitutes an ingredient of consistency within the theory, and reinforces the special form of the energy constraint (39).

3.3.4 q-generalized methods and formulae.

It happens very frequently in BG statistical mechanics that the exact calculation of the thermostatistical properties of nontrivial systems such as interacting many-body ones is mathematically intractable. A variety of methods and formulae have been developed over the years that enable us, at least approximately, to discuss such situations. Many of these methods and formulae are now available in the literature for arbitrary values of q. Among them we count Kubo's linear response theory [172], perturbation method, Bogoliubov inequality (following Feynman' s proof) and variational method [133, 151, 167], many-body Green functions [2, 135, 175], Feynman's path integral [134], uncertainty and entropic bounds [109, 110, 111], Trotter-Lie formula [174], and simulated annealing and related optimization algorithms [27, 28, 94, 106, 107, 156, 163, 189, 220, 246, 241].

3.3.5 Foundations.

The foundations of BG statistical mechanics have a rich and instructive history. There are two central aspects of it, namely the foundation of the thermal equilibrium distribution (BG factor), ubiquitously observed in nature, and the foundation of the BG entropic form, with which it is essentially associated. The main steps associated with the BG factor are the Maxwellian distribution of velocities (Maxwell in 1860 [149]), the Boltzmann kinetic equation (Boltzmann in 1872 [56, 57]) in which the assumption of the *molecular chaos hypothesis, Stosszahlansatz,* leads to a stationary state which is the Boltzmann classical statistical mechanics, the optimization formalism used by Gibbs [103], the steepest descent method (Darwin-Fowler in 1922 [86, 100]), the method based on the laws of large numbers (Khinchin in 1949 [115]), and counting in the microcanonical ensemble (Balian-Balazs in 1987 [46, 118]). The main steps associated with the BG entropic form are the necessary and sufficient conditions for its uniqueness, as proved first by Shannon in 1948 [190] and later on, in a more compact manner, by Khinchin in 1953 [116, 194]. All of these steps have their counterparts for $q \neq 1$ in the literature, as indicated in table 2. It should, however, be stressed that the BG arguments primarily concern thermal equilibrium, whereas the $q \neq 1$ counterparts concern either thermal metaequilibrium or a large class of nonequilibrium stationary states, as illustrated in the next section.

TABLE 2   Entropic form and equilibrium statistics: Foundations. Historical steps of the foundations of Boltzmann-Gibbs statistical mechanics and their nonextensive counterparts. The two lines concerning the conditions of uniqueness refer to the necessary and sufficient conditions for having that particular entropic form (i.e., eq. (2) for $q = 1$ and eq. (4) for $q \neq 1$). The other six lines all refer to the probability distribution which optimizes that particular entropy with appropriate constraints (i.e., exponential for $q = 1$ and q-exponential for $q \neq 1$).

	BG (thermal equilibrium)	$q \neq 1$ (thermal metaequilibrium, nonequilibrium)
Distribution of velocities at equilibrium	Maxwell 1860	[150, 193]
Kinetic equation, molecular chaos hypothesis (Stosszahlansatz)	Boltzmann 1872	[141, 112]
Optimization of entropy with constraints	Gibbs 1902	[85, 210, 227]
Steepest descent	Darwin-Fowler 1922	[11]
Conditions of uniqueness of S	Shannon 1948	[185]
Law of Large Numbers	Khinchin 1949	[10]
Compact conditions of uniqueness of S	Khinchin 1953	[3]
Counting in the microcanonical ensemble	Balian-Balasz 1987 Kubo et al. 1988	[12] [13]

Let us mention at this point that Beck and Cohen [51, 209, 219] have recently extended the q-generalized Boltzmann factor to even more general situations. They coined the word "superstatistics" for that purpose.

3.3.6 Microcanonical ensemble.

As shown in some of the foundational papers (particularly those of Abe and Rajagopal in the preceding subsection) of nonextensive statistical mechanics, the microcanonical ensemble (isolated system) is treated analogously to what is done in BG statistical mechanics. In other words, one has to first evaluate W (the measure of the phase space of the system corresponding to a given total energy, as if it was fully and uniformly occupied, i.e., ergodic occupancy), and then use eq. (5) (with the appropriate value of q) to calculate the entropy S_q as a function of the energy. Its connection with the temperature follows from eq. (46). This procedure can, in principle, be used for both finite ($N < \infty$) or infinite ($N \to \infty$) systems. Only in the $N \to \infty$ limit one expects this procedure to converge onto the result corresponding to the q-canonical ensemble (a very large system in thermal contact with an even much larger thermostat). Although the present microcanonical procedure is well defined once q is known, its ultimate validity for a concrete system (with finite or infinite number of elements N) is to be found in the microscopic dynamics (see next section). More explicitly, if the system covers (as a function of time) the allowed phase space for that total energy in a uniform manner, then $q = 1$. If it covers it (at least after averaging many initial conditions within a given basin of attraction of the space of initial conditions) in the hierarchical manner focused on by the present generalized formalism, then we must use the appropriate value of q (determined by the geometrical structure of occupancy of the phase space). If it covers it in an even more complex manner, then the methods of known statistical mechanics cannot be used, and we are obliged to rely exclusively on dynamics... or to further generalize statistical mechanics! For the hierarchical case addressed by the present theory, if N is finite, one expects a metastable state associated with $q \neq 1$, later on making a terminal crossover to the standard $q = 1$ state. If N is larger and larger, one expects the present $q \neq 1$ description to be the correct one for longer and longer time. If $N \to \infty$, then one expects the system to remain for ever at this anomalous state and never pass onto the standard $q = 1$ microcanonical state.

The above prescription for the microcanonical ensemble closely follows what is usually done for fractals. One first calculates a one-dimensional Euclidean measure L (Lebesgue measure), and then the fractal measure of the system is given by L^{d_f} where d_f is the Hausdorff dimension.

4 CONNECTION WITH THE ELEMENTARY DYNAMICS—THE A PRIORI CALCULATION OF AN ENTROPIC INDEX q

4.1 FROM DYNAMICS TO q

For the formalism presented above to constitute a complete (closed) physical theory, one needs to know *how* to determine (at least in principle) the particular value (or values) of q to be associated with a given physical system susceptible to being described within the present framework. In other words, one needs to determine the particular value of q to be used for calculating (hence predicting) specific statistical-mechanical properties of specific systems. It is these predictions which are to be compared with their experimental counterparts. So, the question arises: *where is the value of q hidden?* The answer is consistent with Einstein's viewpoint [92]: *in the elementary dynamics of the system.* These elementary dynamics can be microscopic (e.g., classical or quantum mechanical models which can be considered as a *first principle* level) or mesoscopic (e.g., Langevin and Fokker-Planck equations, and similar ones).

A variety of such *a priori* calculations of q is already available in the literature. Let us briefly describe a few illustrative ones.

1. *One-dimensional dissipative maps*: See subsection 4.2.
2. *Two-dimensional conservative maps*: See subsection 4.3.
3. *Quantum chaos.* The edge of quantum chaos has recently been exhibited [240] for the first time. A paradigmatic model has been used, namely the quan-

tum kicked top. By exploring the border between the well-known regular and chaotic orbits, it was found (for a typical value of the kicking strength α) that, before entering into the quantum interference region, the *fidelity* relaxes like $e_q^{-(t/\tau)^2}$ ($q \simeq 3.8$ for $\alpha = 3$; τ increases for decreasing perturbation of the Hamiltonian parameter α).

4. *Correlated anomalous diffusion.* In a variety of physical situations [218], it is appropriate to consider the following nonlinear Fokker-Planck-like equation (sometimes referred to as the *Porous Medium Equation*):

$$\frac{\partial p(x,t)}{\partial t} = -\frac{\partial}{\partial x}[F(x)p(x,t)] + D\frac{\partial^2[p(x,t)]^\nu}{\partial x^2} \quad (\nu \in \mathcal{R}) . \tag{57}$$

If we assume that at $t = 0$ we have the paradigmatic (and quite usual) distribution $p(x,0) = \delta(x)$, it can be shown [218] that, for $F(x) = k_1 - k_2 x$ ($k_1 \in \mathcal{R}$; $k_2 \geq 0$) and all (x,t), the (stable) solution is given by

$$p(x,t) \propto e_q^{-\beta(t)[x-x_M(t)]^2} \quad (q < 3) , \tag{58}$$

where $\beta(t)$ and $x_M(t)$ are smooth explicit functions of t, and

$$q = 2 - \nu \quad (q < 3) . \tag{59}$$

5. *Lévy-like anomalous diffusion.* It can be shown [25, 71, 170, 224, 247, 248] that $p(x) \propto e_q^{-\beta x^2}$ ($q < 3$) optimizes

$$S_q = \frac{1 - \int dx[p(x)]^q}{q-1} \tag{60}$$

under appropriate constraints. If we convolute N times $p(x)$ ($N \to \infty$), we approach a Gaussian distribution if $q < 5/3$ and a Lévy $L_{\gamma_L}(x)$ one if $5/3 < q < 3$. The index $\gamma_L < 2$ of this Lévy distribution is related to q as follows:

$$q = \frac{\gamma_L + 3}{\gamma_L + 1} \quad (5/3 < q < 3) . \tag{61}$$

6. *Multiplicative noise.* If we consider the following Langevin-like equation [30]

$$\dot{x} = -\gamma x|x|^{2(s-1)} + x|x|^{s-1}\xi(t) + \eta(t) \quad (s > 0) , \tag{62}$$

where $\xi(t)$ and $\eta(t)$ are independent and Gaussian-distributed zero-mean white noises, satisfying

$$\langle \xi(t)\xi(t') \rangle = 2M\delta(t - t') \quad (M \geq 0) \tag{63}$$

and

$$\langle \eta(t)\eta(t') \rangle = 2A\delta(t - t') \quad (A > 0) . \tag{64}$$

M and A stand for *multiplicative* and *additive*, respectively. If $\gamma \geq M(1-s)$, the distribution corresponding to the stationary state is given by

$$P(x) \propto e_q^{-\frac{(\gamma/s)+M}{2A}|x|^{2s}} , \qquad (65)$$

with

$$q = \frac{(\gamma/s) + 3M}{(\gamma/s) + M} . \qquad (66)$$

7. *Scale-free networks.* A growth model including preferential attachment has been recently introduced [23] (see also Latora and Marchiori [120]) as a prototype of emergence of the ubiquitous scale-free networks (at each time step, m new links are added with probability p, or m existing links are rewired with probability r, or a new node with m links is added with probability $1 - p - r$; all linkings are done with probability $\Pi(k_i) = (k_i + 1)/\sum_j (k_j + 1)$, where k_i is the number of links of the ith node). The exact stationary state distribution of the number k of links at each site can be written as $p(k) \propto e_q^{-k/k_0}$ with

$$q = \frac{2m(2-r) + 1 - p - r}{m(3 - 2r) + 1 - p - r} , \qquad (67)$$

k_0 being an explicit function of (p, r, m).

8. *Many-body long-range classical Hamiltonians*: See subsection 4.4.

Cases (1), (2), and (8) can be considered as particularly instructive and, consistently, have been chosen for more detailed discussion in the following subsections.

4.2 ONE-DIMENSIONAL DISSIPATIVE MAPS

Let us focus on the well-known logisticlike map

$$x_{t+1} = 1 - a|x_t|^z \quad (t = 0, 1, 2, \ldots; \; -1 \leq x_t \leq 1; \; a > 0; \; z > 1) . \qquad (68)$$

For $a < a_c(z)$ ($a_c(2) = 1.401155189\ldots$; $a_c(z)$ increases from 1 to 2 when z increases from 1 to infinity), the attractor is a finite-cycle one and the corresponding Lyapunov exponent λ_1 is negative for almost all values of a. For $a > a_c(z)$, the attractor can be a finite-cycle one (with a typically negative value for λ_1), or a chaotic one (with a typically positive value for λ_1). For all these values of a, the *sensitivity to the initial conditions* $\xi \equiv \lim_{\Delta x(0) \to 0} (\Delta x(t))/(\Delta x(0))$ satisfies the equation

$$\dot{\xi} = \lambda_1 \xi , \qquad (69)$$

hence, as is well known,

$$\xi = e^{\lambda_1 t} \quad (\lambda_1 \neq 0) . \qquad (70)$$

There is, however, an infinite number of values of a for which λ_1 vanishes. For such values of a, we expect [229] the largest possible value of $\xi(t)$ to satisfy

$$\dot{\xi} = \lambda_{q_{\mathrm{sen}}} \xi^{q_{\mathrm{sen}}} \quad (\lambda_1 = 0; \; \lambda_{q_{\mathrm{sen}}} \neq 0; \; q_{\mathrm{sen}} \neq 1) , \qquad (71)$$

hence

$$\xi = e_{q_{sen}}^{\lambda_{q_{sen}} t} \, . \tag{72}$$

This dependence has been proven recently [41, 42, 182] for period-doubling bifurcations, tangent bifurcations, and at the edge of chaos a_c. For the $z = 2$ logistic map it has been found [41, 42, 182, 229]

$$q_{sen} = 5/3 \, ; \quad \lambda_{q_{sen}} = -27/8 \, , \tag{73}$$

at the period-doubling bifurcation occuring at $a = 3/4$;

$$q_{sen} = 3/2 \, ; \quad \lambda_{q_{sen}} = \pm 31.216 \ldots \, , \tag{74}$$

at the tangent bifurcation occuring at $a = 7/4$;

$$q_{sen} = 0.244487701341282 \ldots \, ; \quad \lambda_{q_{sen}} = \frac{1}{1 - q_{sen}} = 1.323605190511561 \ldots \, , \tag{75}$$

at the edge of chaos at $a = a_c(2) = 1.401155189\ldots$. In fact, at the threshold $a_c(z)$, $q_{sen}(z)$ varies from $-\infty$ to $q_{sen}(\infty) \simeq 0.72$ when z varies from unity to infinity [61, 84]. Analogously, for the most chaotic point, namely $a = 2$ ($\forall z$), we find $q_{sen} = 1$ (for $z = 2$ we have $\lambda_1 = \ln 2$). It is in fact $q_{sen} = 1$ for all values of (a, z) for which $\lambda_1 \neq 0$. All maps belonging to the same universality class of the z-logistic one are expected to share the same exponent q_{sen}.

Let us describe now a *second* method which yields the same q_{sen} just described. At the edge of chaos $a_c(z)$, the attractor is well known to be a multifractal. As such it can be characterized with the *multifractal function* $f(\alpha)$ (see, for instance, Beck and Schlogl [52]). This function is always concave, tangent to the bisector, its maximum value coincides with the *fractal* or *Hausdorff dimension* d_f, and it typically vanishes at α_{min} and α_{max}. The following scaling law can be established from simple scaling arguments [143]:

$$\frac{1}{1 - q_{sen}} = \frac{1}{\alpha_{min}} - \frac{1}{\alpha_{max}} \, . \tag{76}$$

This relationship is fascinating: on the right hand, we have a purely geometric quantity (indeed, it can, in principle, be calculated from a photograph of the attractor, without any specification of time ordering) and, on the left hand, we have a dynamical quantity (since it is given by the sensitivity to the initial conditions of the map). Unfortunately, no mathematical proof is yet available for this equality, but it has been numerically verified for a great variety of specific one-dimensional maps [142, 143, 204, 205] (and even two-dimensional ones [202]). These maps include a z-generalization of the circle map, and various others which do not belong to the universality class of the z-logistic one. For the logisticlike map, it is known that $\alpha_{max} = z \, \alpha_{min} = \ln 2 / \ln \alpha_F$, where α_F is the so-called

Feigenbaum constant ($\alpha_F = 2.502907875095892\ldots$ for the $z = 2$ case). These equalities replaced into eq. (76) yield

$$\frac{1}{1 - q_{\text{sen}}(z)} = \frac{(z - 1) \ln \alpha_F(z)}{\ln 2} . \tag{77}$$

It is by using this relationship that we calculated q with great precision for the $z = 2$ logistic map, as indicated in eq. (75).

Let us now address a *third* method for obtaining q_{sen}. This procedure directly refers to the entropy S_q, and consequently provides the justification for using the notation q within the two previous methods, namely the sensitivity-to-initial-conditions and the multifractal ones. We make a partition of the admissible phase space (i.e., $-1 \leq x \leq 1$ in the present case) into W cells. We then locate (uniformly or randomly) N initial conditions within one of these cells, and then run the dynamics of the map for each of those N points. At time t, there will be $N_i(t)$ points in the ith cell ($i = 1, 2, \ldots, W$), and, of course, $\sum_{i=1}^{W} N_i(t) = N$. We can define a probability set through $p_i(t) \equiv N_i(t)/N \ (\forall i)$, hence an entropy $S_q(t)$ through definition (4). Finally, we calculate the quantity $\langle S_q \rangle(t)$ by averaging through all (or almost all) possible initial cells. The representation of $\langle S_q \rangle$ versus t shows curves which depend on the value chosen for q. It is verified that only one special value of q, *precisely* q_{sen}, exists such that, after a relatively short transient, $\langle S_q \rangle$ increases *linearly* with time, thus defining a *finite* entropy production per unit time. For all values of q above q_{sen} the entropy production vanishes, and, for all values below q_{sen}, it diverges. The phenomenon becomes more and more evident with increasing W, which makes the regime, before saturation (at very large times), become longer and longer. Rigorously speaking, we may define a q-generalized Kolmogorov-Sinai-like entropy production K_q as follows:

$$K_q \equiv \lim_{t \to \infty} \lim_{W \to \infty} \lim_{N \to \infty} \frac{\langle S_q \rangle(t)}{t} . \tag{78}$$

What we, therefore, verify [127] is that $0 < K_{q_{\text{sen}}} < \infty$, whereas $K_q = 0 \ (\forall q > q_{\text{sen}})$ and $K_q \to \infty \ (\forall q < q_{\text{sen}})$. The mathematical proof of these statements remains to be done, but these phenomena have been numerically observed in many examples ([127, 142, 204, 205] among others). In practice, we obtain quite good numerical results by using $N = 10 \times W$ with $W \geq 10^5$, the average being done over a few hundreds of cells randomly chosen among the W ones. Moreover, it is known that, whenever $\lambda_1 > 0$, we have $K_1 = \lambda_1$ (Pesin theorem). It has been conjectured [229] that an analogous connection exists in general between $K_{q_{\text{sen}}}$ and $\lambda_{q_{\text{sen}}}$ whenever $\lambda_{q_{\text{sen}}} > 0 \ (q_{\text{sen}} \leq 1)$. This conjecture has been recently numerically verified [43] for the logistic map. For $\lambda_{q_{\text{sen}}} \leq 0$ we expect $K_{q_{\text{sen}}} = 0$.

An important comment should be made at this point. Although for both $q_{\text{sen}} = 1$ and $q_{\text{sen}} < 1$ there is entropy production, the former makes "your finger burns if you touch it" (an expression I learned from E.G.D. Cohen), but the latter does not! This is totally analogous to what happens if we compare the weights of a $(10\ \mathrm{cm})^3$ cubic compact cheese with another one having a Sierpinski sponge structure, in other words, one constructed starting with a $(10\ \mathrm{cm})^3$ compact cheese, then dividing it into $3 \times 3 \times 3$ little cubes, and then eliminating the central one plus the six at the centers of the faces. Then we repeat the operation with each of the 20 little cubes that are left. If we continue this operation indefinitely, at the end we will have a Sierpinski sponge that will measure $(10\ \mathrm{cm})^{d_f}$ with $d_f = \ln 20 / \ln 3 \simeq 2.73$. This interesting "cheese" will have no volume, hence no weight! You can "see" it, you can measure it (its measure is $2^{d_f} \simeq 6.63$ times smaller than that of a $(20\ \mathrm{cm})^{d_f}$ one), but it shows no weight on a balance...nor on your hand! You can sell it per unit, not per weight! Of course, the real object will not strictly coincide with this ideal construction (we cannot eliminate little cubes smaller than the size of the molecules of the cheese), but within some maximal and minimal scales, the concept is what we have described. It is in these terms that one must understand $S_{q_{\mathrm{sen}}}$ and $K_{q_{\mathrm{sen}}}$ if $q_{\mathrm{sen}} < 1$.

So far so good, but the entropic characterization of one-dimensional dissipative maps at the edge of chaos is even more complex. Let us address another special value of q, noted q_{rel}, where *rel* stands for *relaxation*. Suppose that we randomly distribute the $N \gg 1$ initial conditions in not only one of the W cells, but over the entire phase space and let the dynamics of the map act. We observe [87] that the number of occupied cells tends to decrease with time (shrinking of the Lebesgue measure) as $e_{q_{\mathrm{rel}}}^{-t/\tau_{q_{\mathrm{rel}}}} \propto t^{-1/(q_{\mathrm{rel}}-1)}$ ($q_{\mathrm{rel}} > 1$ and $t \to \infty$), which defines q_{rel}, in fact $q_{\mathrm{rel}}(z)$ for the z-logistic map. When z increases from unity to infinity, q_{rel} increases from close to $4/3$ to a value which is close to 4 (perhaps higher) [61]. If we consider not the edge of chaos but values of (a, z) where $\lambda_1 > 0$, then $q_{\mathrm{rel}} = 1$.

Let us now go back to the case where we start with all N initial points in only one window and calculate $S_{q_{\mathrm{sen}}}(t)$. For many such initial windows, this entropy first overshoots and then slowly relaxes toward the large-time plateau (stationary state). It relaxes as $t^{-1/(q_{\mathrm{rel}}(z,W)-1)}$, and $q_{\mathrm{rel}}(z, W)$ slowly increases with increasing W. It exhibits in fact the following interesting finite-graining scaling [61]:

$$q_{\mathrm{rel}}(z, \infty) - q_{\mathrm{rel}}(z, W) \propto \frac{1}{W^{|q_{\mathrm{sen}}(z)|}}\ , \tag{79}$$

where $q_{\mathrm{rel}}(z, \infty)$ coincides with $q_{\mathrm{rel}}(z)$! This relationship connects three important concepts, namely relaxation toward equilibrium, sensitivity to the intial conditions, and the degree of graining in phase space. In addition to this, it shows that two different values of q, namely q_{sen} and q_{rel}, are involved. There is no doubt that the fact that these two entropic indexes happen to coincide ($q_{\mathrm{sen}} = q_{\mathrm{rel}} = 1$) whenever $\lambda_1 > 0$ can easily be at the origin of all types of conceptual confusions.

Before going to conservative maps, let us mention that two-dimensional dissipative ones have also been addressed [202], namely the Henon map: no important difference was found with the one-dimensional dissipative case.

4.3 TWO-DIMENSIONAL CONSERVATIVE MAPS

Let us now address two-dimensional conservative maps. We focus on a paradigmatic one, namely the *standard map*, defined as follows

$$x_{t+1} = x_t + \frac{a}{2\pi} \sin(2\pi x_t) + y_t \quad (\text{mod } 1)$$

$$y_{t+1} = \frac{a}{2\pi} \sin(2\pi x_t) + y_t \qquad (\text{mod } 1) \ (a \in \mathcal{R}) . \tag{80}$$

If $|a|$ is very large, say $|a| > 10$, the system is quickly mixing and ergodic. When $|a|$ decreases, integrable islands start appearing, and for $a = 0$ the system is fully integrable (see fig. 6). Numerical evidence from both the sensitivity to the initial conditions and the entropy production exhibits that the system behaves, in the neighborhood of $a = 1$, like a $q = q_{\text{sen}} \simeq 0.3$ one during intermediate times, eventually making a crossover to a $q = 1$ behavior (see Baldovin [40] and Baldovin et al. [45, 39] for details). In fact, q_{sen} smoothly decreases from 0.3 to 0.1 when a decreases from 1 to 0. The duration of the $q = q_{\text{sen}}$ regime diverges when a decreases. During this regime the system tends to occupy a limited region of the (x, y) phase space, if we start, at $t = 0$, from an ensemble of points in the neighborhood of $y = 0.5$. Only very slowly the occupation diffuses outside of that region, hereafter referred to as the "prison" (see fig. 6). If we start inside the prison, the system lives in a dynamically metastable state for a long time, then escapes from the prison and occupies a good part of the full phase space, thus generating a chaotic see, responsible for the crossover to $q = 1$. If we start outside the prison, it lives there for some time, and gradually enters into the prison as well. The exit from the prison, as well as the entrance into it, occurs through little holes that become smaller and smaller when a decreases. A similar behavior is observed for the $a < 0$ region of the control parameter.

The phenomena that we have observed for the standard map are expected to be ubiquitous, emerging also in a great variety of conservative two- (or higher-) dimensional maps.

4.4 MANY-BODY LONG-RANGE CLASSICAL HAMILTONIANS

Now let us address a many-body conservative system, namely a classical Hamiltonian one, that is assumed to be isolated (i.e., the microcanonical ensemble). Our present choice is a d-dimensional lattice of N localized planar rotators (for another interesting possibility see Andrade et al. [26]). The corresponding Hamiltonian is assumed to be

$$\mathcal{H} = \sum_{i=1}^{N} \frac{p_i^2}{2} + \sum_{i \neq j} \frac{1 - \cos(\theta_i - \theta_j)}{r_{ij}^{\alpha}} \quad (\alpha \geq 0) , \tag{81}$$

where θ_i is the ith angle and p_i the conjugate variable representing the angular momentum (or the rotational velocity since, without loss of generality, unit

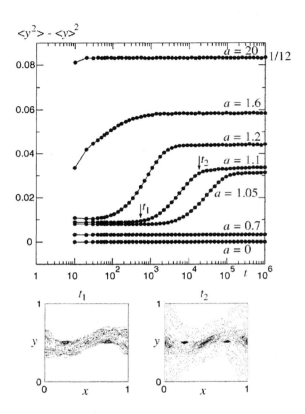

FIGURE 6 Standard map. At $t = 0$ we start [44] with 50×50 replicas with initial locations randomly chosen within $x \in [0, 1]$ and $y \in [0.4995, 0.5005]$. In the upper figure we exhibit, for typical values of a, the time evolution of $\langle y^2 \rangle - \langle y \rangle^2$; $\langle y^2 \rangle$ plays the role of an average kinetic energy (or "temperature"); it is, however, more convenient to plot the "centered temperature" $\langle y^2 \rangle - \langle y \rangle^2$, due to the fact that $(x, y) = (1/2, 1/2)$ is a symmetry point. We generically obtain two plateaus; the duration of the first one increases as a decreases. For a low (high) enough, we only observe the first (second) plateau. We have ergodicity only for large enough a, whose associated "centered temperature" must be $\int_0^1 dy\, y^2 - (\int_0^1 dy\, y)^2 = 1/12$. In the bottom figures we present the ensemble occupancy of the $a = 1.1$ phase space at times t_1 and t_2 indicated by arrows in the upper figure. Practically the same pictures are respectively obtained for all times before t_1 (as soon as a quick transient is overcome) and for all times after t_2. The scenario is that of a "prison" quickly occupied if the points are inside at $t = 0$, but the prison has little "holes," which enable a slow occupation of the outer space.

moment of inertia is assumed). Notice that the summation in the potential is extended to all couples of spins (counted only once) and not restricted to first neighbors; for $d = 1$, $r_{ij} = 1, 2, 3, \ldots$; for $d = 2$, $r_{ij} = 1, \sqrt{2}, 2, \ldots$; for $d = 3$, $r_{ij} = 1, \sqrt{2}, \sqrt{3}, 2, \ldots$. The first-neighbor coupling constant has been assumed, without loss of generality, to be equal to unity. This model is an inertial version of the well-known XY ferromagnet. Although it does not make any relevant difference, we shall assume periodic boundary conditions, in which the distance between a given pair of sites is the smallest one through the 2^d possibilities introduced by the periodicity of the lattice. Notice that the two-body potential term has been written in such a way as to have zero energy for the global fundamental state (corresponding to $p_i = 0$, $\forall i$, and all θ_i equal among them, and equal to say zero). The $\alpha \to \infty$ limit corresponds to only first-neighbor interactions, whereas the $\alpha = 0$ limit corresponds to infinite-range interactions (a typical mean-field situation, frequently referred to as the HMF model [32]).

In the limit $N \to \infty$, the quantity $\tilde{N} \equiv \sum_{i \neq j} r_{ij}^{-\alpha}$ converges to a finite value if $\alpha/d > 1$, and diverges like $N^{1-\alpha/d}$ if $0 \leq \alpha/d < 1$ (like $\ln N$ for $\alpha/d = 1$). In other words, the energy is extensive for $\alpha/d > 1$ and nonextensive otherwise. In the extensive case (*short-range interactions*), the thermal equilibrium (stationary state attained in the $t \to \infty$ limit) is known to be the BG one (see Fisher [95, 96, 97] and Fisher, Lebowitz, and Ruelle [99, 98]). The situation is much more subtle in the nonextensive case (*long-range interactions*). It is this situation that we focus on here. In order to conform to the most usual writing, we shall replace the Hamiltonian $\mathcal{H}$ by the following rescaled one:

$$\mathcal{H}' = \sum_{i=1}^{N} \frac{p_i^2}{2} + \frac{1}{\tilde{N}} \sum_{i \neq j} \frac{1 - \cos(\theta_i - \theta_j)}{r_{ij}^{\alpha}} \qquad (\alpha \geq 0) . \tag{82}$$

The results associated with this Hamiltonian (now artificially transformed into an extensive one for *all* values of α/d) can be trivially transformed into those associated with Hamiltonian $\mathcal{H}$ (see Anteneodo and Tsallis [29]).

The order parameter is the magnetization $\mathbf{M} = 1/N \sum_{i=1}^{N} \mathbf{m}_i$, where $\mathbf{m}_i = [\cos(\theta_i), \sin(\theta_i)]$. The analytical solution of the $\alpha/d \leq 1$ model is possible for the canonical ensemble (equivalent to the microcanonical one in the $N \to \infty$ limit). It predicts a second-order phase transition from a low-energy ferromagnetic phase with magnetization $M \simeq 1$ to a high-energy one, where the spins are homogeneously oriented on the unit circle and $M \simeq 0$. The microcanonical *caloric curve*, that is, the dependence of the energy density $u \equiv U/N$ on the temperature T (U being the fixed total energy), is given by $u = T/2 + 1/2(1 - M^2)$ and shown in the inset (b) of figure 7 as a full curve. The critical point is at $u_c = 0.75$ (reported as a dashed vertical line) corresponding to a critical temperature $T_c = 0.5$ [32, 73, 197].

The dynamical behavior of the model has been investigated in the microcanonical ensemble by starting the system out-of-equilibrium and numerically

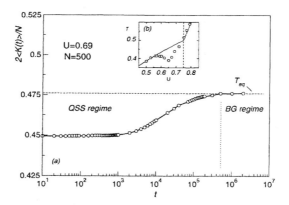

FIGURE 7   The temporal evolution of twice the average kinetic energy per particle is plotted for $N = 500$ and energy density $U = E/N = 0.69$ [125] (a). Two different regimes are clearly visible. The first plateau corresponds to a metastable quasistationary state (QSS) regime, while the second one refers to the Boltzmann-Gibbs (BG) equilibrium regime. We report as a dashed line the final equilibrium temperature which for $U = 0.69$ is equal to $T_{eq} = 0.4756$. We plot, in the inset (b), the equilibrium *caloric curve* (T vs. U) as a full line in comparison with the numerical simulations relative to the QSS regime (open circles). Also in this case $N = 500$.

integrating the equations of motion [122, 123]. For example, one can use *water-bag initial conditions*, in other words, velocities uniformly distributed, and $\theta_i = 0$ for all i ($M = 1$). As shown in the inset (b) of figure 9, microcanonical simulations (open circles) are, in general, in good agreement with the canonical equilibrium prediction, except for a region immediately below u_c, where a dynamics characterized by Lévy walks, anomalous diffusion [124], a negative specific heat [121] (see also Borges and Tsallis [60] and Borges et al. [66]), and aging [154] has been found. In figure 9(a) we report the evolution of twice the average kinetic energy per particle $2\langle K \rangle/N$, a quantity that coincides with the temperature T (brackets denote time averages). The system, started with out-of-equilibrium initial conditions, rapidly reaches a metastable state (quasistationary state; QSS) which is very different from the canonical equilibrium prediction. The main results can be summarized as follows.

For $u > u_c = 3/4$, during the stationary state emerging after the transient:

1. The Lyapunov spectrum collapses to zero in the $N \to \infty$ limit for $\alpha/d \leq 1$, whereas it remains finite for $\alpha/d > 1$. The collapse to zero occurs as A/N^{κ}, where κ decreases from $1/3$ to zero when α/d increases from zero to unity

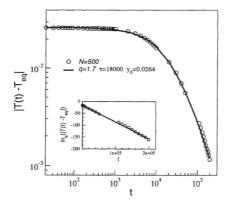

FIGURE 8 We show (open circles) the temporal evolution of the difference between the temperature $T(t) = 2\langle K(t)\rangle/N$ extracted from the numerical simulation with $N = 500$ and the equilibrium value $T_{eq} = 0.4756$. The same data used in figure 9 are plotted. The relaxation is very well reproduced with the q-exponential function $y = y_0\, e_q^{-t/\tau}$ [179]. Inset: The same in $q - log$ versus *linear* representation.

(κ depends on α/d but not on d nor on u [75]); A approaches zero when u diverges.

2. The one-body distribution of velocities is quasi-Maxwellian for finite N and hopefully approaches the Maxwellian distribution in the limit $N \to \infty$. The temperature is practically that of the canonical equilibrium, and hopefully approaches precisely that one in the limit $N \to \infty$.

3. There is no aging (however, the time-dependent correlation functions are of the q-exponential form).

4. Angular diffusion is normal.

5. Specific heat is positive for all values of N.

For $u = 0.69$, during the longstanding plateau which starts immediately after a quick transient:

1. The Lyapunov spectrum collapses to zero in the $N \to \infty$ limit for $\alpha/d \le 1$, whereas it remains finite for $\alpha/d > 1$. The collapse to zero occurs as $N^{-\kappa/3}$ [69, 178]. Although one expects this property to be valid for all d, it has, in fact, been numerically verified only for $d = 1$ [69].

2. The one-body distribution of velocities is far from quasi-Maxwellian for finite N and approaches, for low velocities and in the limit $N \to \infty$, the distribution

(which is time independent along the entire plateau) [125, 126]

$$P(p) = \left[1 - (1 - q) \frac{p^2}{2T_\infty}\right]^{1/(1-q)} , \tag{83}$$

where T_∞ ($= 0.38$) is lower than the canonical one ($\simeq 0.48$). For $\alpha = 0$, we find $q \simeq 7$. We expect this value to decrease to a value close to say 1.5, if we divide the system into two parts, in such a way that the one under study is quite large but still much smaller than the rest, which would act as a thermostat. Under these circumstances we expect the numerical results to coincide with the above analytical expression for a range which could diverge in the $N \to \infty$ limit. Let us mention that this value of q might or might not coincide with that of the distribution of energies in the full phase space (however, if it is not the same, it must be directly related to it).

3. There is aging. The data can be collapsed in such a way that a q-exponential dependence on time emerges.
4. Angular diffusion is superdiffusive.
5. Specific heat is negative for all values of N.
6. The duration of the plateau diverges with N, possibly like $\tilde{N}$ ([69, 74]). Its temperature T_N decreases with N, and approaches $T_\infty > 0$ through a power law.

For $u = 0.69$, after the longstanding plateau has relaxed into the terminal one:

1. The Lyapunov spectrum remains finite.
2. The one-body distribution of velocities is quasi-Maxwellian for finite N and hopefully approaches, in the limit $N \to \infty$, the Maxwellian distribution at the canonical temperature.
3. There is probably no aging, although this remains to be verified.
4. Angular diffusion is normal.
5. Specific heat is positive for all values of N.
6. The temperature relaxation from the longstanding plateau to the terminal plateau occurs through a q-exponential function (see fig. 8).

The scenario which emerges is that, analogously to what occurs with the standard map and other two-dimensional conservative maps, the phase space of this many-body Hamiltonian contains, for $u < u_c$, a "prison" whose holes decrease in size and/or number when N diverges. If, at $t = 0$, the system is started inside the prison (waterbag or double waterbag initial conditions), then it quickly occupies (possibly in a hierarchical, fractal-like way) the interior of the prison and stays there for a long time, eventually also occupying the outside with a chaotic sea, for which the equiprobable occupation typical of BG systems applies. During the longstanding metastable state, the system is in a thermal metaequilibrium which for sure is non-BG, and moreover might be described by the q-exponential distribution obtained within the present formalism. After the system has fully achieved the terminal equilibrium, one reasonably expects this to be the BG one. It is worth stressing at this point that, in the numerical simulations we have described here, not a single thermostatistical concept has been introduced *a priori*. We are exclusively using the laws of Newtonian mechanics, namely *force= mass × acceleration*.

5 APPLICATIONS

We present a nonexhaustive list of applications of or connections to nonextensive statistical mechanics that have been explored in the literature. Each case merits a separate analysis. Indeed, some of them present strong evidence due to powerful experimental and/or theoretical results. Others constitute phenomenological observations of validity, where q is obtained through direct fitting, basically because of uncertainty about the pertinent microscopic dynamics. Some, finally, can only be considered as suggestive arguments for plausibility, typically due to the sparseness of the observational data. Also, the reader should be aware that, in some of the following examples, q refers to the distribution law, whereas in others it refers to the cumulative probability; the connection among them is given by eq. (22). In others, finally, it refers to the concepts of sensitivity to the initial conditions, multifractality, and entropy production.

5.1 PHYSICS

In astrophysics and cosmology, connections have been established with polytropic and other models for self-gravitating systems [165, 195, 200], the distribution of peculiar velocities of spiral galaxies [130], the solar neutrinos problem [22, 83, 113, 114, 128, 129, 171], the cosmic background microwave radiation [68, 131, 147, 148, 169, 225], the primordial helium production and related matters [203, 208], the Robertson-Walker universe [104, 105], and inflationary cosmology and related matters [89, 90].

In astronomy, for the planets of the solar system, it has been recently suggested $q = 0.60$ [192].

In high-energy physics, connections have been established with the transverse momenta distribution in hadronic jets produced by electron-positron annihilation [54, 49], with heavy nuclei collisions [21, 117, 157, 234, 242, 243, 244], with the diffusion of a charm quark in quark-gluon soup [238], and with the energy distribution of cosmic rays [221].

In condensed matter physics, connections have been suggested with manganites [180, 181], high T_c superconductivity [160, 235], Bose-Einstein condensation [81, 93, 184, 199], and electron tight-binding [64, 65, 158], among others.

In nonlinear dynamics, there are numerous studies of low-dimensional dissipative and conservative maps, as well as of noise-induced phenomena [245].

Also, important applications have been done to three-dimensional Eulerian and Lagrangian turbulence [33, 34, 35, 36, 37, 38, 50, 53, 176, 191], a one-dimensional model mimicking turbulence [164], plasma phenomena [55, 140] (e.g., two-dimensional turbulence in non-neutral electron plasma), among others. In fact, the applications to turbulence are, at present, among the most successful ones of the present theory.

In the area of quantum entanglement, a variety of connections have been suggested [6, 9, 16, 236, 237]. In particular, a quite efficient bound of the frontier between separable and entangled states has been proposed [8, 14] and applied [24, 223, 230, 231].

5.2 CHEMISTRY

The Arrhenius law has been generalized to anomalous diffusion [132] within the present formalism. A connection with the lattice Lotka-Volterra model for conservative three-component reactive dynamics has been recently exhibited [232] ($q = 1 - 1/d$). Also, many applications to various chemical systems use the optimization algorithm referred to as generalized simulated annealing (GSA) [27, 28, 94, 106, 107, 156, 163, 189, 220, 246, 241], to be briefly presented in subsection 5.9.

5.3 BIOLOGY

Experimental curves of reassociation of CO in heme-proteins have been quite satisfactorily fitted [222] by solving the ordinary differential equations which appear naturally within the present theory. The velocity distribution and the anomalous diffusion of *Hydra viridissima* have been connected [233] to the present formalism ($q \simeq 1.5$).

The phenomenon of self-organized criticality (SOC) appears to be closely related to the concepts emerging within the present nonextensive formalism. In particular, the connection has been established in two biological models, namely the Bak-Sneppen model for biological evolution [198, 226], and a model for imitation games [162].

5.4 ECONOMICS

A theory of risk aversion in economic trading has been developed using biased averages (p_i^q) [31]. The Black-Scholes differential equation for pricing options and its closed-form solutions have been generalized to arbitrary values of q, thus enabling a remarkable agreement with the market [62, 63] (see also Osorio et al. [161]). The geographical distribution of wealth and its time evolution have been discussed recently [59].

5.5 LINGUISTICS

The well-known Zipf law for linguistics was generalized by Mandelbrot, thus taking the form $p(r) \propto e_q^{-r/r_0}$ ($q > 1$ and $r_0 > 0$), where $p(r)$ is the frequency at which words with rank r appear in a given text or set of texts. Denisov showed [88] that this can be obtained within the present formalism. And, recently, it was shown [152, 153] that a great variety of literary texts can be very well fitted with this law, or with its generalizations along the lines of Tsallis et al. [222].

5.6 MEDICINE

Several works [77, 78, 101, 168, 183] have addressed the problem of processing the electroencephalographic (EEG) signal of epileptic crisis in humans (and also in turtles), as well as in EEG signals following brain injury from cardiac arrest [206]. Similar implementation has been made with electrocardiographic (ECG) signals [196]. The nonextensive entropy has been used to substantially improve the "readability" of the signals. Also, a model for interacting epithelial cell migration has been developed [201].

5.7 GEOPHYSICS

The present nonextensive concepts have been used in studies of mare-highland contacts [138], earthquakes [18, 19, 20], and in the analysis of fine-scale canopy turbulence within and above an Amazon forest [58].

5.8 COGNITIVE SCIENCES

A specific intellectual task, namely the learning/memorization of a few $n \times n$ matrix ($n = 5, 7$) with binary symbols (circles and crosses) has been quantitatively studied. The time evolution of the number of errors (at most n^2) has been measured for a population of 120 university students. The results have been compared with those obtained with a simple perceptron using a q-generalized Langevin dynamics. The comparison suggests that, for humans accomplishing this specific task, q is slightly above unity [76].

5.9 COMPUTER SCIENCES

The problem of finding the global minimum (or minima) of a complex cost function (typically having a large number of local minima) usually becomes computationally very heavy as soon as the dimension of the space of the independent variables overcomes a not so high value. In order to perform calculations that are otherwise very difficult, a variety of optimization algorithms have been introduced over the years. One of the most popular is the so-called *simulated annealing* (SA). It uses BG statistics at two different steps, namely at the *visitation* step (which uses a Gaussian distribution) and at the *acceptance* step (which uses the Boltzmann factor). By q-generalizing both the Gaussian distribution and the Boltzmann factor, the speed, the precision, and the success rate can be sensibly improved. This procedure is usually referred to as *generalized simulated annealing* (GSA). This and related algorithms are commonly used for a great variety of practical problems, including the Travelling Salesman Problem, Lennard-Jones clusters, and other hard classical or quantum problems [27, 28, 94, 106, 107, 156, 163, 189, 220, 246, 241].

From a different perspective, several new laws have recently been discovered for the statics and dynamics of the Internet network. They exhibit a strong connection with nonextensive statistical mechanics (see Abe and Suzuki [17] for the dynamics and references therein for the statics).

5.10 SOCIAL SCIENCES

A great variety of laws emerging in social activities appear to suggest connection to the present theory. This is true for citations of scientific publications [228] ($q_{ISI} = 1.53$ and $q_{PRD} = 1.64$, where *ISI* and *PRD* stand for *Institute for Scientific Information* and *Physical Review D*, respectively); scientific co-authorships [159] (the data for biomedicine are well fitted with a q-exponential with $q = 1.33$); co-actorships [47] ($q = 1.4$); goal distribution in football championships in Italy, England, Spain, and Brazil [144] ($q = 1.33$); city population distributions in Brazil and USA [145] ($q = 1.7$ for both); sexual partnerships in Sweden [139] ($q_F = 1.40$ and $q_M = 1.58$, where F and M stand for female and male, respectively); and teenage mothers in Texas [186, 187, 188].

6 CONCLUSIONS

We have briefly summarized the present status of nonextensive statistical mechanics, its foundations, and its connection to thermodynamics. Particular emphasis has been given to the *a priori* calculation of the entropic index(es) q. Such calculation can be done through various properties (e.g., sensitivity to the initial conditions, multifractality, entropy production, shrinking of the Lebesgue measure), always directly related to and determined by the microscopic (or mesoscopic) dynamics of the system. A variety of natural, social, and artificial systems exhibit relevant properties, that follow the distributions that emerge within the present theory. We believe that the central reason for so many different phenomena sharing the same type of (power-) laws is a scale-free-like structure in the phase space of the system. This is probably the ultimate explanation of the validity of this formalism. In contrast, when the dynamics are so chaotic that the occupied phase space eventually becomes translationally (rotationally) invariant, ergodicity holds, and, of course, the Boltzmann-Gibbs statistical mechanical concepts are applied, as is widely known by physicists, chemists, and others.

To conclude, let us mention some of the relevant questions that, to the best of our knowledge, remain open or not completely understood at present.

1. The zeroth principle of thermodynamics, or whether systems (associated with the same or different values of q) in "equilibrium" or "metaequilibrium" share some kind of temperature (possibly just the usual one). There are several papers in the literature that claim to prove or disprove its validity. In our opinion, the matter is very deep and delicate, and still in progress. Hopefully, the answer to this question will explain why the energy constraint is to be written with the escort distribution (as indicated in eq. (39)), instead of with the original distribution (as in eq. (38)). See Moyano et al. [155] and Tsallis [217] for very recent results on this topic.
2. The dependence on α/d of the index q characterizing the distribution of energies in the canonical ensemble associated with long-range many-body classical Hamiltonian systems. Numerical studies in microcanonical ensembles composed of two subsystems, one of them being the thermostat, and the other one being the system under observation, are welcome. Such studies are also expected to clarify the previous point (1).
3. The characterization of q in geometrical (multi)fractal terms of the phase space of a (few- or many-) body conservative system. And the analogous discussion of dissipative systems.

The understanding of these points will surely clarify "why" the theory works. If the present review provides some insight about this "why," and helps us to understand "how" it works, we will feel fully rewarded. In Galileo's (freely translated) words, *"Knowing with certainty a conclusion is not negligible at all if one wants to discover its proof."*

ACKNOWLEDGMENTS

I am grateful to S. Abe, C. Anteneodo, F. Baldovin, M. Baranger, E. P. Borges, E. Brigatti, E. G. D. Cohen, A. Coniglio, E. M. F. Curado, F. Family, M. Gell-Mann, V. Latora, B. Lesche, L. G. Moyano, F. D. Nobre, A. K. Rajagopal, A. Rapisarda, and A. M. C. de Souza for many and fruitful remarks.

REFERENCES

[1] Abe, S. "A Note on the q-Deformation Theoretic Aspect of the Generalized Entropies in Nonextensive Physics." *Phys. Lett. A* **224** (1997): 326.
[2] Abe, S. "The Thermal Green Functions in Nonextensive Quantum Statistical Mechanics." *Eur. Phys. J. B* **9** (1999): 679.

[3] Abe, S. "Axioms and Uniqueness Theorem for Tsallis Entropy." *Phys. Lett. A* **271** (2000): 74.

[4] Abe, S. "General Pseudoadditivity of Composable Entropy Prescribed by Existence of Equilibrium." *Phys. Rev. E* **63** (2001): 061105.

[5] Abe, S. "Macroscopic Thermodynamics based on Composable Nonextensive Entropies." *Physica A* **305** (2002): 62.

[6] Abe, S. "Nonadditive Entropies and Quantum Entanglement." *Physica A* **306** (2002): 316;

[7] Abe, S. "Stability of Tsallis Entropy and Instabilities of Renyi and Normalized Tsallis Entropies: A Basis for q-Exponential Distributions." *Phys. Rev. E* **66** (2002): 046134.

[8] Abe, S. "Generalized Nonadditive Information Theory and Quantum Entanglement." This volume.

[9] Abe, S., and A. K. Rajagopal. "Quantum Entanglement Inferred by the Principle of Maximum Nonadditive Entropy." *Phys. Rev. A* **60** (1999): 3461.

[10] Abe, S., and A. K. Rajagopal. "Justification of Power-Law Canonical Distributions based on Generalized Central Limit Theorem." *Europhys. Lett.* **52** (2000): 610.

[11] Abe, S., and A. K. Rajagopal. "Microcanonical Foundation for Systems with Power-Law Distributions." *J. Phys. A* **33** (2000): 8733.

[12] Abe, S., and A. K. Rajagopal. "Nonuniqueness of Canonical Ensemble Theory Arising from Microcanonical Basis." *Phys. Lett. A* **272** (2000): 341.

[13] Abe, S., and A. K. Rajagopal. "Macroscopic Thermodynamics of Equilibrium Characterized by Power-Law Canonical Distributions." *Europhys. Lett.* **55** (2001): 6.

[14] Abe, S., and A. K. Rajagopal. "Nonadditive Conditional Entropy and Its Significance for Local Realism." *Physica A* **289** (2001): 157.

[15] Abe, S., and A. K. Rajagopal. "Reexamination of Gibbs' Theorem and Nonuniqueness of Canonical Ensemble Theory." *Physica A* **295** (2001): 172.

[16] Abe, S., and A. K. Rajagopal. "Towards Nonadditive Quantum Information Theory." In *Classical and Quantum Complexity and Nonextensive Thermodynamics*, edited by P. Grigolini, C. Tsallis, and B. J. West, 431. Amsterdam: Pergamon-Elsevier, 2002.

[17] Abe, S., and N. Suzuki. "Itineration of the Internet over Nonequilibrium Stationary States in Tsallis Statistics." April 2002. arXiv e-Print Archive, Condensed Matter, Cornell University. January 2003. ⟨http://lanl.arXiv.org/abs/cond-mat/0204336⟩.

[18] Abe, S., and N. Suzuki. "Law for the Distance between Successive Earthquakes." *J. Geophys. Res. (Solid Earth)* **108** (2003): B2, 2113.

[19] Abe, S., and N. Suzuki. "Scale-Free Network of Earthquakes." October 2002. arXiv e-Print Archive, Condensed Matter, Cornell University. January 2003. ⟨http://lanl.arXiv.org/abs/cond-mat/0210289⟩.

[20] Abe, S., and N. Suzuki. "Time Interval Distribution of Earthquakes." July 2002. arXiv e-Print Archive, Condensed Matter, Cornell University. January 2003. ⟨http://lanl.arXiv.org/abs/cond-mat/0207657⟩.

[21] Aguiar, C. E., and T. Kodama. "Nonextensive Statistics and Multiplicity Distribution in Hadronic Collisions." *Physica A* **320** (2003): 371.

[22] Alberico, W. M., A. Lavagno, and P. Quarati. "Nonextensive Statistics, Fluctuations and Correlations in High Energy Nuclear Collisions." *Eur. Phys. J C* **12** (1999): 499.

[23] Albert, R., and A. L. Barabasi. "Topology of Evolving Networks: Local Events and Universality." *Phys. Rev. Lett.* **85** (2000): 5234.

[24] Alcaraz, F. C., and C. Tsallis. "Frontier between Separability and Quantum Entanglement in a Many Spin System." *Phys. Lett. A* **301** (2002): 105.

[25] Alemany, P. A., and D. H. Zanette. "Fractal Random Walks from a Variational Formalism for Tsallis Entropies." *Phys. Rev. E* **49** (1994): R956.

[26] Andrade, J. S., Jr., M. P. Almeida, A. A. Moreira, A. B. Adib, and G. A. Farias. "A Hamiltonian Approach for Tsallis Thermostatistics." This volume.

[27] Andricioaei, I., and J. E. Straub. "Generalized Simulated Annealing Algorithms using Tsallis Statistics: Application to Conformational Optimization of a Tetrapeptide." *Phys. Rev. E* **53** (1996): R3055.

[28] Andricioaei, I., and J. E. Straub. "On Monte Carlo and Molecular Dynamics Inspired by Tsallis Statistics: Methodology, Optimization and Applications to Atomic Clusters." *J. Chem. Phys.* **107** (1997): 9117.

[29] Anteneodo, C., and C. Tsallis. "Breakdown of Exponential Sensitivity to Initial Conditions: Role of the Range of Interactions." *Phys. Rev. Lett.* **80** (1998): 5313–5316.

[30] Anteneodo, C., and C. Tsallis. "Multiplicative Noise: A Mechanism Leading to Nonextensive Statistical Mechanics." May 2002. arXiv e-Print Archive, Condensed Matter, Cornell University. January 2003. ⟨http://lanl.arXiv.org/abs/cond-mat/0205314⟩.

[31] Anteneodo, C., C. Tsallis, and A. S. Martinez. "Risk Aversion in Economic Transactions." *Europhys. Lett.* **59** (2002): 635.

[32] Antoni, M., and S. Ruffo. "Clustering and Relaxation in Hamiltonian Long-Range Dynamics." *Phys. Rev. E* **52** (1995): 2361–2374.

[33] Arimitsu, N., and T. Arimitsu. "Analysis of Velocity Derivatives in Turbulence based on Generalized Statistics." *Europhys. Lett.* **60** (2002): 60.

[34] Arimitsu, N., and T. Arimitsu. "Analysis of Velocity Fluctuation in Turbulence based in Generalized Statistics." *J. Phys.: Condens. Matter* **14** (2002): 2237.

[35] Arimitsu, T., and N. Arimitsu. "Analysis of Fully Developed Turbulence in Terms of Tsallis Statistics." *Phys. Rev. E* **61** (2000): 3237.

[36] Arimitsu, T., and N. Arimitsu. "Tsallis Statistics and Fully Developed Turbulence." *J. Phys. A* **33** (2000): L235 [Corrigenda: **34** (2001): 673.]

[37] Arimitsu, T., and N. Arimitsu. "Analysis of Fully Developed Turbulence by a Generalized Statistics." *Prog. Theor. Phys.* **105** (2001): 355.

[38] Arimitsu, T., and N. Arimitsu. "Analysis of Turbulence by Statistics based on Generalized Entropies." *Physica A* **295** (2001): 177.

[39] Baldovin, F. "Numerical Analysis of Conservative Maps: A Possible Foundation of Nonextensive Phenomena." This volume.

[40] Baldovin, F. "Mixing and Approach to Equilibrium in the Standard Map." *Physica A* **305** (2002): 124.

[41] Baldovin, F., and A. Robledo. "Sensitivity to Initial Conditions at Bifurcations in One-Dimensional Nonlinear Maps: Rigorous Nonextensive Solutions." *Europhys. Lett.* **60** (2002): 518.

[42] Baldovin, F., and A. Robledo. "Universal Renormalization-Group Dynamics at the Onset of Chaos in Logistic Maps and Nonextensive Statistical Mechanics." *Phys. Rev. E* **66** (2002): (R)045104.

[43] Baldovin, F., and A. Robledo. "Nonextensive Pesin Identity. Exact Renormalization Group Analytical Results for the Dynamics at the Edge of Chaos of the Logistic Map." April 2002. arXiv e-Print Archive, Condensed Matter, Cornell University. January 2003. ⟨http://lanl.arXiv.org/abs/cond-mat/0304410⟩.

[44] Baldovin, F., and C. Tsallis. (2002): in preparation.

[45] Baldovin, F., C. Tsallis, and B. Schulze. "Nonstandard Entropy Production in the Standard Map." *Physica A* **320** (2003): 184.

[46] Balian, R., and N. L. Balazs. *Ann. Phys. (NY)* **179** (1987): 97.

[47] Barabasi, A.-L., and R. Albert. "Emergence of Scaling in Random Networks." *Science* **286** (1999): 509–512.

[48] Barlow, H. "Conditions for Versatile Learning, Helmholtz's Unconscious Inference, and the Task of Perception." *Vision. Res.* **30** (1990): 1561.

[49] Beck, C. "Nonextensive Statistical Mechanics and Particle Spectra in Elementary Interactions." *Physica A* **286** (2000): 164.

[50] Beck, C. "Dynamical Foundation of Nonextensive Statistical Mechanics." *Phys. Rev. Lett.* **87** (2001): 180601.

[51] Beck, C., and E. G. D. Cohen. "Superstatistics." *Physica A* **321** (2003): 267.

[52] Beck, C., and F. Schlogl. *Thermodynamics of Chaotic Systems.* Cambridge: Cambridge University Press, 1993.

[53] Beck, C, G. S. Lewis, and H. L. Swinney. "Measuring Nonextensitivity Parameters in a Turbulent Couette-Taylor Flow." *Phys. Rev. E* **63** (2001): R 035303.

[54] Bediaga, I., E. M. F. Curado, and J. Miranda. "A Nonextensive Thermodynamical Equilibrium Approach in $e^+e^- \to hadrons$." *Physica A* **286** (2000): 156.

[55] Boghosian, B. M. "Thermodynamic Description of the Relaxation of Two-Dimensional Turbulence using Tsallis Statistics." *Phys. Rev. E* **53** (1996): 4754.

[56] Boltzmann, L., and Ber Wien. **66** (1872): 275.

[57] Boltzmann, L., and Ber Wien. *Lectures on Gas Theory*. Translated by S. Brush. Berkeley: University California Press, 1964.

[58] Bolzan, M. J. A., F. M. Ramos, L. D. A. Sa, C. Rodrigues Neto, and R. R. Rosa. "Analysis of Fine-Scale Canopy Turbulence within and above an Amazon Forest using Tsallis' Generalized Thermostatistics." *J. Geophys. Res.* (2002): in press.

[59] Borges, E. P. "Empirical Nonextensive Laws for the Geographical Distribution of Wealth." May 2002. arXiv e-Print Archive, Condensed Matter, Cornell University. January 2003. ⟨http://lanl.arXiv.org/abs/cond-mat/0205520⟩.

[60] Borges, E. P., and C. Tsallis. "Negative Specific Heat in a Lennard-Jones-like Gas with Long-Range Interactions." *Physica A* **305** (2002): 148–151.

[61] Borges, E. P., C. Tsallis, G. F. J. Ananos, and P. M. C. Oliveira. "Nonequilibrium Probabilistic Dynamics at the Edge of Chaos of the Logistic Map." *Phys. Rev. Lett.* **89** (2002): 254103.

[62] Borland, L. "Closed Form Option Pricing Formulas based on a Non-Gaussian Stock Price Model with Statistical Feedback." *Phys. Rev. Lett.* **89** (2002): 098701.

[63] Borland, L. "The Pricing of Stock Options." This volume.

[64] Borland, L., and J. G. Menchero. "Nonextensive Effects in Tight-Binding Systems with Long-Range Hopping." *Braz. J. Phys.* **29** (1999): 169.

[65] Borland, L., J. G. Menchero, and C. Tsallis. "Anomalous Diffusion and Nonextensive Scaling in a One-Dimensional Quantum Model with Long-Range Interactions." *Phys. Rev. B* **61** (2000): 1650.

[66] Boyer, L. L. *Application of Vector and Parallel Machines to Molecular Dynamics of Large Clusters*, edited by L. P. Kartashev and S. I. Kartashev. *Supercomputing Projects, Applications and Artificial Intelligence*. Proc. Third International Conference on Supercomputing, International Supercomputing Institute, Inc., 1988.

[67] Budde, C., D. Prato, and M. Re. "Superdiffusion in Decoupled Continuous Time Random Walks." *Phys. Lett. A* **283** (2001): 309.

[68] Buyukkilic, F., I. Sokmen, and D. Demirhan. "Nonextensive Thermostatistical Investigation of the Blackbody Radiation." *Chaos, Solitons and Fractals* **13** (2002): 749.

[69] Cabral, B. J. C., and C. Tsallis. "Metastability and Weak Mixing in Classical Long-Range Many-Rotator System." *Phys. Rev. E* **66** (2002): 065101.

[70] Caceres, M. O. "Irreversible Thermodynamics in the Framework of Tsallis Entropy." *Physica A* **218** (1995): 471.

[71] Caceres, M. O., and C. E. Budde. "Comment on 'Thermodynamics of Anomalous Diffusion.'" *Phys. Rev. Lett.* **77** (1996): 2589.

[72] Caceres, M. O., and C. Tsallis. Private discussion, 1993.

[73] Campa, A., A. Giansanti, and D. Moroni. "Canonical Solution of a System of Long-Range Interacting Rotators on a Lattice." *Phys. Rev E* **62** (2000): 303–306.

[74] Campa, A., A. Giansanti, and D. Moroni. "Metastable States in a Class of Long-Range Hamiltonian Systems." *Physica A* **305** (2002): 137–143.

[75] Campa, A., A. Giansanti, D. Moroni, and C. Tsallis. "Classical Spin Systems with Long-Range Interactions: Universal Reduction of Mixing." *Phys. Lett. A* **286** (2001): 251.

[76] Cannas, S. A., D. Stariolo, and F. A. Tamarit. "Learning Dynamics of Simple Perceptrons with Nonextensive Cost Functions." *Network: Comp. Neural Sci.* **7** (1996): 141.

[77] Capurro, A., L. Diambra, D. Lorenzo, O. Macadar, M. T. Martin, C. Mostaccio, A. Plastino, E. Rofman, M. E. Torres, and J. Velluti. "Tsallis Entropy and Cortical Dynamics: The Analysis of EEG Signals." *Physica A* **257** (1998): 149.

[78] Capurro, A., L. Diambra, D. Lorenzo, O. Macadar, M. T. Martins, C. Mostaccio, A. Plastino, J. Perez, E. Rofman, M. E. Torres, and J. Velluti. "Human Brain Dynamics: The Analysis of EEG Signals with Tsallis Information Measure." *Physica A* **265** (1999): 235.

[79] Chame, A., and E. V. L. de Mello. "The Fluctuation-Dissipation Theorem in the Framework of the Tsallis Statistics." *J. Phys. A* **27** (1994): 3663.

[80] Chame, A., and E. V. L. de Mello. "The Onsager Reciprocity Relations within Tsallis Statistics." *Phys. Lett. A* **228** (1997): 159.

[81] Chen, J., Z. Zhang, G. Su, L. Chen, and Y. Shu. "q-Generalized Bose-Einstein Condensation based on Tsallis Entropy." *Phys. Lett. A* **300** (2002): 65.

[82] Cohen, E. G. D. "Statistics and Dynamics." *Physica A* **305** (2002): 19–26.

[83] Coraddu, M., G. Kaniadakis, A. Lavagno, M. Lissia, G. Mezzorani, and P. Quarti. "Thermal Distributions in Stellar Plasmas, Nuclear Reactions and Solar Neutrinos." *Braz. J. Phys.* **29** (1999): 153.

[84] Costa, U. M. S., M. L. Lyra, A. R. Plastino, and C. Tsallis. "Power-Law Sensitivity to Initial Conditions within a Logistic-like Family of Maps: Fractality and Nonextensivity." *Phys. Rev. E* **56** (1997): 245.

[85] Curado, E. M. F., and C. Tsallis. "Generalized Statistical Mechanics: Connection with Thermodynamics." *J. Phys. A* **24** (1991): L69. [Corrigenda: **24** (1991): 3187 and **25** (1992): 1019].

[86] Darwin, C. G., and R. H. Fowler. *Phil. Mag. J. Sci.* **44** (1922): 450.

[87] de Moura, F. A. B. F., U. Tirnakli, and M. L. Lyra. "Convergence to the Critical Attractor of Dissipative Maps: Log-Periodic Oscillations, Fractality and Nonextensivity." *Phys. Rev. E* **62** (2000): 6361.

[88] Denisov, S. "Fractal Binary Sequences: Tsallis Thermodynamics and Zipf Law." *Phys. Lett. A* **235** (1997): 447.

[89] de Oliveira, H. P., S. L. Sautu, I. D. Soares, and E. V. Tonini. "Chaos and Universality in the Dynamics of Inflationary Cosmologies." *Phys. Rev. D* **60** (1999): 121301-121304.

[90] de Oliveira, H. P., I. D. Soares, and E. V. Tonini. "Universality in the Chaotic Dynamics Associated with Saddle-Centers Critical Points." *Physica A* **295** (2001): 348.

[91] de Souza, A. M. C., E. Brigatti, and C. Tsallis. (2002): to be published.

[92] Einstein, A. *Annalen der Physik* **33** (1910): 1275. Translation: Abraham Pais, *Subtle is the Lord...*, Oxford University Press, 1982.

[93] Fa, K. S., R. S. Mendes, P. R. B. Pedreira, and E. K. Lenzi. "q-Gaussian Trial Function and Bose-Einstein Condensation." *Physica A* **295** (2001): 242.

[94] Fachat, A., K. H. Hoffmann, and A. Franz. "Simulated Annealing with Threshold Accepting or Tsallis Statistics." *Comput. Phys. Commun.* **132** (2000): 232.

[95] Fisher, M. E. *Arch. Rat. Mech. Anal.* **17** (1964): 377.

[96] Fisher, M. E. *J. Chem. Phys.* **42** (1965): 3852.

[97] Fisher, M. E. *J. Math. Phys.* **6** (1965): 1643.

[98] Fisher, M. E., and J. L. Lebowitz. *Commun. Math. Phys.* **19** (1970): 251.

[99] Fisher, M. E., and D. Ruelle. *J. Math. Phys.* **7** (1966): 260.

[100] Fowler, R. H. *Phil. Mag. J. Sci.* **45** (1923): 497.

[101] Gamero, L. G., A. Plastino, and M. E. Torres. "Wavelet Analysis and Nonlinear Dynamics in a Nonextensive Setting." *Physica A* **246** (1997): 487.

[102] Gell-Mann, M. *The Quark and the Jaguar.* New York: W. H. Freeman, 1994.

[103] Gibbs, J. W. *Elementary Principles in Statistical Mechanics.* New York: C. Scribner's Sons, 1902; New Haven: Yale University Press, 1948.

[104] Hamity, V. H., and D. E. Barraco. "Generalized Nonextensive Thermodynamics Applied to the Cosmic Background Radiation in a Robertson-Walker Universe." *Phys. Rev. Lett.* **76** (1996): 4664.

[105] Hamity, V. H., and D. E. Barraco. "Relativistic Nonextensive Kinetic Theory." *Physica A* **282** (2000): 203.

[106] Hansmann, U. H. E. "Simulated Annealing with Tsallis Weights: A Numerical Comparison." *Physica A* **242** (1997): 250.

[107] Hansmann, U. H. E., and Y. Okamoto. "Generalized-Ensemble Monte Carlo Method for Systems with Rough Energy Landscape." *Phys. Rev. E* **56** (1997): 2228.

[108] Hardy, G., J. E. Littlewood, and G. Polya. *Inequalities*. Cambridge, MA: Cambridge University Press, 1952.

[109] Ion, D. B., and M. L. D. Ion. "Entropic Lower Bound for the Quantum Scattering of Spinless Particles." *Phys. Rev. Lett.* **81** (1998): 5714.

[110] Ion, D. B., and M. L. D. Ion. "Angle-Angular-Momentum Entropic Bounds and Optimal Entropies for Quantum Scattering of Spinless Particles." *Phys. Rev. E* **60** (1999): 5261.

[111] Ion, D. B., and M. L. D. Ion. "Optimal Bounds for Tsallis-like Entropies in Quantum Scattering." *Phys. Rev. Lett.* **83** (1999): 463.

[112] Kaniadakis, G. "Nonlinear Kinetics Underlying Generalized Statistics." *Physica A* **296** (2001): 405.

[113] Kaniadakis, G., A. Lavagno, and P. Quarati. "Generalized Statistics and Solar Neutrinos." *Phys. Lett. B* **369** (1996): 308.

[114] Kaniadakis, G., A. Lavagno, and P. Quarati. "Nonextensive Statistics and Solar Neutrinos." *Astrophys. & Space Sci.* **258** (1998): 145.

[115] Khinchin, A. I. *Mathematical Foundations of Statistical Mechanics*. New York: Dover, 1949.

[116] Khinchin, A. I. *Uspekhi Matem. Nauk* **8** (1953): 3.

[117] Korus, R., St. Mrowczynski, M. Rybczynski, and Z. Wlodarczyk. "Transverse Momentum Fluctuations due to Temperature Variation in High-Energy Nuclear Collisions." *Phys. Rev. C* **64** (2001): 054908.

[118] Kubo, R., H. Ichimura, T. Usui, and N. Hashitsume. *Statistical Mechanics*. Amsterdam: North-Holland, 1988.

[119] Landsberg, P. T., and V. Vedral. "Distributions and Channel Capacities in Generalized Statistical Mechanics." *Phys. Lett. A* **247** (1998): 211.

[120] Latora, V., and M. Marchiori. "The Architecture of Complex Systems." This volume.

[121] Latora, V., and A. Rapisarda. "Microscopic Dynamics of a Phase Transition: Equilibrium vs. Out-of-Equilibrium Regime." *Nucl. Phys. A* **681** (2001): 406c–413c.

[122] Latora, V., A. Rapisarda, and S. Ruffo. "Lyapunov Instability and Finite Size Effects in A System with Long-Range Forces." *Phys. Rev. Lett.* **80** (1998): 692–695.

[123] Latora, V., A. Rapisarda, and S. Ruffo. "Chaos and Statistical Mechanics in the Hamiltonian Mean Field Model." *Physica D* **131** (1999): 38–54.

[124] Latora, V., A. Rapisarda, and S. Ruffo. "Superdiffusion and Out-of-Equilibrium Chaotic Dynamics with Many Degrees of Freedoms." *Phys. Rev. Lett.* **83** (1999): 2104–2107.

[125] Latora, V., A. Rapisarda, and C. Tsallis. "Non-Gaussian Equilibrium in a Long-Range Hamiltonian System." *Phys. Rev. E* **64** (2001): 056134.

[126] Latora, V., A. Rapisarda, and C. Tsallis. "Fingerprints of Nonextensive Thermodynamics in a Long-Range Hamiltonian System." *Physica A* **305** (2002): 129–136.

[127] Latora, V., M. Baranger, A. Rapisarda, and C. Tsallis. "Generalization to Nonextensive Systems of the Rate of Entropy Increase: The Case of the Logistic Map." *Phys. Lett. A* **273** (2000): 97–103.

[128] Lavagno, A., and P. Quarati. "Nonextensive Statistics in Stellar Plasma and Solar Neutrinos." *Nucl. Phys. B, Proc. Suppl.* **87** (2000): 209.

[129] Lavagno, A., and P. Quarati. "Solar Reaction Rates, Nonextensivity and Quantum Uncertainty." February 2001. arXiv e-Print Archive, Nuclear Theory, Cornell University. January 2003. ⟨http://lanl.arXiv.org/abs/nucl-th/0102016⟩.

[130] Lavagno, A., G. Kaniadakis, M. Rego-Monteiro, P. Quarati, and C. Tsallis. "Nonextensive Thermostatistical Approach of the Peculiar Velocity Function of Galaxy Clusters." *Astrophys. Lett. & Comm.* **35** (1998): 449.

[131] Lenzi, E. K., and R. S. Mendes. "Blackbody Radiation in Nonextensive Tsallis Statistics: The Exact Solution." *Phys. Lett. A* **250** (1998): 270.

[132] Lenzi, E. K., C. Anteneodo, and L. Borland. "Escape Time in Anomalous Diffusive Media." *Phys. Rev. E* **63** (2001): 051109.

[133] Lenzi, E. K., L. C. Malacarne, and R. S. Mendes, "Perturbation and Variational Methods in Nonextensive Tsallis Statistics." *Phys. Rev. Lett.* **80** (1998): 218.

[134] Lenzi, E. K., L. C. Malacarne, and R. S. Mendes. "Path Integral Approach to the Nonextensive Canonical Density Matrix." *Physica A* **278** (2000): 201.

[135] Lenzi, E. K., R. S. Mendes, and A. K. Rajagopal. "Quantum Statistical Mechanics for Nonextensive Systems." *Phys. Rev. E* **59** (1999): 1398.

[136] Lenzi, E. K., R. S. Mendes, and L. R. da Silva. "Statistical Mechanics based on Renyi Entropy." *Physica A* **280** (2000): 337.

[137] Lesche, B. "Instabilities of Renyi Entropies." *J. Stat. Phys.* **27** (1982): 419–422.

[138] Li, L., and J. F. Mustard. "Compositional Gradients Across Mare-Highland Contacts: Importance and Geological Implication of Lateral Transport." *J. Geophys. Res.—Planet* **105** (2000): 20431.

[139] Liljeros, F., C. R. Edling, L. A. N. Amaral, H. E. Stanley, and Y. Aberg. "The Web of Human Sexual Contacts." *Nature* **411** (2001): 907–908.

[140] Lima, J. A. S., R. Silva Jr., and J. Santos. "Plasma Oscillations and Nonextensive Statistics." *Phys. Rev. E* **61** (2000): 3260.

[141] Lima, J. A. S., R. Silva, and A. R. Plastino. "Nonextensive Thermostatistics and the *H*-Theorem." *Phys. Rev. Lett.* **86** (2001): 2938.

[142] Lyra, M. L. "Nonextensive Entropies and Sensitivity to Initial Conditions of Complex Systems." This volume.

[143] Lyra, M. L., and C. Tsallis. "Nonextensivity and Multifractality in Low-Dimensional Dissipative Systems." *Phys. Rev. Lett.* **80** (1998): 53.

[144] Malacarne, L. C., and R. S. Mendes. "Regularities in Football Goal Distributions." *Physica A* **286** (2000): 391.

[145] Malacarne, L. C., R. S. Mendes, and E. K. Lenzi. "q-Exponential Distribution in Urban Agglomeration." *Phys. Rev. E* **65** (2001): 017106.

[146] Mariz, A. M. "On the Irreversible Nature of the Tsallis and Renyi Entropies." *Phys. Lett. A* **165** (1992): 409.

[147] Martinez, S., F. Pennini, A. Plastino, and C. Tessone. "Blackbody Radiation in a Nonextensive Scenario." *Physica A* **295** (2001): 224.

[148] Martinez, S., F. Pennini, A. Plastino, and C. J. Tessone. "q-Thermostatistics and the Black-Body Radiation Problem." *Physica A* **309** (2002): 85.

[149] Maxwell, J. C. *Philos. Mag. (Ser. 4)* **19** (1860): 19.

[150] Mendes, R. S., and C. Tsallis. "Renormalization Group Approach to Nonextensive Statistical Mechanics." *Phys. Lett. A* **285** (2001): 273.

[151] Mendes, R. S., K. Fa, E. K. Lenzi, and J. N. Maki. "Perturbation Expansion, Bogoliubov Inequality and Integral Representations in Nonextensive Tsallis Statistics." *Eur. Phys. J. B* **10** (1999): 353.

[152] Montemurro, M. A. "Beyond the Zipf-Mandelbrot Law in Quantitative Linguistics." *Physica A* **300** (2001): 567,

[153] Montemurro, M. A. "A Generalization of the Zipf-Mandelbrot Law in Linguistics." This volume.

[154] Montemurro, M. A., F. Tamarit, and C. Anteneodo. "Aging in an Infinite-Range Hamiltonian System of Coupled Rotators." *Phys. Rev. E* **67** (2003): 031106.

[155] Moyano, L. G., F. Baldovin, and C. Tsallis. "Zeroth Principle of Theremodynamics in Aging Quasi-stationary States." ⟨http://lanl.arXiv.org/abs/cond-mat/0305070⟩.

[156] Mundim, K. C., and C. Tsallis. "Geometry Optimization and Conformational Analysis through Generalized Simulated Annealing." *Int. J. Quantum Chem.* **58** (1996): 373.

[157] Navarra, F. S., O. V. Utyuzh, G. Wilk, and Z. Wlodarczyk. "Violation of the Feynman Scaling Law as a Manifestation of Nonextensivity." *N. Cimento C* **24** (2001): 725.

[158] Nazareno, H. N., and P. E. de Brito. "Long-Range Interactions and Nonextensivity in One-Dimensional Systems." *Phys. Rev. B* **60** (1999): 4629.

[159] Newman, M. E. J. "Random Graphs as Models of Networks." In *Handbook of Graphs and Networks*, edited by S. Bornholdt and H. G. Schuster. Berlin: Wiley-VCH, 2002. Also published at arXiv e-Print Archive, Condensed Matter, Cornell University. February 2002. ⟨http://lanl.arXiv.org/abs/cond-mat/0202208⟩.

[160] Nunes, L. H. C. M., and E. V. L. de Mello. "BCS Model in Tsallis' Statistical Framework." *Physica A* **296** (2001): 106.

[161] Osorio, R., L. Borland, and C. Tsallis. "Distributions of High-Frequency Stock-Market Observables." This volume.

[162] Papa, A. R. R., and C. Tsallis. "Imitation Games: Power-Law Sensitivity to Initial Conditions and Nonextensivity." *Phys. Rev. E* **57** (1998): 3923.

[163] Penna, T. J. P. "Traveling Salesman Problem and Tsallis Statistics." *Phys. Rev. E* **51** (1995): R1.

[164] Peyrard, M., and I. Daumont. "Statistical Properties of One-Dimensional 'Turbulence."' *Europhys. Lett.* **59** (2002): 834.

[165] Plastino, A. R., and A. Plastino. "Stellar Polytropes and Tsallis' Entropy." *Phys. Lett. A* **174** (1993): 384.

[166] Plastino, A. R., and A. Plastino. "Tsallis' Entropy, Ehrenfest Theorem and Information Theory." *Phys. Lett. A* **177** (1993): 177.

[167] Plastino, A., and C. Tsallis. "Variational Method in Generalized Statistical Mechanics." *J. Phys. A* **26** (1993): L893.

[168] Plastino, A., M. T. Martin, and O. A. Rosso. "Generalized Information Measures and the Analysis of Brain Electrical Signals." This volume.

[169] Plastino, A. R., A. Plastino, and H. Vucetich. "A Quantitative Test of Gibbs' Statistical Mechanics." *Phys. Lett. A* **207** (1995): 42;

[170] Prato, D., and C. Tsallis. "Nonextensive Foundation of Levy Distributions." *Phys. Rev. E* **60** (1999): 2398.

[171] Quarati, P., A. Carbone, G. Gervino, G. Kaniadakis, A. Lavagno, and E. Miraldi. "Constraints for Solar Neutrinos Fluxes." *Nucl. Phys. A* **621** (1997): 345c.

[172] Rajagopal, A. K. "Dynamic Linear Response Theory for a Nonextensive System based on the Tsallis Prescription." *Phys. Rev. Lett.* **76** (1996): 3469.

[173] Rajagopal, A. K., and S. Abe. "Implications of Form Invariance to the Structure of Nonextensive Entropies." *Phys. Rev. Lett.* **83** (1999): 1711.

[174] Rajagopal, A. K., and C. Tsallis. "Generalization of the Lie-Trotter Product Formula for q-Exponential Operators." *Phys. Lett. A* **257** (1999): 283.

[175] Rajagopal, A. K., R. S. Mendes, and E. K. Lenzi. "Quantum Statistical Mechanics for Nonextensive Systems—Prediction for Possible Experimental Tests." *Phys. Rev. Lett.* **80** (1998): 3907.

[176] Ramos, F. M., R. R. Rosa, C. R. Neto, M. J. A. Bolzan, and L. D. A. Sa. "Nonextensive Thermostatistics Description of Intermittency in Turbulence and Financial Markets." *Nonlinear Analysis: Theory, Methods and Applications* **47** (2001): 3521.

[177] Ramshaw, J. D. "H-Theorems for the Tsallis and Renyi Entropies." *Phys. Lett. A* **175** (1993): 169.

[178] Rapisarda, A., and V. Latora. "Nonextensive Effects in Hamiltonian Systems." This volume.

[179] Rapisarda, A., C. Tsallis, and A. Giansanti. (2002): to be published.

[180] Reis, M. S., J. P. Araújo, V. S. Amaral, E. K. Lenzi, and I. S. Oliveira. "Magnetic Behavior of a Nonextensive S-spin System: Possible Connections to Manganites." *Phys. Rev. B* **66** (2002): 134417.

[181] Reis, M. S., J. C. C. Freitas, M. T. D. Orlando, E. K. Lenzi, and I. S. Oliveira. "Evidences for Tsallis Non-extensivity on CMR Manganites." *Europhys. Lett.* **58** (2002): 42;

[182] Robledo, A. "Unifying Laws in Multidisciplinary Power-Law Phenomena: Fixed-Point Universality and Nonextensive Entropy." This volume.

[183] Rosso, O. A., M. T. Martin, and A. Plastino. "Brain Electrical Activity Analysis using Wavelet-Based Informational Tools." *Physica A* **313** (2002): 587.

[184] Salasnich, L. "BEC in Nonextensive Statistical Mechanics." *Int. J. Mod. Phys. B* **14** (2000): 405.

[185] Santos, R. J. V. "Generalization of Shannon's Theorem for Tsallis Entropy." *J. Math. Phys.* **38** (1997): 4104.

[186] Scafetta, N., P. Hamilton, and P. Grigolini. "The Thermodynamics of Social Processes: The Teen Birth Phenomenon." *Fractals* **9** (2001): 193–208.

[187] Scafetta, N., P. Grigolini, P. Hamilton, and B. J. West. "Nonextensive Diffusion Entropy Analysis: Nonstationarity in Teen-Birth Phenomena." May 2002. arXiv e-Print Archive, Condensed Matter, Cornell University. January 2003. ⟨http://lanl.arXiv.org/abs/cond-mat/0205524⟩.

[188] Scafetta, N., P. Grigolini, P. Hamilton, and B. J. West. "Nonextensive Diffusion Entropy Analysis: Nonstationarity in Teen Birth Phenomena." This volume.

[189] Serra, P., A. F. Stanton, S. Kais, and R. E. Bleil. "Comparison Study of Pivot Methods for Global Optimization." *J. Chem. Phys.* **106** (1997): 7170.

[190] Shannon, C. E. *Bell System Tech. J.* **27** (1948): 379 and 623.

[191] Shibata, H. "Statistics of Phase Turbulence II." *Physica A* (2002): in press.

[192] Siekman, W. H. "The Entropic Index of the Planets of the Solar System." *Chaos, Solitons and Fractals* **16** (2003): 119–124.

[193] Silva, R., A. R. Plastino, and J. A. S. Lima. "A Maxwellian Path to the *q*-Nonextensive Velocity Distribution Function." *Phys. Lett. A* **249** (1998): 401.

[194] Silverman, R. A., M. D. Friedman, trans., *Mathematical Foundations of Information Theory*. New York: Dover, 1957.

[195] Sota, Y., O. Iguchi, M. Morikawa, T. Tatekawa, and K. Maeda. "Origin of Scaling Structure and Non-Gaussian Velocity Distribution in a Self-Gravitating Ring Model." *Phys. Rev. E* **64** (2001): 056133.

[196] Sun, Y. K. L. Chan, S. M. Krishnan, and D. N. Dutt. "Tsallis' Multiscale Entropy for the Analysis of Nonlinear Dynamical Behavior of ECG Signals." In *Medical Diagnostic Techniques and Procedures*, edited by Megha Singh et al., 49. London: Narosa Publishing House, 1999.

[197] Tamarit, F., and C. Anteneodo. "Rotators with Long-Range Interactions: Connection with the Mean-Field Approximation." *Phys. Rev. Lett.* **84** (2000): 208–211.

[198] Tamarit, F. A., S. A. Cannas, and C. Tsallis. "Sensitivity to Initial Conditions and Nonextensivity in Biological Evolution." *Eur. Phys. J. B* **1** (1998): 545.

[199] Tanatar, B. "Trapped Interacting Bose Gas in Nonextensive Statistical Mechanics." *Phys. Rev. E* **65** (2002): 046105.

[200] Taruya, A., and M. Sakagami. "Gravothermal Catastrophe and Tsallis' Generalized Entropy of Self-Gravitating Systems." *Physica A* **307** (2002): 185.

[201] Thurner, S., N. Wick, R. Hanel, R. Sedivy, and L. Huber. "Anomalous Diffusion on Dynamical Networks: A Model for Interacting Epithelial Cell Migration." *Physica A* **320** (2003): 475.

[202] Tirnakli, U. "Two-Dimensional Maps at the Edge of Chaos: Numerical Results for the Henon Map." *Phys. Rev. E* **66** (2002): 066212.

[203] Tirnakli, U., and D. F. Torres. "Quantal Distribution Functions in Nonextensive Statistics and an Early Universe Test Revisited." *Physica A* **268** (1999): 225.

[204] Tirnakli, U., C. Tsallis, and M. L. Lyra. "Circular-like Maps: Sensitivity to the Initial Conditions, Multifractality and Nonextensivity." *Eur. Phys. J. B* **11** (1999): 309.

[205] Tirnakli, U., C. Tsallis, and M. L. Lyra. "Asymmetric Unimodal Maps at the Edge of Chaos." *Phys. Rev. E* **65** (2002): 036207.

[206] Tong, S., A. Bezerianos, J. Paul, Y. Zhu, and N. Thakor. "Nonextensive Entropy Measure of EEG Following Brain Injury from Cardiac Arrest." *Physica A* **305** (2002): 619.

[207] Toral, R. "On the Definition of Physical Temperature and Pressure for Nonextensive Thermostatistics." *Physica A* **317** (2003): 209.

[208] Torres, D. F., H. Vucetich, and A. Plastino. "Early Universe Test of Nonextensive Statistics." *Phys. Rev. Lett.* **79** (1997): 1588 [Erratum: **80** (1998): 3889.]

[209] Touchette, H. "Temperature Fluctuations and Mixtures of Equilibrium States in the Canonical Ensemble." This volume.

[210] Tsallis, C. "Possible Generalization of Boltzmann-Gibbs Statistics." *J. Stat. Phys.* **52** (1988): 479–487.

[211] Tsallis, C. "Nonextensive Thermostatistics: Brief Review and Comments." *Physica A* **221** (1995): 277.

[212] Tsallis, C. "Some Comments on Boltzmann-Gibbs Statistical Mechanics." *Chaos, Solitons and Fractals* **6** (1995): 539.

[213] Tsallis, C. "Nonextensive Thermostatistics: Brief Review and Comments." *Braz. J. Phys.* **29(1)** (1999).

[214] Tsallis, C. "Nonextensive Thermostatistics: Brief Review and Comments." In *Nonextensive Statistical Mechanics and Its Applications*, edited by S. Abe and Y. Okamoto. Series Lecture Notes in Physics. Berlin: Springer, 2001.

[215] Tsallis, C. "Nonextensive Thermostatistics: Brief Review and Comments." *Chaos, Solitons and Fractals* **13(3)** (2002).

[216] Tsallis, C. "Nonextensive Thermostatistics: Brief Review and Comments." *Physica A* **305(1/2)** (2002).

[217] Tsallis, C. "Comment on 'A Critique of q-Entropy for Thermal Statistics' by M. Nauenberg." ⟨http://lanl.arXiv.org/abs/cond-mat/0304696⟩.

[218] Tsallis, C., and D. J. Bukman. "Anomalous Diffusion in the Presence of External Forces: Exact Time-Dependent Solutions and Their Thermostatistical Basis." *Phys. Rev. E* **54** (1996): R2197.

[219] Tsallis, C., and A. M. C. Souza. "Constructing a Statistical Mechanics for Beck-Cohen Superstatistics." *Phys. Rev. E* **67** (2003): 026106.

[220] Tsallis, C., and D. A. Stariolo. "Generalized Simulated Annealing." *Physica A* **233** (1996): 395. Also published as a preprint, Centro Brasileiro de Pesquisas Fisicas , 1994.

[221] Tsallis, C., J. C. Anjos, and E. P. Borges. "Fluxes of Cosmic Rays: A Delicately Balanced Stationary State." *Phys. Lett. A* **310** (2003): 372.

[222] Tsallis, C., G. Bemski, and R. S. Mendes, "Is Re-association in Folded Proteins a Case of Nonextensivity?" *Phys. Lett. A* **257** (1999): 93.

[223] Tsallis, C., P. W. Lamberti, and D. Prato. "A Nonextensive Critical Phenomenon Scenario for Quantum Entanglement." *Physica A* **295** (2001): 158.

[224] Tsallis, C., S V. F. Levy, A. M. C. de Souza, and R. Maynard. "Statistical-Mechanical Foundation of the Ubiquity of Levy Distributions in Nature." *Phys. Rev. Lett.* **75** (1995): 3589 [Erratum: **77** (1996): 5442].

[225] Tsallis, C., F. C. Sa Barreto, and E. D. Loh. "Generalization of the Planck Radiation Law and Application to the Microwave Background Radiation." *Phys. Rev. E* **52** (1995): 1447.

[226] Tsallis, A. C., C. Tsallis, A. C. N. de Magalhães, and F. Tamarit. "Human and Computer Learning: An Experimental Study." (2003): in preparation.

[227] Tsallis, C., R. S. Mendes, and A. R. Plastino. "The Role of Constraints within Generalized Nonextensive Statistics." *Physica A* **261** (1998): 534.

[228] Tsallis, C., and M. P. de Albuquerque. "Are Citations of Scientific Papers a Case of Nonextensivity?" *Eur. Phys. J. B* **13** (2000): 777.

[229] Tsallis, C., A. R. Plastino, and W.-M. Zheng. "Power-Law Sensitivity to Initial Conditions—New Entropic Representation." *Chaos, Solitons and Fractals* **8** (1997): 885.

[230] Tsallis, C., S. Lloyd, and M. Baranger. "Peres Criterion for Separability through Nonextensive Entropy." *Phys. Rev. A* **63** (2001): 042104.

[231] Tsallis, C., D. Prato, and C. Anteneodo. "Separable-Entangled Frontier in a Bipartite Harmonic System." *Europhys. J. B* **29** (2002): 605.

[232] Tsekouras, G. A., A. Provata, and C. Tsallis. "Nonextensivity of the Cyclic Lattice Lotka Volterra Model." ⟨http://lanl.arXiv.org/abs/cond-mat/0303104⟩.

[233] Upadhyaya, A., J.-P. Rieu, J. A. Glazier, and Y. Sawada. "Anomalous Diffusion and Non-Gaussian Velocity Distribution of Hydra Cells in Cellular Aggregates." *Physica A* **293** (2001): 549.

[234] Utyuzh, O. V., G. Wilk, and Z. Wlodarczyk. "The Effects of Nonextensive Statistics on Fluctuations Investigated in Event-by-Event Analysis of Data." *J. Phys. G* **26** (2000): L39.

[235] Uys, H., H. G. Miller, and F. C. Khanna. "Generalized Statistics and High T_c Superconductivity." *Phys. Lett. A* **289** (2001): 264.

[236] Vidiella-Barranco, A. "Entanglement and Nonextensive Statistics." *Phys. Lett. A* **260** (1999): 335.

[237] Vidiella-Barranco, A., and H. Moya-Cessa. "Nonextensive Approach to Decoherence in Quantum Mechanics." *Phys. Lett. A* **279** (2001): 56.

[238] Walton, D. B., and J. Rafelski. "Equilibrium Distribution of Heavy Quarks in Fokker-Planck Dynamics." *Phys. Rev. Lett.* **84** (2000): 31.

[239] Watanabe, S. *Knowing and Guessing*. New York: Wiley, 1969.

[240] Weinstein, Y. S., S. Lloyd, and C. Tsallis. "Border between Regular and Chaotic Quantum Dynamics." *Phys. Rev. Lett.* **89** (2002): 214101.

[241] Whitfield, T. W., L. Bu, and J. E. Straub. "Generalized Parallel Sampling." *Physica A* **305** (2002): 157.

[242] Wilk, G., and Z. Wlodarczyk. "Interpretation of the Nonextensivity Parameter q in Some Applications of Tsallis Statistics and Levy Distributions." *Phys. Rev. Lett.* **84** (2000): 2770.

[243] Wilk, G., and Z. Wlodarczyk. "Application of Nonextensive Statistics to Particle and Nuclear Physics." *Physica A* **305** (2002): 227.

[244] Wilk, G., and Z. Wlodarczyk. "The Imprints of Nonextensive Statistical Mechanics in High-Energy Collisions." In *Classical and Quantum Complexity and Nonextensive Thermodynamics*, edited by P. Grigolini, C. Tsallis, and B. J. West, 547. Amsterdam: Pergamon-Elsevier, 2002.

[245] Wio, H. "On the Role of Non Gaussian Noises on Noise Induced Phenomena. This volume.

[246] Xiang, Y., D. Y. Sun, W. Fan, and X. G. Gong. "Generalized Simulated Annealing Algorithm and Its Aplication to the Thomson Model." *Phys. Lett. A* **233** (1997): 216.

[247] Zanette, D. H., and P. A. Alemany. "Reply to Comment on 'Thermodynamics of Anomalous Diffusion.'" *Phys. Rev. Lett.* **77** (1996): 2590.

[248] Zanette, D. H., and P. A. Alemany. "Thermodynamics of Anomalous Diffusion." *Phys. Rev. Lett.* **75** (1995): 366.

Generalized Nonadditive Information Theory and Quantum Entanglement

Sumiyoshi Abe

Nonadditive classical information theory is developed in the axiomatic framework and then translated into quantum theory. The nonadditive conditional entropy associated with the Tsallis entropy indexed by q is given in accordance with the formalism of nonextensive statistical mechanics. The theory is applied to the problems of quantum entanglement and separability of the Werner-Popescu-type mixed state of a multipartite system, in order to examine if it has any points superior to the additive theory with the von Neumann entropy realized in the limit $q \to 1$. It is shown that the nonadditive theory can lead to the necessary and sufficient condition for separability of the Werner-Popescu-type state, whereas the von Neumann theory can give only a much weaker condition.

1   INTRODUCTION

Tsallis' nonextensive generalization of Boltzmann-Gibbs statistical mechanics [3, 15, 16] and its success in describing behaviors of a large class of complex systems naturally lead to the question of whether information theory can also admit an analogous generalization. If the answer is affirmative, then that will be of particular importance in connection with the problem of quantum entanglement and quantum theory of measurement [6, 8], in which necessities of a nonadditive information measure and an information content are suggested. One should also remember that there exists a conceptual similarity between a complex system and an entangled quantum system. In these systems, a "part" is indivisibly connected with the rest. An external operation on any part drastically influences the whole system, in general. Thus, the traditional reductionistic approach to an understanding of the nature of such a system may not work efficiently.

In this chapter, we report a recent development in nonadditive quantum information theory based on the Tsallis entropy indexed by q [15] and its associated nonadditive conditional entropy [1]. This theory includes the ordinary additive theory with the von Neumann entropy in a special limiting case: $q \to 1$. To see if it has points superior to the additive theory, we apply it to the problems of separability and quantum entanglement. Employing the Werner-Popescu-type state of an N^n-system (i.e., n-partite N-level system), we show that the present nonadditive information theory with $q > 1$ can, in fact, yield a limitation on separability which is stronger than the one derived from the additive theory. We find that in particular the strongest limitation (i.e., the necessary and sufficient condition) on separability can be obtained in the limit $q \to \infty$.

2   AXIOMS AND UNIQUENESS THEOREM FOR TSALLIS ENTROPY

Before considering quantum theory, we wish to devote this section to the axiomatic foundation of the classical nonadditive information theory. This is regarded as the nonadditive counterpart of the Shannon-Khinchin axiomatic framework for the ordinary additive theory.

The set of axioms presented here is the following: (i) the quantity $S_q(p_1, p_2, \cdots, p_W)$ is continuous with respect to its arguments and takes its maximum for the equiprobability distribution $p_i = 1/W$ ($i = 1, 2, \cdots, W$), (ii) $S_q(A, B) = S_q(A) + S_q(B|A) + (1 - q)S_q(A)S_q(B|A)$ for a composite bipartite system (A, B), where $S_q(B|A)$ is the quantity under consideration given in terms of the conditional probability, and (iii) $S_q(p_1, p_2, \ldots, p_W, p_{W+1} = 0) = S_q(p_1, p_2, \cdots, p_W)$. In Abe [1], it is shown that S_q satisfying (i)–(iii) is, up to a multiplicative constant, uniquely given by

$$S_q(p_1, p_2, \cdots, p_W) \equiv S_q[p] = \frac{1}{1-q} \left[\sum_{i=1}^{W} (p_i)^q - 1 \right],$$ (1)

which is precisely the Tsallis entropy (provided that the conditional entropy is defined in eq. (2) below). The one and only difference between the above set of axioms and Khinchin's [9] is in (ii), and the latter is recovered in the limit $q \to 1$. This correspondence is consistent with the fact that S_q in eq. (1) converges to the Boltzmann-Gibbs-Shannon entropy, $S[p] = -\sum_{i=1}^{W} p_i \ln p_i$, in the limit $q \to 1$. The nonadditive conditional entropy, $S_q(B|A)$, is given, in conformity with the formalism of nonextensive statistical mechanics [16], by the normalized q-expectation value:

$$S_q(B|A) = \langle S_q(B|A_i) \rangle_q^{(A)},$$ (2)

where $S_q(B|A_i)$ is the Tsallis entropy of the conditional probability $p_{ij}(B|A) = p_{ij}(A, B)/p_i(A)$ with the joint probability $p_{ij}(A, B)$ and the marginal probability $p_i(A) = \sum_j p_{ij}(A, B)$. The symbol $\langle Q \rangle_q^{(A)}$ stands for the above-mentioned normalized q-expectation value: $\langle Q \rangle_q^{(A)} = \sum_i Q_i P_i^{(q)}(A)$, where $P_i^{(q)}(A) = [p_i(A)]^q / \sum_k [p_k(A)]^q$ is the escort distribution associated with $p_i(A)$ [5].

From (ii), it immediately follows that

$$S_q(B|A) = \frac{S_q(A, B) - S_q(A)}{1 + (1-q)S_q(A)}.$$ (3)

In a particular case when A and B are statistically independent, $p_{ij}(B|A) = p_j(B)$ and accordingly (ii) becomes $S_q(A, B) = S_q(A) + S_q(B) + (1-q)S_q(A)S_q(B)$, which is termed pseudoadditivity of the Tsallis entropy.

3 NONADDITIVE QUANTUM CONDITIONAL ENTROPY OF WERNER-POPESCU-TYPE STATES

Classical theory in the previous section may be translated into quantum theory by replacing the probability distributions with the density matrices. The quantum Tsallis entropy is written as

$$S_q[\hat{\rho}] = \frac{1}{1-q} \left(\mathrm{Tr} \hat{\rho}^q - 1 \right).$$ (4)

This quantity converges to the von Neumann entropy, $S[\hat{\rho}] = -\mathrm{Tr}(\hat{\rho} \ln \hat{\rho})$, in the limit $q \to 1$. The nonadditive conditional entropy in eq. (3) is also translated in an obvious manner:

$$S_q(B|A) = \frac{S_q(A, B) - S_q(A)}{1 + (1-q)S_q(A)}.$$ (5)

A very important point is that classically it is nonnegative, whereas it can take negative values in quantum theory. For example, suppose $\hat{\rho}(A, B)$ is a pure state. Then, $S_q(A, B) = 0$ but, on the other hand, $S_q(A)$ of $\hat{\rho}(A) = \text{Tr}_B \hat{\rho}(A, B)$ can be positive, leading to negative $S_q(B|A)$. This simple example may indicate that a negative value of $S_q(B|A)$ is a signal of quantum entanglement.

However, it is known that the above-mentioned naïve information-theoretic characterization of quantum entanglement is not valid, in general. For detailed discussions about this point, see Rajagopal [14] and the references cited therein. We shall not consider the problem of quantum entanglement in generic mixed states. Instead, we shall limit our discussion to a special class of mixed states, since our purpose here is to see if the present nonadditive theory has some points which are superior to the ordinary additive theory. The density matrices we consider here are those of the Werner-Popescu-type states.

As the simplest case, let us first discuss a (2×2)-system (i.e., a bipartite spin-1/2 system). Its Werner-Popescu-type state [13, 18] is given by

$$\hat{\rho}(A, B) = \frac{1 - x}{4} \hat{I}_2(A) \otimes \hat{I}_2(B) + x |\Psi^-\rangle\langle\Psi^-|, \tag{6}$$

where $x \in [0, 1]$, $\hat{I}_2$ is the (2×2)-unit matrix, and $|\Psi^-\rangle$ is the maximally entangled state

$$|\Psi^-\rangle = \frac{1}{\sqrt{2}} (|\uparrow\rangle_A|\downarrow\rangle_B - |\downarrow\rangle_A|\uparrow\rangle_B). \tag{7}$$

The density matrix is called separable (or, classically correlated) if it can be recast into the following form:

$$\hat{\rho}(A, B) = \sum_\lambda w_\lambda \hat{\rho}_\lambda(A) \otimes \hat{\rho}_\lambda(B), \tag{8}$$

where $w_\lambda \in [0, 1]$ and $\sum_\lambda w_\lambda = 1$. Note that this is not a product state and, therefore, there is correlation between the subsystems A and B. It is known [18] that such correlation can be modeled by using locally realistic hidden variable theories. On the other hand, genuine quantum entanglement cannot accept descriptions by hidden variable theories. In Peres [10] and Horodecki [7], an algebraic method (termed the method of partial transpose) has been devised to determine when $\hat{\rho}(A, B)$ in eq. (6) becomes separable. The result found is that the state is separable if

$$x \in \left[0, \frac{1}{3}\right]. \tag{9}$$

This condition is now known to be the necessary and sufficient condition. It is also known [7] that the method of partial transposition can yield the necessary and sufficient condition for separability of arbitrary density matrices of (2×2) and (2×3)-systems.

Let us examine our generalized information theory for the state in eq. (6). The nonadditive conditional entropy is immediately calculated to be

$$S_q(B|A) = \frac{1}{1-q}\left[\frac{3}{2}\left(\frac{1-x}{2}\right)^q + \frac{1}{2}\left(\frac{1+3x}{2}\right)^q - 1\right]. \tag{10}$$

As mentioned before, the negative values of this quantity may indicate existence of quantum entanglement, and, accordingly, the condition $S_q(B|A) = 0$ should be analyzed. This condition leads to the function $x = x(q)$, below which the state becomes separable. As can be seen, it is actually a *monotonically decreasing function* with respect to the entropic index, q. The von Neumann value is $x(q \to 1) = 0.748 \cdots$. On the other hand, the asymptotic value is found to be $x(q \to \infty) = 1/3 = 0.333 \cdots$, which precisely gives the necessary and sufficient condition in eq. (9). (One may also recall that the celebrated Bell inequality leads to $x = 1/\sqrt{2} = 0.707 \cdots$.) Therefore, we clearly see how the present nonadditive theory with $q > 1$ is superior to the ordinary additive theory corresponding to the limit $q \to 1$ [4, 17].

Now, let us consider a more general system, which is an N^n-system (i.e., an n-partite N-level system). Its Werner-Popescu-type state is given by

$$\hat{\rho}(A_1, A_2, \cdots, A_n) = \frac{1-x}{N^n}\hat{I}_N(A_1) \otimes \hat{I}_N(A_2) \otimes \cdots \otimes \hat{I}_N(A_n)$$
$$+ x|\Psi_N^{(n)}\rangle\langle\Psi_N^{(n)}|, \tag{11}$$

where $x \in [0, 1]$, $\hat{I}_N$ is the $(N \times N)$-unit matrix, and $|\Psi_N^{(n)}\rangle$ is defined by

$$|\Psi_N^{(n)}\rangle = \frac{1}{\sqrt{N}}\sum_{k=0}^{N-1}|k\rangle_{A_1}|k\rangle_{A_2}\cdots|k\rangle_{A_n}, \tag{12}$$

where k labels the levels of each system. This is a pure state of the nonproduct form and, therefore, is an entangled state. In this case, there may exist a series of the conditional entropies of different kinds, corresponding to a variety of the marginal density matrices. In Abe [2], it is shown that the most important one for characterizing separability is $S_q(A_1|A_2, A_3, \cdots, A_n)$. To calculate this quantity, it is necessary to clarify the spectral structures of $\hat{\rho}(A_1, A_2, \cdots, A_n)$ and $\hat{\rho}(A_2, A_3, \cdots, A_n)$, where the latter is given by

$$\hat{\rho}(A_2, A_3, \cdots, A_n) = \mathrm{Tr}_{A_1}\,\hat{\rho}(A_1, A_2, A_3, \cdots, A_n)$$
$$= \frac{1-x}{N^{n-1}}\hat{I}_N(A_2) \otimes \hat{I}_N(A_3) \otimes \cdots \otimes \hat{I}_N(A_n) \tag{13}$$
$$+ \frac{x}{N}\sum_{k=0}^{N-1}|k\rangle_{A_2\,A_2}\langle k| \otimes |k\rangle_{A_3\,A_3}\langle k| \otimes \cdots \otimes |k\rangle_{A_n\,A_n}\langle k|.$$

Analyzing these density matrices, their eigenvalues are found to be

$$\hat{\rho}(A_1, A_2, \cdots, A_n) : \quad \frac{1-x}{N^n} \ [(N^n - 1) - \text{fold degenerate}],$$

$$\frac{1 + (N^n - 1)x}{N^n}, \tag{14}$$

$$\hat{\rho}(A_2, A_3, \cdots, A_n) : \quad \frac{1-x}{N^{n-1}} \ [(N^{n-1} - N) - \text{fold degenerate}],$$

$$\frac{1 + (N^{n-2} - 1)x}{N^{n-1}} \ [N - \text{fold degenerate}]. \tag{15}$$

Consequently, the nonadditive conditional entropy under consideration is calculated to be

$$S_q(A_1 | A_2, A_3, \cdots, A_n)$$

$$= \frac{1}{1-q} \left\{ \frac{(N^n - 1)\left(\frac{1-x}{N^n}\right)^q + \left[\frac{1+(N^n-1)x}{N^n}\right]^q}{(N^{n-1} - N)\left(\frac{1-x}{N^{n-1}}\right)^q + N\left[\frac{1+(N^{n-2}-1)x}{N^{n-1}}\right]^q} - 1 \right\}. \tag{16}$$

Our interest is in the zero of this quantity, which gives rise to a function $x = x(q)$, for fixed n and N. As in the previous example of the (2×2)-system, this function also turns out to be a monotonically decreasing function of q. In the limit $q \to \infty$, $x(q \to \infty) = 1/(1+N^{n-1})$, indicating that the state in eq. (11) may be separable if

$$x \in \left[0, \frac{1}{1 + N^{n-1}}\right]. \tag{17}$$

Actually, this is known to be the necessary and sufficient condition for the state to be separable, as shown in Pittenger [11, 12] in which this condition is derived by an algebraic method. Thus, we again see that the present nonadditive theory with $q > 1$ is in fact superior to the ordinary additive theory.

4   CONCLUSION

We have presented nonadditive generalization of quantum information theory by introducing the nonadditive conditional entropy associated with the Tsallis entropy. We have applied this theory to the problems of quantum entanglement and separability of a class of the Werner-Popescu-type states. We have shown how the present nonadditive approach can, in general, be superior to the ordinary additive theory with the von Neumann entropy.

REFERENCES

[1] Abe, S. "Axioms and Uniqueness Theorem for Tsallis Entropy." *Phys. Lett. A* **271** (2000): 74.

[2] Abe, S. "Nonadditive Information Measure and Quantum Entanglement in a Class of Mixed States of an N^n System." *Phys. Rev. A* **65** (2002): 052323.

[3] Abe, S., and Y. Okamoto, eds. *Nonextensive Statistical Mechanics and Its Applications.* Heidelberg: Springer-Verlag, 2001.

[4] Abe, S., and A. K. Rajagopal. "Nonadditive Conditional Entropy and Its Significance for Local Realism." *Physica A* **289** (2001): 157.

[5] Beck, C., and F. Schlögl. *Thermodynamics of Chaotic Systems: An Introduction.* Cambridge: Cambridge University Press, 1993.

[6] Brukner, Č., and A. Zeilinger. "Conceptual Inadequacy of the Shannon Information in Quantum Measurements." *Phys. Rev. A* **63** (2001): 022113.

[7] Horodecki, M., P. Horodecki, and R. Horodecki. "Separability of Mixed States: Necessary and Sufficient Conditions." *Phys. Lett. A* **223** (1996): 1.

[8] Horodecki, M., P. Horodecki, and R. Horodecki. "Mixed-State Entanglement and Distillation: Is There a 'Bound' Entanglement in Nature?" *Phys. Rev. Lett.* **80** (1998): 5239.

[9] Khinchin, A. I. *Mathematical Foundations of Information Theory.* New York: Dover, 1957

[10] Peres, A. "Separability Criterion for Density Matrices." *Phys. Rev. Lett.* **77** (1996): 1413.

[11] Pittenger, A. O., and M. H. Rubin. "Separability and Fourier Representaions of Density Matrices." *Phys. Rev. A* **62** (2000): 032313.

[12] Pittenger, A. O., and M. H. Rubin. "Note on Separability of the Werner States in Arbitrary Dimensions." Jan. 2000. arXiv.org e-Print Archive, Quantum Physics, Cornell University. ⟨http://arXiv.org/abs/quant-ph/0001110⟩.

[13] Popescu, S. "Bell's Inequality Versus Teleportation: What is Nonlocality?" *Phys. Rev. Lett.* **72** (1994): 797.

[14] Rajagopal, A. K., and R. W. Rendell. "Separability and Correlations in Composite States Based on Entropy Methods." *Phys. Rev. A.* **66** (2002): 022104.

[15] Tsallis, C. "Possible Generalization of Boltzmann-Gibbs Statistics." *J. Stat. Phys.* **52** (1988): 479.

[16] Tsallis, C., R. S. Mendes, and A. R. Plastino. "The Role of Constraints within Generalized Nonextensive Statistics." *Physica A* **261** (1998): 534.

[17] Tsallis, C., S. Lloyd, and M. Baranger. "Peres Criterion for Separability through Nonextensive Entropy." *Phys. Rev. A* **63** (2001): 042104.

[18] Werner, R. F. "Quantum States with Einstein-Podolsky-Rosen Correlations Admitting a Hidden-Variable Model." *Phys. Rev. A* **40** (1989): 4277.

Unifying Laws in Multidisciplinary Power-Law Phenomena: Fixed-Point Universality and Nonextensive Entropy

Alberto Robledo

1 INTRODUCTION

Critical, power-law behavior in space and/or time manifests in a large variety of complex systems [12] within physics and, nowadays, more conspicuously in other fields, such as biology, ecology, geophysics, and economics. Universality, the same power law holding for completely different systems, is a consequence of the characteristic self-similar, scale-invariant property of criticality, and can be understood in terms of basins of attraction of the renormalization-group (RG) fixed points. However, the guiding quality of a variational approach has been seemingly lacking in the theoretical studies of critical phenomena. Here we give an account of entropy extrema associated with fixed points of RG transformations. As illustrations, we consider simple one-dimensional models of random walks and nonlinear dynamical systems. In describing these systems we consider distribution and/or time relaxation functions with power-law decay that may have infinite first- or second- and higher-order moments. When these moments diverge, we observe the emergence of nonexponential or non-Gaussian fractal

Nonextensive Entropy—Interdisciplinary Applications
edited by Murray Gell-Mann and Constantino Tsallis, Oxford University Press

properties that can be measured by the nonextensive Tsallis entropy index q. We note that the presence of nonextensive properties may signal situations of hindered movement among the system's possible configurations. Some representative applications within physics, but with suggested or recognized connections to other fields, are critical behavior in fluids and magnets, anomalous diffusion processes, transitions to chaos in nonlinear systems, and relaxation properties of supercooled liquids near the glass formation.

Two prototypical model systems serve to illustrate the development of critical states characterized by power laws from generic states described by exponential behavior. These are random walks and nonlinear iterated maps that we discuss below in some detail. Random walks [18] are suitable, for example, for representing Brownian motion (molecular thermal motion under the microscope), but also for many types of data originating from diverse disciplines. One type is that which comes in the form of a "time series," a temporal sequence of measured values, for instance, stock market prices in economics or electroencephalographic potentials in medicine. Nonlinear iterated maps [26] like the logistic map, a simple quadratic equation that models restrained growth, have widespread applications in many fields. The logistic map exhibits both the famous period-doubling and intermittency routes to deterministic chaos that are observed in physical phenomena such as hydrodynamic instabilities in Bénard convection cells. The paradigmatic properties of nonlinear maps such as the logistic map are observed in many ecological and economical systems, for example, predator/prey population dynamics and investments with self-limiting rates of interest. In mathematical terms, the power-law scale-invariant properties possessed by random walks are linked to the properties of Weierstrass functions—everywhere continuous but nowhere differentiable. On the other hand, the scale-invariant properties of the dynamics of iterates, or trajectories, in the logistic map at the so-called strange attractor at the edge of chaos relate to those of Cantor sets, sets with a vanishing measure but, unaccountably, with many members. The fractal or multifractal nature of these two mathematical objects is basic to the properties of critical states we describe here.

The renormalization group (RG) method was designed to study systems with scale-invariant properties. In physics, the RG has been the leading application over the last decades in phase transitions and other problems in statistical physics, in nonlinear dynamics, and in other fields [10]. RG studies have shed much light on critical phenomena taking place, for example, in fluids, magnets, and spin glasses, and have played a central role in understanding the onset of chaos in nonlinear maps and the marginal stability of self-organized structures. The RG is generally applied to systems with many degrees of freedom, and the approach consists of devising problem-specific transformations that combine partial elimination of degrees of freedom with rescaling, leaving central expressions unchanged in form, but not the set of parameters in them. Such parameter changes indicate that the RG transformation has the effect of switching from one system into another within a set called a universality class. Repeated application of the RG transformation produces a flow within this set either toward or away from special systems referred to as fixed points. For systems under conditions of criticality, the flow is attracted by a "nontrivial" fixed point. The power laws of the scale-invariant critical states of interest are determined by the properties of the nontrivial fixed point, and these properties are universal—they hold for all systems that belong to the same basin of attraction. We describe below two types of RG transformation for random walks, one involving the range of the step probability distribution and the other the convolution of this function with itself that generates distributions for multiple-step walks. In both cases we point out the meaning of the fixed points. For the logistic map, the RG transformation is that due to Feigenbaum and consists of functional composition and rescaling. When this is applied to the critical points of the map, the self-similar features of the dynamics generated by the number of iterations are revealed, and we determine the corresponding power laws from the associated fixed points. For both walks and maps we point out an important connection that appears between RG fixed points and extremal properties of entropy expressions. Scaling approaches, like the RG method, have proven valuable in obtaining universal properties involving probability distributions of nonequilibrium systems [28].

Also, properties of some systems described by distributions or dynamical response functions with asymptotic power laws for large arguments appear to be connected to a generalized entropy with a nonextensive feature [29, 30, 31]. This entropy expression was proposed more than a decade ago as the basis of a generalization of the canonical, or Boltzmann-Gibbs (BG), statistical mechanics. A growing collection of studies has provided experimental and numerical evidence for both the probable inapplicability of BG statistics and the plausible competence of the generalized theory in describing various types of phenomena and circumstances. Many of the contributions in this volume illustrate the current outlook of the nonextensive theory and its applications. One common condition under which the generalized theory is believed to be applicable is that of restricted or stalled accessibility to otherwise permissible configurations in phase space. This fraction of phase space often has a fractal or multifractal geometrical structure [30, 31]. We illustrate this condition here through both random walk and iterated map models. The nonextensive statistical mechanics [30, 31] stems from the following expression for the entropy

$$S_q = - \sum_j [p(j)]^q \ln_q p(j) \,, \tag{1}$$

where $\ln_q x \equiv (x^{1-q}-1)/(1-q)$ is the q-logarithmic function, $p(j)$ is a distribution representative of the problem at hand, and the index q stands for the degree of nonextensivity (a proportionality constant has been omitted in S_q). In the limit $q \to 1$ the standard expression $S_1 = - \sum_j p(j) \ln p(j)$ is recovered, as are all other generalized properties derived from it [30, 31]. A central property of the generalized theory is that the form of the distribution that maximizes S_q under a given constraint $\sum_j j^\zeta p^q(j) = \text{const}$ is

$$p(j) = \mathcal{N}^{-1} \exp_q(-\beta j^\zeta),\qquad(2)$$

where $\mathcal{N}$ is a normalization constant, $\exp_q(x) \equiv [1 + (1-q)x]^{1/(1-q)}$ is the q-exponential function, and β is a Lagrange multiplier. Also, $\ln_q(\exp_q(x)) = x$.

Our presentation takes the form of short introductions to the topics addressed, along with concise reviews and accounts of recent results with relevant references. In the following two sections we illustrate the above-mentioned concepts and theoretical developments in terms of the two models mentioned. Weierstrass random walks and critical attractors of logisticlike iterated maps. We conclude by giving a brief outline about the possible application of these ideas to a problem in physics that is currently a major challenge. The problem is that of understanding the formation of a glass by the fast cooling of liquids. Disordered substances, such as most liquids, in which the atoms are randomly distributed, can be thought of as chaotic behavior implanted in physical space. Similarly, most solid substances encountered in nature are amorphous, like glasses, just like a liquid, except that they share properties of rigid crystals. Within physics, the word "glass" has become the generic term for a disordered system, but the concept finds application in other domains.

2   RANDOM WALKS

The Weierstrass walk is a symmetric random walk on an infinite lattice [14, 18, 19] that exhibits a scaling property such that under appropriate conditions the trajectories consist of a hierarchy of self-similar clusters of visited sites. As in this chapter, the presentation details are often given for walks on a one-dimensional lattice, but the treatment can be generalized straightforwardly to d-dimensional simple cubic lattices [14, 18, 19]. The walks are generated by a distribution for single steps $p^\mu(l)$ of the form

$$p^\mu(l) = \frac{A}{2} \sum_{n=0}^{\infty} a^{-n}(\delta_{l,-b^n} + \delta_{l,b^n}),\qquad(3)$$

where $a > 1$ and $b > 1$. The allowed steps have step lengths $l = b^n$ and occur with probabilities proportional to a^{-n}. Equation (3) can be rewritten as the power law $p^\mu(l) = A|l|^{-\mu}$, $A = 1 - a^{-1}$, $\mu \equiv \ln a / \ln b$, and it can be seen that when $\mu < 2$, the mean-squared displacement $\langle l^2 \rangle$ per jump diverges. The structure function $\lambda^\mu(k) = \sum_l p^\mu(l) \exp(ikl)$ is the continuous nondifferentiable function of Weierstrass

$$\lambda^\mu(k) = A \sum_{n=0}^{\infty} a^{-n} \cos(b^n k),\qquad(4)$$

and its nonanalytic small-k behavior was demonstrated [14, 27] to arise from an infinite sum of regular terms obtained by iteration of the scaling equation

$$\lambda^\mu(k) = a^{-1}\lambda^\mu(bk) + A\cos k\,. \tag{5}$$

When $\mu \le 2$, the singular part of $\lambda^\mu(k)$ is of the form $Q(k)\,|k|^\mu$ with $Q(k)$ period in $\ln|k|$ with period $\ln b$.

Walks on a lattice are termed persistent when the walker is certain to return to any site, and in this case the set of distinct sites visited by an infinite walk covers the entire lattice. On the other hand, a walk is said to be transient if the probability of return for the walker is less than unity, in which case the set of distinct sites in an infinite walk does not fill the lattice. Polya's theorem [18, 20] asserts that a d-dimensional lattice walk with finite $\langle l^2 \rangle$ is persistent if $d = 1$ or 2 and transient if $d \ge 3$. The consideration of sufficiently long-ranged steps such that $\langle l^2 \rangle \to \infty$, as in the Weierstrass walk, can induce behavior consistent with an effective dimension $d_{\text{eff}} > d$. For the $d = 1$ Weierstrass walk it is found $d_{\text{eff}} = 3-\mu$ [14, 27], since these walks are persistent when $\mu \le 1$ and transient when $\mu > 1$. When the Weierstrass walk is transient, trajectories display a self-similar cluster structure and the exponent μ is identified with the fractal dimension of the set of sites visited by the walks [14, 27].

Let $P_n(l)$ be the probability that the walker is at site l after n steps if the origin is the starting point $P_0(l) = \delta_{l,0}$. For a memory-less walk with single-step distribution $p(l)$, this set of probabilities satisfies

$$P_n(l) = \sum_{l'} p(l - l')P_{n-1}(l')\,, \tag{6}$$

which, in terms of the generating function $P(l; z) \equiv \sum_{n=0}^{\infty} P_n(l)z^n$, becomes

$$P(l; z) = z\sum_{l'} p(l - l')P(l'; z) + \delta_{l,0}\,. \tag{7}$$

In Fourier space one has

$$\widetilde{P}_n(k) = [\lambda(k)]^n \text{ and } \widetilde{P}(k; z) = [1 - z\lambda(k)]^{-1}\,, \tag{8}$$

where $\lambda(k) = \sum_l p(l)e^{ikl}$, $\widetilde{P}_n(k) = \sum_l P_n(l)e^{ikl}$, and $\widetilde{P}(k; z) = \sum_l P(l; z)e^{ikl}$ [20]. The first-passage probabilities can be obtained as follows. Introduce the generating function $F(l; z) \equiv \sum_{n=1}^{\infty} F_n(l)z^n$ where $F_n(l)$ is the probability that the walker steps on l for the first time after n steps. One obtains [20]

$$F(l; z) = \frac{[P(l; z) - \delta_{l,0}]}{P(0; z)}\,. \tag{9}$$

The probability of return to the origin irrespective of the number of steps is $F(0; 1) = 1-1/P(0; 1)$, so it can be seen that Polya's theorem and its generalization follows from the divergence or convergence of the inverse Fourier transform of $\widetilde{P}(k; z)$ at $l = 0$ and $z = 1$ [14, 20].

According to the central limit theorem and its generalization [9], the probabilities $P_n(l)$ for large n approximate the form of either a Gaussian or a Lévy distribution. This can readily be seen by considering the small k form for the Weierstrass $\lambda^\mu(k)$, since then, for $\mu > 2$, large n, and $k = O(n^{-1/2})$ [14], one has

$$\widetilde{P}_n(k) \approx \left[1 - \frac{1}{n} \frac{\langle l^2 \rangle}{2} nk^2 \right]^n \approx \exp\left(-\frac{\langle l^2 \rangle}{2} nk^2 \right), \tag{10}$$

whereas for $\mu \leq 2$, large n, and $k = O(n^{-1/\mu})$, one obtains instead

$$\widetilde{P}_n(k) \approx \left[1 - \frac{1}{n} Qn |k|^\mu \right]^n \approx \exp\left(-Qn |k|^\mu \right). \tag{11}$$

The exponential function in eq. (10) can be seen to be the Fourier space expression for the Gaussian distribution,

$$P_n(l) \approx \left[2\pi \langle l^2 \rangle n \right]^{-1/2} \exp\left(-\frac{l^2}{2\langle l^2 \rangle n} \right), \tag{12}$$

and that in eq. (11) to be the Fourier space expression for the symmetric stable Lévy distribution of order μ [9, 14]. Explicit analytical forms for the Lévy distributions are not obtainable in general. In continuum space where these distributions are defined, they exhibit power-law decay of the form $|x|^{-1-\mu}$ for large x. The repeated convolutions of the step distribution $p(l)$ implicated in the construction of the $P_n(l)$ in eq. (6) can be thought of as an example of repeated RG transformations that lead to a fixed point. In this case, the fixed point is a stable distribution obtained independently of the form of the initial $p(l)$. When $\langle l^2 \rangle$ is finite ($\mu > 2$ in the Weierstrass $p^\mu(l)$), the result is the Gaussian distribution but, when $\langle l^2 \rangle$ is infinite (and $\mu \leq 2$ in $p^\mu(l)$), the fixed point is the Lévy distribution of order μ. The well-known maximum entropy property associated with the Gaussian distribution, however, does not seem to appear in a simple manner for the Lévy distributions [19]. The self-similar properties of the Weierstrass walks $P_n(l)$ derived from the power-law tail when $\mu \leq 2$ make the limiting stable distribution $\lim_{n \to \infty} P_n(l)$ and the single-step distribution $p^\mu(l)$ similar in form, the distribution for the complete trajectory and that for its building block are alike. In relation to this, the single-step distribution for continuum-space Lévy flights has been determined in terms of the nonextensive theory [30, 31]. A step distribution in the form of a q-exponential has been obtained and studied by exploiting a maximum entropy procedure for S_q analogous to that in which the Gaussian distribution is obtained from the canonical extensive entropy [30, 31].

Interestingly, the step distribution for the Weierstrass walk can be written in the form of a q-exponential by a shift $|l| \to |l| + 1$ of the step-length variable, and one obtains

$$p^\mu(l) = A \exp_q(-\mu |l|), \tag{13}$$

where $q = 1 + 1/\mu$. In relation to the n-step walk $P_n(l)$, we notice the following property in Fourier space. According to eqs. (10) and (11), $\widetilde{P}_n(k)$ can also be written as a q-exponential, this time with $q = 1 - 1/n$ and, as it can be observed in the same two equations, these transform into an exponential as $n \to \infty$ and $q \to 1$. When $\mu > 2$ the limiting $\widetilde{P}_n(k)$ acquires the ordinary Gaussian form $\exp(-(n^{1/2}k)^2)$ but, when $\mu \le 2$, this becomes a "stretched" Gaussian with $(n^{1/2}k)^2$ replaced by $\left|n^{1/\mu}k\right|^\mu$. The difference between the Lévy and the distributions $P_n(l)$ with finite n becomes more pronounced when the step distribution $p^\mu(l)$ broadens its range as $\mu \to 0$. This can be more easily investigated in Fourier space by considering

$$\Delta\widetilde{P}(k, \mu, n) \equiv \exp(-Qn\left|k\right|^\mu) - \left[1 - \frac{1}{n}Qn\left|k\right|^\mu\right]^n, \quad k = O(n^{-1/\mu}). \tag{14}$$

The difference $\Delta\widetilde{P}$ vanishes for $n = 0$ and $n \to \infty$, and is maximum at

$$n_{\max} = \frac{\ln Q\left|k\right|^\mu - \ln\ln(1 - Q\left|k\right|^\mu)^{-1}}{Q\left|k\right|^\mu + \ln(1 - Q\left|k\right|^\mu)}, \tag{15}$$

for small k, $n_{\max} \simeq (Q\left|k\right|^\mu)^{-1}$, and $\Delta\widetilde{P}(k, \mu, n_{\max}) = e^{-1} - (1 - Q\left|k\right|^\mu)^{1/Q\left|k\right|^\mu}$. When $\mu > 1$, the difference $\Delta\widetilde{P}(k, \mu, n_{\max})$ is small almost everywhere, since $n_{\max}$ remains large ($q_{\max} = 1 - 1/n_{\max} \simeq 1$) within a broad neighborhood of $k = 0$. On the contrary, when $\mu < 1$ the difference $\Delta\widetilde{P}(k, \mu, n_{\max})$ is large almost everywhere, since $n_{\max}$ now remains small ($q_{\max} = 1 - 1/n_{\max} < 1$) in a wide neighborhood of $k = 0$. Notice, too, that the validity of the approximations in eq. (11) requires $k = O(n^{-1/\mu})$ with n large.

The Weierstrass walk in eq. (3) has been enlarged to an infinite family of walks $p_r(l)$ such that a class of them with an asymptotic power-law decay $\left|l\right|^{-\mu}$ would be attracted under an RG transformation to the Weierstrass walk as a nontrivial fixed point [21]. The expression for the distribution of single steps $p_r(l)$ for the walks in this family is

$$p_r(l; \{a_n\}) = \frac{A_r}{2} \sum_{n=0}^{r} a_n(\delta_{l,-b^n} + \delta_{l,b^n}), \tag{16}$$

where the set of step lengths b^n, $b > 1$, has been maintained, but now the probabilities assigned to them are proportional to arbitrary numbers $0 < a_n < 1$. Also, a range for the step lengths b^r has been introduced, but the possibility $r \to \infty$ is included; A_r normalizes $p_r(l)$, i.e., $A_r^{-1} = \sum_{n=0}^{r} a_n$. The elementary RG transformation $a_n' = R[a_n] \equiv aa_{n+1}$ can be applied to our family of walks. This transformation maps the sites $l = b^{n+1}$ into the sites $l' = b^n$ (eliminating intermediate lattice space between allowed step lengths) and renormalizes the step probability by a restoring factor, a. It is found that the Weierstrass walk $p^\mu(l)$ and the simple nearest-neighbor step walk $p_0(l)$ are both fixed points of R.

The first one is nontrivial in the sense that it is associated with an infinite-ranged step distribution that can be reached via the RG transformation only from other infinite-ranged step distributions $p_\infty(l)$ required to approach asymptotically the condition $a_n = a^{-n}$, $n \to \infty$. The distributions $p_\infty(l)$ make up the "critical hypersurface" and the quantities $\alpha_n \equiv a_n - a^{-n}$ are the irrelevant variables that vanish as R is repeatedly applied. The other fixed point $p_0(l)$ is trivial, since it is generated by the application of the RG transformation to any "noncritical" finite-ranged $p_r(l)$, $r < \infty$. It has been shown [21] that along the critical RG transformation flows, over which the irrelevant variables decrease in value, the entropy evolves toward either a maximum or a minimum (depending on the nature of an imposed constraint [17, 21]) at the fixed point where these variables vanish. The existing analogy [22, 23] between a random walk and the Ornstein-Zernike relation for the pair correlation functions in a fluid or magnet is employed to describe the critical phenomena in a lattice gas or Ising model. We find that the anomalous dimension η at criticality is simply related to the index μ of the Lévy lattice walk, and the parameter q in the Tsallis entropy is used as a measure of the nonextensivity associated to the non-Gaussian fixed point [21].

3   NONLINEAR MAPS

As has become manifest [31], the critical points of one-dimensional nonlinear dissipative maps are instances for which the predictions of the nonextensive generalization of the BG statistical mechanics can be examined [5, 16, 32] and, more recently, unambiguously corroborated [1, 2, 25]. (For a description of related issues in conservative maps see Baldovin [3].) The one-dimensional logistic map $f_\nu(x) = 1 - \nu |x|^2$, $-1 \le x \le 1$, and its generalization to nonlinearity of order $\jmath > 1$, exhibit several types of infinite sequences of such critical points as the control parameter ν which varies across the interval $0 \le \nu \le 2$. These sequences of critical points correspond to pitchfork bifurcations, chaotic-band splittings, and tangent bifurcations [26]. The pitchfork bifurcation is the mechanism for the successive doubling of periods of orbits, while the tangent bifurcation is a different mechanism linking periodic orbits with chaos in which intermittency is a precursor to periodic behavior. The accumulation point of the pitchfork bifurcations and band splittings is the Feigenbaum attractor that marks the dividing state between periodic and chaotic orbits, at $\nu_\infty = 1.40115\ldots$. The locations for the pitchfork bifurcations $\nu_n < \nu_\infty$ approach their limit according to $\nu_n - \nu_\infty \sim \delta^{-n}$ (n large) where $\delta = 0.46692\ldots$ is one of the two Feigenbaum universal constants. Similarly, the locations for band-splitting transitions $\widehat{\nu}_n > \nu_\infty$ follow $\nu_\infty - \widehat{\nu}_n \sim \delta^{-n}$ (n large) [26]. A measure of the amplitudes of the periodic orbits is defined by the diameters d_n of the "bifurcation forks" at the "super stable" periodic orbits of lengths 2^n that contain the point $x = 0$. These superstable orbits occur at $\overline{\nu}_n < \nu_\infty$ and also gather as $\overline{\nu}_n - \nu_\infty \sim \delta^{-n}$ (n large). The diameter $d_n \equiv f_{\overline{\nu}_n}^{(2^{n-1})}(0)$ is the iterate position in such 2^n-cycle that is closest to $x = 0$ and for large n these distances have constant ratios $d_n/d_{n+1} = -\alpha$, where $\alpha = 2.50290...$ is the second of the Feigenbaum constants [26]. A set of diameters with scaling properties similar to those of d_n can also be defined for the band-splitting sequence [26]. For clarity, we use here only absolute values of positions so that below d_n means $|d_n|$. At the tangent bifurcations the chaotic states for $\nu > \nu_\infty$ become interrupted at certain values ν_m by windows of periodic orbits of lengths $m \neq 2^n$ that, in turn, double, $m \cdot 2^n$, as ν varies. Also, periodic triplings $p \cdot 3^n$, quadruplings $p \cdot 4^n$, and so on, occur [26].

The scaling laws described above with indexes δ and α are universal, independent of the details of the map, and are, therefore, shared by all maps of the same order $\mathfrak{z}$. As with other statistical-mechanical systems, an explanation of this universality is provided via the RG method, which has been successfully applied to this type of iterated map. The Feigenbaum RG doubling transformation, consisting of functional composition and rescaling, was first devised to study the cascade of period-doubling transitions and its accumulation point, and was subsequently applied to the intermittency transition. It has been pointed out [26] that the latter case is one of the rare examples where the RG equations can be solved exactly. We have recently found [1, 25] that this analytic solution holds generally for both tangent *and* pitchfork bifurcations. By studying the time (the number of iterations) evolution of trajectories at these bifurcations, we have concluded that the physical property expressed by the RG solution is, in fact, the nonextensive entropy extrema associated with the fixed-point map. We have also analyzed [2] the universal properties related to the dynamics of iterates at the onset of chaos $\nu = \nu_\infty$ and found previously unrevealed scaling properties that are a factual confirmation of the generalized nonextensive theory. At this state, the most singular of the map critical points, the trajectories of the iterates, exhibit an elaborate structure that is governed by the Feigenbaum RG transformation.

The solution to the Feigenbaum RG recursion relation for the case of the tangent bifurcation [13, 26] is briefly stated as follows. For the transition to periodicity of order m, the mth composition $f^{(m)}$ of the original map f is considered and the neighborhood of one of the m points tangent to the line with unit slope is studied. A shift of the origin of coordinates to that point and expansion yields

$$f^{(m)}(x) = x + u\,|x|^z + O(|x|^z)\,, \qquad (17)$$

where $u > 0$ is the expansion coefficient. The exact RG fixed-point map $x' = f^*(x)$ was found to be

$$x' = x[1 - (z - 1)ux^{z-1}]^{-1/(z-1)}\,, \qquad (18)$$

as it satisfies $f^*(f^*(x)) = \alpha^{-1}f^*(\alpha x)$ with $\alpha = 2^{1/(z-1)}$ and has a power-series expansion in x that coincides with eq. (17) in the two lowest-order terms. (We have used $x^{z-1} \equiv |x|^{z-1}\,\mathrm{sgn}(x)$.) From eq. (18) we notice that the ratio x'/x is a q-exponential, with $q = z$; that is, $x'/x = \exp_z(ux^{z-1})$. In Baldovin and Robledo [1] and Robledo [25], it is observed that the same fixed-point solution is also applicable to the pitchfork bifurcations

$$df^{(2^{k-1})}(x)/dx\Big|_{x=0} = -1$$

of order $n = 2^k$, $k = 1, 2, \ldots$, provided that the sign of u is changed for $x > 0$. It is also indicated there that the power z appearing in eqs. (17) and (18) is different from the power $_3$ in the original map and the relationship between them is determined.

The dynamical properties at the pitchfork and tangent bifurcations follow from those of the static fixed-point map via the use of the following scaling property [1]

$$f^{*(l)}(x) = l^{-\frac{1}{z-1}} f^*(l^{\frac{1}{z-1}} x), \quad l = 1, 2, \ldots . \tag{19}$$

This property implies that, for a total number of iterations $t = ml$, $l = 1, 2, \ldots$, with sufficiently small initial x_0, and for m and, therefore, t large, the fixed-point map can be written as

$$x_t \equiv \left[f^{(m)} \right]^{(l)} (x_0) = x_0 \exp_z(ax_0^{z-1}), \tag{20}$$

where $a \equiv u/m$. It can be readily verified that eq. (20) satisfies the properties $dx_t/dt = ax_t^z$ and $dx_t/dx_0 = (x_t/x_0)^z$. Similarly, the effect of a perturbation of the form εx^{-p} to lowest order in ε [13, 26], can be expressed as a dynamical property. As shown previously [25], this is

$$x_t = x \exp_z(ax_0^{z-1}) - \frac{\varepsilon x_0^{z-p}}{z-1} \left[\exp_z(ax_0^{z-1}) \right]^z$$
$$+ \frac{\varepsilon x_0^{z-p}}{z-1} \left[\exp_z(ax_0^{z-1}) \right]^{z-p} . \tag{21}$$

The recurrence relation in eq. (21) remains invariant under iteration and rescaling by α when ε is multiplied by the factor $\gamma = 2^{(p-z+1)/(z-1)}$. By considering an ensemble of iterates, it is possible to evaluate the generalized entropy S_q along the RG flow. It is found [25] that the entropy S_q^* for the fixed-point map is maximum as the required value of p implies $\gamma > 1$ and the RG flow is away from the fixed-point, making the entropy decrease in value as the RG transformation is applied.

The sensitivity to initial conditions ξ_t is an all-important quantity in characterizing the dynamical behavior of nonlinear maps. It is defined as $\xi_t \equiv \lim_{\Delta x_0 \to 0}(\Delta x_t/\Delta x_0)$ where Δx_t is the position difference of two trajectories at time t with initial position difference Δx_0. Generically, $\xi_t = \exp(\lambda_1 t)$ where λ_1 is the so-called Lyapunov exponent that measures either periodic ($\lambda_1 < 0$) or chaotic behavior ($\lambda_1 > 0$). At each of the map critical points, the Lyapunov λ_1 exponent vanishes, and ξ_t for large iteration time t no longer obeys exponential behavior, displaying instead power-law behavior. In order to describe the dynamics at such critical points, the q-exponential expression

$$\xi_t = \exp_q(\lambda_q t) \tag{22}$$

containing a generalized Lyapunov exponent λ_q has been proposed [32] based on the nonextensive entropy of Tsallis S_q, and clearly, the standard exponential is recovered when $q \to 1$.

For the tangent and pitchfork bifurcations we [1] find that eq. (22) is obeyed exactly with the identifications [4]

$$q = 2 - \frac{1}{z} \text{ and } \lambda_q(x_0) = zax_0^{z-1}. \tag{23}$$

Moreover, by using the known form $\rho(x) \sim \left|x^{-(z-1)}\right|$ [11] for the invariant distribution of $f^{(m)}$ in eq. (17), we have that the average $\bar{\lambda}_q$ of $\lambda_q(x_0)$ over x_0 yields the piecewise constant

$$\bar{\lambda}_q = za \, \text{sgn}(x_0). \tag{24}$$

It is interesting to notice [25] that this average corresponds to the q-extension of the customary expression for the Lyapunov exponent λ_1, obtained as the average of $\ln|df(x)/dx|$ over $\rho(x)$. Here we have

$$\bar{\lambda}_q = \int dx \rho(x) \ln_q \left| \frac{df^{(m)}(x)}{dx} \right|. \tag{25}$$

For all the pitchfork bifurcations of any nonlinearity $\mathfrak{z}$ it is found that $z = 3$, $q = 5/3$, and $\lambda_q < 0$; therefore, ξ_t displays "weak" (power law) insensitivity [32] to initial conditions at both sides of the critical point. For all the tangent bifurcations of any nonlinearity $\mathfrak{z}$ it is found that $z = 2$ and $q = 3/2$. At the left-hand side of the tangent bifurcation points $\lambda_q < 0$, implying, in an analogy to the pitchfork bifurcations, a weak insensitivity to initial conditions. However, at the right-hand side of the bifurcation $\lambda_q > 0$, and this, together with $q > 1$, results in a "super strong" sensitivity to initial conditions, a sensitivity that grows faster than exponentially [1].

The celebrated RG fixed-point map static solution at the edge of chaos $\nu = \nu_\infty$ for $\mathfrak{z} = 2$ is obtained as a power series of a smooth unimodal transcendental function $g(x)$, the universal limit of $(-\alpha)^n f_{\bar{\nu}_{n+1}}^{(2^n)}(x/(-\alpha)^n)$ as $n \to \infty$ [26]. Here again, the dynamical properties can be obtained [2] from the same RG transformation. In Baldovin and Robledo [2] it is shown that the iterates at ν_∞ follow trajectories that proceed according to the entire period-doubling cascade that takes place for $\nu < \nu_\infty$. It was found that the positions of the trajectories are given by the diameters $d_n(\bar{\nu}_n)$ and other related iterate positions in the 2^n-supercycles. For example, the trajectory starting at $x_0 = 0$ is seen to consist of interwoven position subsequences that have power-law decay. Some of these subsequences are given by times of the form $t = 2^n + 2^{n-k}$, $k = 0, 1, \ldots$. In particular, the position values for the trajectory subsequence $t = 2^n$ are asymptotically given by $x_t = d_n$, $n \geq 0$. The entire trajectory approaches the origin $x_0 = 0$ progressively as n increases every time that $t = 2^n$, but in between the values 2^n and 2^{n+1}, it returns in an oscillatory manner toward $x_1 = 1$, repeating twice the positions visited in the previous cycle between 2^{n-1} and 2^n and introducing a new position between these two subcycles [2]. At the edge of chaos the sensitivity ξ_t is rigorously obtained as the q-exponential

$$\xi_t = \exp_q(\lambda_q t), \tag{26}$$

where $q = 1 - \ln 2 / \ln \alpha$ and $\lambda_q = \ln \alpha / \ln 2$. This corroborates the previously known value of q at ν_∞ for $\mathfrak{z} = 2$ [32, 16] and also determines for the first time λ_q at ν_∞ for $\mathfrak{z} = 2$. Both x_t/x_0 and ξ_t were found to satisfy the dynamical fixed-point relations $h(t) = \alpha h(h(t/\alpha))$, with $\alpha = 2^{-1/(q-1)}$ and $\alpha = 2^{1/(q-1)}$ respectively. The results are universal for all one-dimensional maps with a single maximum of order $\mathfrak{z} = 2$.

For the pitchfork and tangent bifurcations, the noncanonical dynamics displayed by ξ_t appear to arise from the tangency shape of the map at these critical points, causing trajectories to be either effectively trapped or expelled, causing a deficient sampling of the phase space $-1 \leq x \leq 1$. At the onset of chaos, the reduced phase subspace is represented by the strange attractor, a Cantor subset of the interval $-1 \leq x \leq 1$. In this case, the permissible positions are asymptotically captured by the attractor and this acts as a blockage for access to other locations.

4   GLASS FORMATION

Experimentally, the glass transition is observed as a dramatic slowing down in which the characteristic time for liquid structural relaxation increases many orders of magnitude within a small window of temperatures [6]. The energy landscape [6], which is also called the multidimensional surface that represents the dependence of the potential energy of a liquid system on the positions of all its N molecules, has become a key concept in the study of glass formation. This is because information on the landscape structure facilitates interpretations and predictions on dynamic behavior. A precise picture of the statistical properties of the landscape is required for its use as a means for determining the relaxation phenomena. The energy landscape is visualized as made of shallow potential energy minima or shallow basins that are vastly more numerous than deeper basins, and these rarer minima exhibit higher energy barriers as their depth increases. The way in which the system samples its landscape as a function of temperature provides information on its dynamical properties. Upon cooling, molecules rearrange so slowly that they cannot sample configurations in the time allowed by the process, suggesting that phase-space mixing is decelerated and that it is possibly only partially fulfilled during glass-forming dynamics.

An informative preview of our analysis [24] follows. Some relevant aspects of relaxation dynamics associated with supercooling and structural arrest in liquids can be predicted via simple assumptions for the topography and topology of the energy landscape at low energies. First, if the landscape is assumed to acquire a scaling property at low energies, we obtain a power-law (therefore non-Gaussian) distribution for molecular velocities. The scaling property develops when particles at the wells of, say, a Lennard-Jones potential, approximate an array of oscillators, whereas the resulting dynamical heterogeneity is observed in experiments and simulations [33]. Second, in line with recent numerical simulations [8] on small systems, the connectivity of the landscape minima is assumed to increase as energy decreases, leading to a network with scale-free properties. As energy lowers, the evolution of the system exhibits increasingly hindered accessibility to configurations, as this is a transient random walk in which only a decreasing fraction of the minima is visited. Third, if the barrier heights between minima increase above a threshold as energy decreases, the induced distribution of trapping times has a power-law decay and the associated first-passage time probabilities obey power-law or stretched-exponential forms. We argue that the predicted properties act in accordance with those expected by nonextensive statistical mechanics for systems with blocked phase-space accessibility.

As we have illustrated in the two previous sections by considering universality classes of random walks and iterated maps, when conditions arise so that there is accessibility to only a fraction of total phase space, a set that often has a fractal or multifractal geometrical structure, we make contact with nonextensive statistical properties [30, 31].

ACKNOWLEDGMENTS

I would like to thank F. Baldovin, C. Tsallis, and P. Tartaglia for useful discussions and comments. I gratefully acknowledge the hospitality of the Dipartimento di Fisica de la Università di Roma, "La Sapienza," where this assignment was carried out. This work was partially supported by INFM (Italy), CONACyT grant 34572-E (Mexico), and by DGAPA-UNAM grant IN110100 (Mexico).

REFERENCES

[1] Baldovin, F., and A. Robledo. "Sensitivity to Initial Conditions at Bifurcations in One-Dimensional Nonlinear Maps: Rigorous Nonextensive Solutions." *Europhys. Lett.* **60** (2002): 518.

[2] Baldovin, F., and A. Robledo. "Universal Renormalization-Group Dynamics at the Onset of Chaos in Logistic Maps and Nonextensive Statistical Mechanics." *Phys. Rev. E* **66** (2002): 045104-1 (R).

[3] Baldovin, F. "Numerical Analysis of Conservative Maps: A Possible Foundation of Nonextensive Phenomena." This volume.

[4] Buiatti, M., P. Grigolini, and A. Montagnini. "Dynamic Approach to the Thermodynamics of Superdiffusion." *Phys. Rev. Lett.* **82** (1999): 3383.

[5] Costa, U. M. S., M. L. Lyra, A. R. Plastino, and C. Tsallis. "Power-Law Sensitivity to Initial Conditions within a Logisticlike Family of Maps: Fractality and Nonextensivity." *Phys. Rev. E* **56** (1997): 245.

[6] De Benedetti, P. G., and F. H. Stillinger. "Supercooled Liquids and the Glass Transition." *Nature* **410** (2001): 267.

[7] de Moura, F. A. B. F., U. Tirnakli, and M. L. Lyra. "Convergence to the Critical Attractor of Dissipative Maps: Log-Periodic Oscillations, Fractality, and nonextensivity." *Phys. Rev. E* **62** (2000): 6361.

[8] Doye, J. P. K. Network Topoloty of a Potential Energy Landscape: A Static Scale-Free Network." *Phys. Rev. Lett.* **88** (2002): 238701-1.

[9] Feller, W. *An Introduction to Probability Theory and Its Applications*, vol. 2. 2d ed. New York: Wiley, 1971.

[10] Fisher, M. E. "Renormalization Group Theory: Its Basis and Formulation in Statistical Physics." *Rev. Mod. Phys.* **70** (1998): 653.

[11] Gaspard, P., and X. J. Wang. "Sporadicity: Between Periodic and Chaotic Dynamical Behaviors." *Proc. Natl. Acad. Sci. USA* **85** (1988): 4591.

[12] Gell-Mann, M. "Effective Complexity." This volume.

[13] Hu, B., and J. Rudnick. "Exact Solutions to the Feigenbaum Renormalization-Group Equations for Intermittency." *Phys. Rev. Lett.* **48** (1982): 1645.

[14] Hughes, B. D., M. F. Shlesinger, and E. W. Montroll. "Random Walks with Self-Similar Clusters." *Proc. Natl. Acad. Sci. USA* **78** (1981): 3287.

[15] Lyra, M. L. "Nonextensive Entropies and Sensitivity to Initial Conditions of Complex Systems." This volume.

[16] Lyra, M. L., and C. Tsallis. "Nonextensivity and Multifractality in Low-Dimensional Dissipative Systems." *Phys. Rev. Lett.* **80** (1998): 53.

[17] Mayoral-Villa, E. Principios Variacionales y Grupo de Renormalización en el Estudio de la Percolación en una Dimensión." M.Sc. Thesis, Universidad Nacional Autonóma de México, 2001, unpublished.

[18] Montroll, E. W., and M. F. Shlesinger. "On the Wonderful World of Random Walks." In *Studies in Statistical Mechanics*, edited by J. L. Lebowitz and E. W. Montroll, vol. 11, 1–121. Amsterdam: North Holland, 1984.

[19] Montroll, E. W., and M. F. Shlesinger. "Maximum Entropy Formalism,Fractals, Scaling Phenomena, and1/fnoise: A Tale of Tails." *J. Stat. Phys.* **32** (1983): 209.

[20] Montroll, E. W., and G. H. Weiss. "Random Walks on Lattices. II." *J. Math. Phys.* **6** (1965): 167.

[21] Robledo, A. "Renormalization Group, Entropy Optimization, and Nonextensivity at Criticality." *Phys. Rev. Lett.* **83** (1999): 2289.

[22] Robledo, A., and I. E. Farquhar. "Random-Walk Theory and Ornstein-Zernike Systems with Extended-Core Potentials." *J. Chem. Phys.* **61** (1974): 1594.

[23] Robledo, A., and I. E. Farquhar. "Random-Walk Theory and Correlation Functions in Classical Statistical Mechanics." *Physica A* **84** (1976): 435.

[24] Robledo, A. "Nonextensive Properties along the super-Cooled Liquid Route to the Vitreous State." Workshop on Relaxation in Complex Systems, Anacapri, Italy, May 23–24, 2002.

[25] Robledo, A. "The Renormalization Group and Optimization of Nonextensive Entropy: Criticality in nonlinear One-Dimensional Maps." *Physica A* **314** (2002): 437.

[26] Schuster, H. G. *Deterministic Chaos. An Introduction.* Revised, 2d ed. Weinheim: VCH Publishers, 1988.

[27] Shlesinger, M. F., and B. D. Hughes. "Analogs or Renormalization Group Transformations in Random Processes." *Physica A.*

[28] Stinchcombe, R. "Nonequilibrium Systems." This volume.

[29] Tsallis, C. "Nonextensive Statistical Mechanics: Construction and Physical Interpretation." This volume.

[30] Tsallis, C. "Possible Generalization of Boltzmann-Gibbs Statistics." *J. Stat. Phys.* **52** (1988): 479.

[31] Tsallis, C. "Nonextensive Statistical Mechanics and Thermodynamics: Historical Background and Present Status." In *Nonextensive Statistical Mechanics and Its Applications*, edited by S. Abe and Y. Okamoto, 3. Lecture Notes in Physics, vol. 3. Berlin: Springer, 2001.

[32] Tsallis, C., A. R. Plastino, and W.-M. Zheng. "Power-Law Sensitivity to Initial Conditions—New Entropic Representation." *Chaos, Solitons and Fractals* **8** (1997): 885.

[33] Weeks, E. R. , J. C. Crocker, A. C. Levitt, A. Schofield, and D. A. Weitz. "Three-Dimensional Direct Imaging of Structural Relazation near the Colloidal Glass Transition." *Science* **287** (2000): 627.

Nonextensive Entropies and Sensitivity to Initial Conditions of Complex Systems

Marcelo L. Lyra

Tsallis generalized statistics has been successfully applied to describe some relevant features of several natural systems exhibiting a nonextensive character. It is based on an extended form for the entropy, namely $S_q = (1 - \sum_i p_i^q)/(q-1)$, where q is a parameter that measures the degree of nonextensivity ($q \rightarrow 1$ for the traditional Boltzmann-Gibbs statistics). A series of recent works have shown that the power-law sensitivity to initial conditions in a complex state provides a natural link between the q-entropic parameter and the scaling properties of dynamical attractors. These results contribute to the growing set of theoretical and experimental evidences that Tsallis statistics can be a natural frame for studying systems with a fractal-like structure in phase space. Here, the main ideas underlying this relevant aspect are reviewed. The starting point is the weak sensitivity to initial conditions exhibited by low-dimensional dynamical systems at the onset of chaos. It is shown how general scaling arguments can provide a direct relation between the entropic index q and the scaling exponents associated with the multifractal critical attractor.

Nonextensive Entropy—Interdisciplinary Applications
edited by Murray Gell-Mann and Constantino Tsallis, Oxford University Press

These works shed light in the elusive problem concerning the connection between the q-entropic parameter of Tsallis statistics and the underlying microscopic dynamics of nonextensive systems.

1   INTRODUCTION

Inspired on the scaling properties of multifractals, Tsallis introduced a generalized entropy with the aim of extending the usual statistical mechanics and thermodynamics [37]. The postulated generalized form for entropy reads:

$$S_q = k\frac{1 - \sum_{i=1}^{W} p_i^q}{q - 1} \;\; ; \;\; \sum_{i=1}^{W} p_i = 1 \; , \tag{1}$$

where k is a positive constant depending on the particular units to be used (we will consider $k = 1$ hereafter) and $q \in \mathcal{R}$ is the parameter which characterizes a particular statistics (the usual Boltzmann-Gibbs entropy is recovered in the limit of $q \to 1$). The above generalization has been shown to retain most of the formal structure of the standard theory as, for example, Legendre thermodynamic structure, H-theorem, Nyquist and Onsager reciprocity theorem, Kramers and Wannier relations, thermodynamic stability, among others [9, 10, 12, 13, 14, 27, 31, 38]. Further, for equiprobability, it is a continuous monotonic increasing function of W and therefore can still be used as a measure of disorder. However, it satisfies a pseudo-additivity rule $S_q(A + B) = S_q(A) + S_q(B) + (1 - q)S_q(A)S_q(B)$ and in this sense is intrinsically nonextensive with the entropic index q controlling the degree of nonextensivity.

Tsallis statistics has been successfully applied to a variety of systems with underlying nonextensivity. These belong to two main classes: (i) systems with long-range interactions, for example gravitational and Coulomb forces; and (ii) systems with long-range correlations such as systems at thermodynamic equilibrium at the vicinity of a second-order phase transition and open dynamical systems exhibiting self-organized criticality. The distributions derived from the generalized entropy postulate display power-law asymptotic behaviors. This property gives support for arguments suggesting that the q-generalized statistics may refer to a scale-invariant statistical mechanics; i.e., it would be the proper frame for studying systems with a fractal-like structure in phase space [1]. Such conjecture is corroborated by the success of Tsallis' prescription in describing features such as Hurst exponents for anomalous diffusion and Lévy flight distributions [40]. However, a full and general understanding of the precise relationship between the entropic index q and the underlying fractality in phase space is still lacking. This task is quite difficult to tackle even numerically because most of the physical systems of interest have many degrees of freedom and, consequently, a high-dimensional phase space.

A simple way to investigate possible relations between fractal scaling exponents in phase space and the nonextensive entropic index q can be provided by studying the so-called thermodynamic behavior of low-dimensional nonlinear maps. These models can describe a great variety of systems with few degrees of freedom. Their typical behavior include the occurrence of bifurcation instabilities, long-range correlated sequences, fractal structures, and chaos, which are commonly observed in a great variety of systems ranging from fluids, magnetism, biology, social sciences, and many others [28]. The underlying nonlinearity can induce these systems to exhibit complicated behaviors with quite structured paths in phase space. Usually they present a transition between phases of periodic and chaotic orbits. At the onset of chaos these maps present features quite similar to those of high-dimensional systems at criticality, namely, space-time scale invariance, critical slowing down, and fractality with universal exponents that are independent of the detailed model and determined only by some relevant symmetries.

This chapter will review some recent works which have explored the simplicity of low-dimensional maps poised at criticality to reveal some interesting relations connecting the index q of nonextensive dynamical entropies and geometric exponents characterizing the scaling properties of critical attractors. This chapter is organized as follows. Section 2 will review the main characteristics of the dynamical attractor of the logistic map, which will be used as a prototype map to the subsequent analysis, with a discussion of the sensitivity to the initial conditions problem. In section 3, it will be shown how the sensitivity to initial conditions can be represented in an entropic form and how a generalized nonextensive entropy emerges as the natural representation of the critical slowing down at the onset of chaos. In section 4, it will be shown how the index q of the dynamical nonextensive entropy characterizing the mixing dynamics can be related to geometric scaling exponents. Section 5 extends the above analysis to the equilibration dynamics. In section 6 some remarks concerning recent similar studies in Hamiltonian maps are discussed before we conclude giving some perspective on possible future extensions to higher-dimensional complex systems.

2   SENSITIVITY TO INITIAL CONDITIONS OF ONE-DIMENSIONAL DISSIPATIVE MAPS

The study of nonlinear systems' sensitivity to initial conditions is one of the most important tools used to investigate the nature of the phase-space attractor. It is usually characterized by the Lyapunov exponent λ, defined for the simple case of a one-dimensional dynamical variable x as

$$\xi(t) \equiv \lim_{\Delta x(0) \to 0} \frac{\Delta x(t)}{\Delta x(0)} \equiv \frac{dx(t)}{dx(0)} = e^{\lambda t} \quad , \tag{2}$$

where the sensitivity function $\xi(t)$ measures how the distance $\Delta x(0)$ between two initially nearby orbits, or in an equivalent point of view, the uncertainty on the precise initial condition evolves in time. If $\lambda > 0$, the system is said to be strongly sensitive on the initial condition with the uncertainty on the dynamical variable growing exponentially in time and this characterizes a chaotic motion in the phase space. On the other hand, if $\lambda < 0$, the system becomes strongly insensitive to the initial condition that is expected for any state whose dynamical attractor is an orbit with a finite period. The above features can be well illustrated using the standard logistic map

$$x_{t+1} = 1 - ax_t^2 \tag{3}$$

with $x_t \in [-1, 1]$; $a \in [0, 2]$; $t = 0, 1, 2, ...$ The dynamical attractor as a function of a is shown in figure 1(a). For small a it exhibits periodic orbits that bifurcate as a increases and the bifurcation points accumulate at a critical value $a_c = 1.40115518909...$ above which chaotic orbits emerge. The Lyapunov exponent λ as a function of the parameter a is displayed in figure 1(b). The predicted trend, i.e., $\lambda < 0$ ($\lambda > 0$), for periodic (chaotic) orbits is clearly observed. Notice that $\lambda = 0$ describes indistinctly the bifurcation points and chaos threshold.

3   THE ENTROPIC REPRESENTATION OF THE SENSITIVITY TO INITIAL CONDITIONS

A quite instructive approach for the problem of sensitivity to initial conditions is the one introduced by Kolmogorov and Sinai. They quantified the intuitive concept that the sensitivity to initial conditions is closely related to a process of loss of information about the system as time goes on. To characterize this loss of information, KS introduced the so-called Kolmogorov-Sinai entropy K which basically is the increase per unit time of the Boltzmann-Gibbs-Sinai entropy $S = -\sum_{i=1}^{W} p_i \ln p_i$, where W is the total number of possible configurations and $\{p_i\}$ their associated probabilities. A precise expression for K can be obtained by considering the evolution of an ensemble of identical copies of the system. The probabilities p_i can be defined as the fractional number of points of the ensemble that are in the ith cell of a suitable partition of the phase space. The K-entropy can therefore be written as

$$K \equiv \lim_{\tau \to 0} \lim_{l \to 0} \lim_{N \to \infty} \frac{1}{N\tau}[S(N) - S(0)] \ , \tag{4}$$

where $S(0)$ and $S(N)$ are the entropies of the system evaluated at times $t = 0$ and $t = N\tau$ (for maps $\tau = 1$). With the simplifying assumption that at time t there are $W(t)$ occupied cells with the same occupation number, we have from eq. (4) that

$$W(t) = W(0)e^{Kt} , \tag{5}$$

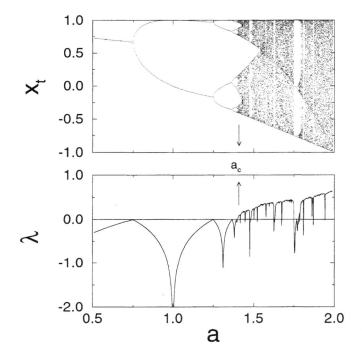

FIGURE 1   (a) The dynamical attractor of the logistic map as a function of the parameter a. The attractor exhibits a series of bifurcations as a increases that accumulate at
$a_c = 1.40115518909...$, above which chaotic orbits emerge; (b) The Lyapunov exponent
λ versus a. Notice that $\lambda < 0$ for periodic orbits, $\lambda > 0$ for chaotic orbits and $\lambda = 0$ at
bifurcation and critical points. Strong fluctuations of λ for $a > a_c$ reflects the presence
of periodic windows at all scales.

which is equivalent to eq. (2) for the sensitivity to initial condition and provides
the well-known Pesin equality $K = \lambda$ [22].

However, figure 1 does not suitably describe the sensitivity to initial conditions at bifurcation points and at the threshold to chaos which are the marginal
cases where $\lambda = 0$. At these points, the BG entropy does not vary at a constant
rate and therefore does not provide a useful tool to characterize the rhythm of
information loss or recovery. The failure of the above prescription to characterize these points is related to the fact that the extensive BG entropy can not
properly deal with the underlying fractality (and therefore nonextensivity) of
the phase-space attractor.

In figure 2, the sensitivity function at bifurcation and chaos threshold points
are shown to exhibit a power-law evolution in time. Such power-law sensitivity

has been shown to be naturally derived from the assumption that a proper nonextensive entropy exhibits a constant variation rate at these points [11, 41]. Namely, using Tsallis entropy form

$$S_q = \frac{1 - \sum_{i=1}^{W} p_i^q}{q-1} , \tag{6}$$

a generalized Kolmogorov-Sinai entropy can be defined as

$$K_q \equiv \lim_{\tau \to 0} \lim_{l \to 0} \lim_{N \to \infty} \frac{1}{N\tau} [S_q(t) - S_q(0)] . \tag{7}$$

Assuming equiprobability and a constant K_q, it can be readily obtained that, starting with an initial ensemble concentrated in a single box, the volume on the phase space shall evolve in time as

$$W(t) = [1 + (1-q)K_q t]^{1/(1-q)} \tag{8}$$

consistent with the asymptotic power-law behavior at marginal points where $\lambda = 0$. Assuming a generalized Pesin equality $K_q = \lambda_q$, we can also write the sensitivity function within the present formalism as

$$\xi(t) = [1 + (1-q)\lambda_q t]^{1/(1-q)} . \tag{9}$$

The above relation provides a direct relationship between the entropic index q and the sensitivity power-law exponent. For $q > 1$ the system becomes weakly insensitive to the initial conditions once the visited volume on the phase space slowly shrinks as the system evolves in time. This is the case for period-doubling bifurcation points of the logistic map where $1/(1-q) = -3/2$ and therefore $q = 5/3$. On the other hand, for $q < 1$ the system becomes weakly sensitive to the initial conditions as $W(t)$ slowly grows with time. This is observed at the onset of chaos of the standard logistic map, where it was obtained $1/(1-q) = 1.325$ and therefore $q = 0.2445$ [41].

The close relationship between the entropic index q of Tsallis entropies and the sensitivity to initial conditions at the onset of chaos of such nonlinear low-dimensional dissipative maps provides a useful recipe to estimate the proper entropic index from the system dynamical rules. This relationship has been further used to investigate a recent conjecture that the nonextensive Tsallis statistics is the natural framework for studying systems with a fractal-like structure in the phase space [1]. The critical dynamical attractor of such nonlinear dissipative systems can be associated with a multifractal measure whose scaling exponents can be obtained from traditional methods. Therefore, both the entropic index q and the scaling properties of the critical attractor can be estimated independently and their relation revealed.

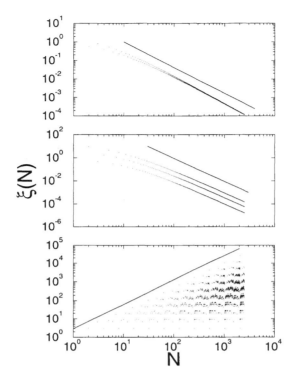

FIGURE 2 $\xi(N)$ versus N. From top to bottom $a = 3/4$ (period-doubling point); $a = 7/4$ (tangent bifurcation point); and $a = 1.4011...$ (onset of chaos). Straight lines give the asymptotic slopes $1/(1 - q)$. These imply in $q = 5/3$ for the period-doubling point and $q = 3/2$ for the tangent bifurcation point (power-law insensitivity). Power-law sensitivity takes place at the onset of chaos with $q = 0.2445$.

4   MIXING NONEXTENSIVE ENTROPY AND MULTIFRACTALITY

In order to completely describe the scaling behavior of critical dynamical attractors it is necessary to introduce a multifractal formalism [20]. The partition function $\chi_Q(N) = \sum_{i=1}^{N} p_i^Q$ is a central quantity within this formalism, where p_i represents the probability (integrated measure) on the ith box among the N boxes of the measure (we use Q instead of the standard notation q in order to avoid confusion with the entropic index q).

In chaotic systems p_i is the fraction of times the trajectory visits the box i. In the $N \to \infty$ limit, the contribution to $\chi_Q(N) \propto N^{-\tau(Q)}$, with a given Q, comes from a subset of all possible boxes, whose number scales with N as $N_Q \propto$

$N^{f(Q)}$, where $f(Q)$ is the fractal dimension of the subset ($f(Q = 0)$ is the fractal dimension d_f of the support of the measure). The content on each contributing box is roughly constant and scales as $P_Q \propto N^{-\alpha(Q)}$. These exponents are all related by a Legendre transformation

$$\tau(Q) = Q\alpha(Q) - f(Q), \tag{10}$$

$$\alpha(Q) = \frac{d}{dQ}\tau(Q). \tag{11}$$

The multifractal object is then characterized by the continuous function $f(\alpha)$, which reflects the different dimensions of the subsets with singularity strength α. $f(\alpha)$ is usually shaped like an asymmetric $\cap$. The α values at the end points of the $f(\alpha)$ curve are the singularity strengths associated with the regions in the set where the measure is most concentrated ($\alpha_{\min} = \alpha(Q = +\infty)$) and most rarefied ($\alpha_{\max} = \alpha(Q = -\infty)$). The end points of the $f(\alpha)$ curves can be inferred theoretically from well-known scaling properties related to the most concentrated and most rarefied intervals in the attractor. Feigenbaum has shown that, after $N = \omega^n$ iterations (ω is a natural scaling factor inherent to the recursive relations), the size of these intervals scale respectively as $l_{-\infty} \sim [\alpha_F]^{-n}$ and $l_{+\infty} \sim [\alpha_F(z)]^{-zn}$, where α_F is a universal scaling factor [18] and z is the inflexion at the vicinity of the extremal point of the map. Since the measures in each box are simply $p_i = 1/N = \omega^{-n}$, the end points are expected to be

$$\alpha_{\max} = \ln p_i / \ln l_{+\infty} = \frac{\ln \omega}{\ln \alpha_F(z)}, \tag{12}$$

$$\alpha_{\min} = \ln p_i / \ln l_{-\infty} = \frac{\ln \omega}{z \ln \alpha_F(z)}. \tag{13}$$

Scaling argumentsindexnonextensive entropy applied to the most rarefied and most concentrated regions of the attractor provide a precise relationship between the singularity spectrum extremals and the entropic index q [25, 26]. Considers an ensemble of identical systems whose initial conditions spread over a region of the order of the typical box size in the most concentrated region $l_{+\infty}$. In other words, considers that the uncertainty on the precise initial conditions is $\Delta x(t = 0) \sim l_{+\infty}$. After N time steps these systems will be spread over a region whose size is at most of the order of the typical size of the boxes in the most rarefied region ($\Delta x(N) \sim l_{-\infty}$). Therefore, assuming power-law sensitivity on the initial conditions on the critical state, we can write for large N

$$\xi(N) \equiv \lim_{\Delta x()) \to 0} \frac{\Delta x(N)}{\Delta x(0)} = \frac{l_{-\infty}}{l_{+\infty}} \sim N^{1/(1-q)}, \tag{14}$$

and using eqs. (12) and (13), it follows immediately that

$$\frac{1}{1-q} = \frac{1}{\alpha_{\min}} - \frac{1}{\alpha_{\max}}. \tag{15}$$

The above relation indicates that the proper nonextensive statistics can be inferred from the knowledge of the scaling properties associated with the extremal sets of the dynamical attractor. This relation follows from very usual and general scaling arguments and therefore shall be applicable for a large class of nonlinear dynamical systems irrespective of the underlying topological and metrical properties.

The above relation has been numerically observed to hold with very high accuracy for the critical attractors of the family of generalized logistic maps [26]

$$x_{t+1} = 1 - a|x_t|^z \; ; (z > 1 \; ; \; 0 < a < 2 \; ; \; t = 0, 1, 2, \dots \; ; \; x_t \in [-1, 1]), \qquad (16)$$

where z is the inflexion of the map in the neighborhood of the extremal point $\bar{x} = 0$. These maps are well known [3, 21] to have the topological properties (such as the sequence of bifurcations while varying the parameter a) independent of z, but the metric properties (such as Feigenbaum's exponents and multifractal singularity spectra of the attractors) do depend on z.indexnonextensive entropy

The scaling relation has also been checked and holds for the family of circular maps [35]

$$\theta_{t+1} = \Omega + \left[\theta_t - \frac{K}{2\pi} \sin(2\pi\theta_t)\right]^{z/3}, \qquad \text{mod}(1). \qquad (17)$$

with $0 < \Omega < 1 \; ; \; 0 < K < \infty$. These maps describes dynamical systems possessing a natural frequency ω_1 which are driven by an external force of frequency ω_2 ($\Omega \equiv \omega_1/\omega_2$ is known as the *bare* winding number). These systems tend to mode lock at a frequency ω_1^*, and $\omega \equiv \omega_1^*/\omega_2$ is known as the *dressed* winding number. For $K < 1$ these maps are linear at the vicinity of the extremal point and exhibit only periodic orbits. And $K = 1$ is the onset value above which chaotic orbits exist. Once mode-locked, the *dressed* winding number remains constant and rational for a small range of *bare* winding numbers. Quasi-periodic orbits are produced at special irrational *dressed* winding number. The best studied one equals the golden mean $\omega_{GM} = (\sqrt{5} - 1)/2$ [23].

It is worth mentioning that the above two family of maps belong to distinct universality classes and therefore exhibit distinct scaling behavior for the same value of the inflexion z. The multifractal singularity spectra for these two families were numerically obtained and the extremal values of the singularity strength α_{min} and α_{max} estimated for a wide range of z values (see fig. 3). From the power-law exponent of the sensitivity function the value of $1/(1-q)$ could be independently estimated. In table 1, the numerically obtained values of the extremal values of the singularity spectrum together with the q_{mix} value obtained from the sensitivity function power-law behavior are reported and show that the scaling relation (17) holds irrespective to the value of the inflexion z. In table 2, the numerically computed values of the corresponding critical *bare* winding number, the extremal values of the singularity spectrum and q_{mix} from the sensitivity function. Again, the scaling relation (17) holds with high accuracy for a wide range of inflexions z. Recently, it has been shown that the proposed scaling

TABLE 1   Mixing dynamics and fractality of the z-generalized family of logistic maps. Numerical values correspond to: (i) map inflexion z; (ii) the critical parameter at the onset of chaos a_c; (iii) α_{min}; (iv) α_{max}; (v) a_{mix} as predicted by the scaling relation; and (vi) a_{mix} from the sensitivity function.

z	a_c	α_{min}	α_{max}	q_{mix} (scaling)	q_{mix} (sensitivity)
1.10	1.1249885...	0.302	0.332	-2.34 ± 0.02	-2.33 ± 0.02
1.25	1.2095137...	0.355	0.443	-0.79 ± 0.01	-0.78 ± 0.01
1.50	1.2955099...	0.380	0.568	-0.15 ± 0.01	-0.15 ± 0.01
1.75	1.3550607...	0.383	0.667	0.10 ± 0.01	0.11 ± 0.01
2.00	1.4011551...	0.380	0.755	0.23 ± 0.01	0.24 ± 0.01
2.50	1.4705500...	0.367	0.912	0.39 ± 0.01	0.39 ± 0.01
3.00	1.5218787...	0.354	1.054	0.47 ± 0.01	0.47 ± 0.01
5.00	1.6455339...	0.315	1.561	0.61 ± 0.01	0.61 ± 0.01

TABLE 2   Mixing dynamics and fractality of the z-generalized family of circle maps. Numerical values correspond to (i) map inflexions z; (ii) the critical parameter at the onset of chaos a_c; (iii) α_{min}; (iv) α_{max}; (v) q_{mix} as predicted by the scaling relation, and (vi) q_{mix} from the sensitivity function.

z	Ω_c	α_{min}	α_{max}	q_{mix} (scaling)	q_{mix} (sensitivity)
3.0	0.606661063469...	0.632	1.895	0.05 ± 0.01	0.05 ± 0.01
3.5	0.629593799039...	0.599	2.097	0.16 ± 0.01	0.15 ± 0.01
4.0	0.648669091983...	0.572	2.289	0.24 ± 0.01	0.24 ± 0.01
4.5	0.664861001064...	0.542	2.440	0.30 ± 0.01	0.30 ± 0.01
5.0	0.678831756505...	0.516	2.581	0.36 ± 0.01	0.36 ± 0.01
5.5	0.691048981515...	0.491	2.701	0.40 ± 0.01	0.40 ± 0.01
6.0	0.701853340894...	0.473	2.838	0.43 ± 0.01	0.44 ± 0.01
7.0	0.720182442561...	0.438	3.065	0.49 ± 0.01	0.50 ± 0.01
8.0	0.735233625356...	0.410	3.280	0.53 ± 0.01	0.53 ± 0.01

relation also holds for asymmetric maps with distinct inflexions on the left and right side of the inflexion point [36].indexnonextensive entropy

5   EQUILIBRATION NONEXTENSIVE ENTROPY AND FRACTALITY

It is well known that the temporal evolution of critical systems can present distinct regimes. At the usual second-order phase transition, for example, short-time

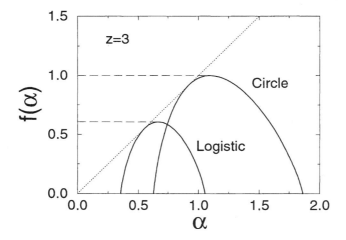

FIGURE 3 Multifractal singularity spectra of the critical attractor of generalized logistic and circle maps with $z = 3$. The maximum of the $f(\alpha)$ curves gives the fractal dimension of the support ($d_f = 0.605$ for the ($z = 3$)-logistic map and $d_f = 1$ for the circle map). The curves were obtained after $N = 2048$ (logistic) and $N = 2584$ (circle) iterations starting from the extremal point $\bar{x} = 0$ and following Halsey et al. prescription.

and long-time dynamics are described by different and independent power-law exponents. Similarly, the dynamics of critical maps, within an entropic representation, can be dependent on the particular initial ensemble. Although some scaling laws were found for an ensemble of initial conditions concentrated around the map inflexion point, these are usually not universal with respect to a general ensemble.

This section deals with the critical temporal evolution of the volume of the phase space occupied by an ensemble of initial conditions spread over the entire phase space of the one-dimensional maps described in the previous section. This ensemble is expected to contract towards the critical attractor as the systems relax to the equilibrium distribution in phase space and we are going to explore the parametric dependence of the dynamical exponent on the fractal dimension of the critical attractor. In practice, a partition of the phase space on N_{box} cells of equal size is performed and a set of N_c identical copies of the system is followed whose initial conditions are uniformly spread over the phase space. The ratio $r = N_c/N_{\text{box}}$ is a control parameter giving the degree of sampling of the phase space.

Within the nonextensive Tsallis statistics, there shall exist a proper entropy S_q evolving at a constant rate. Assuming again equiprobability, the phase-space

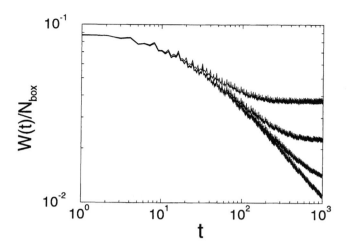

FIGURE 4   The volume occupied by the ensemble $W(t)$ (number of occupied boxes) as a function of discrete time in the standard logistic map ($z = 2$) and with sampling ratio $r = 0.1$. From top to bottom $N_{\text{box}} = 2000, 8000, 32000, 128000$.

volume occupied by the ensemble decreases as

$$W(t) = [W(0)^{1-q} + (1-q)K_q t]^{1/(1-q)} \tag{18}$$

where $K_q < 0$ and the exponent $\mu = 1/(q-1) > 0$ governs the asymptotic power-law decay.

In figure 4, the results are shown for $W(t)/N_{\text{box}}$ in the standard logistic map with inflexion $z = 2$ and from distinct partitions of the phase space with sampling ratio $r = 0.1$. Observe that, after a short transient period when $W(t)$ is nearly constant, a power-law contraction of the volume occupied by the ensemble sets up. $W(t)$ saturates at a finite fraction corresponding to the phase-space volume occupied by the critical attractor on a given finite partition. The saturation is postponed when a finer partition is used once the fraction occupied by the critical attractor vanishes in the limit $N_{\text{box}} \to \infty$.

In figure 5, the behavior of $W(t)/N_{\text{box}}$ is illustrated for a given fine partition of the phase space and distinct sampling ratios r. Notice that the crossover regime to the power-law scaling is quite short for large values of r so that a clear power-law scaling regime sets up even at early times. This feature is consistent with eq. (18) which states that the crossover time τ scales as $\tau \sim 1/W(0)^{q-1}$. Further, the scaling regime exhibits log-periodic oscillations once the multifractal nature of the critical attractor is closely probed by such dense ensemble. A general form for $W(t)$ reflecting the discrete scale invariance of the attractor can be written as

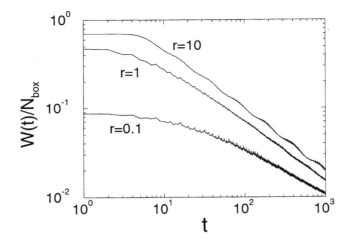

FIGURE 5   The volume occupied by the ensemble $W(t)$ as a function of discrete time in the standard logistic map ($z = 2$) and for a partition containing $N_{box} = 128000$ cells. Notice the emergence of log-periodic oscillations for large sampling ratios.

$$W(t) = t^{-\mu} P\left(\frac{\ln t}{\ln \Lambda}\right) \tag{19}$$

where P is a function of period unity and Λ is the characteristic scaling factor between the periods of two consecutive oscillations. Log-periodic modulations correcting a pure power law have been found in several systems exhibiting discrete scale invariance [32], for example, diffusion-limited aggregation [33], crack growth [7], earthquakes [29], and financial markets [16]. It has also been observed in thermodynamic systems with a fractal-like energy spectrum [39, 42].

The critical exponent μ has been measured as a function of the map inflexion z [15]. The results are summarized in table 3. It is a decreasing function of z. The volume occupied by the ensemble depicts a fast contraction for $z \sim 1$ where the fractal dimension is small. On the other side, a very slow contraction is observed for large values of z, pointing towards a saturation or at most to a logarithmic decrease of $W(t)$ in the limit of dense attractors. The exponent governing the mixing dynamics of an ensemble of initial conditions concentrated around the inflexion point exhibits a reversed trend. Although scaling arguments have shown that the mixing exponent can be written in terms of scaling exponents characterizing the extremal sets in the attractor, a similar scaling relation between the equilibration exponent μ and the multifractal singularity spectrum is still missing. However, it was observed that, when plotted against the fractal dimension of the attractor as shown in figure 6, the equilibration dynamical exponent μ is

TABLE 3   Equilibration Dynamics and Fractality of logistic and circle maps. Numerical values, within the z-generalized family of logistic maps, of: (i) the dynamic exponent μ governing the contraction towards the critical attractor of the uniform ensemble; (ii) the entropic index q_{eq} of the proper Tsallis entropy decreasing at a constant rate; and (iii) the fractal dimension d_f of the critical attractor. These values also hold for the generalized periodic maps. The last line represents our results for the z-generalized circle maps.

z	$\mu = 1/(q_{eq} - 1)$	q_{eq}	d_f
1.10	1.62 $\pm$ 0.02	1.62 $\pm$ 0.01	0.32 $\pm$ 0.02
1.25	1.23 $\pm$ 0.01	1.81 $\pm$ 0.01	0.40 $\pm$ 0.01
1.5	0.95 $\pm$ 0.01	2.05 $\pm$ 0.01	0.47 $\pm$ 0.01
1.75	0.80 $\pm$ 0.01	2.25 $\pm$ 0.015	0.51 $\pm$ 0.01
2.0	0.71 $\pm$ 0.01	2.41 $\pm$ 0.02	0.54 $\pm$ 0.01
2.5	0.59 $\pm$ 0.01	2.70 $\pm$ 0.02	0.58 $\pm$ 0.01
3.0	0.515 $\pm$ 0.005	2.94 $\pm$ 0.02	0.60 $\pm$ 0.01
5.0	0.395 $\pm$ 0.005	3.53 $\pm$ 0.03	0.66 $\pm$ 0.01
z-circular maps			
	0.0	∞	1.0

very well fitted by $\mu \propto (1 - d_f)^2$, which indicates d_f as the relevant geometric exponent coupled to the dynamics of the uniform ensemble. The same dynamic exponents were obtained for the generalized periodic maps which belong to the same universality class of logisticlike maps.

The above result indicates that a slow convergence to the critical attractor shall be expected for dense critical attractors. However, it is not clear how this convergence takes place when the dynamical attractor fills the phase space with a multifractal probability density as occurs for the one-dimensional critical circle map. In figure 7 we show results for the temporal evolution of the phase-space volume occupied by an ensemble of initial conditions uniformly spread over the circle. The ensemble $W(t)$ exhibits a rich pattern which resembles the one observed for the sensitivity function associated to the expansion of the phase space from initial conditions concentrated around the inflexion point. However, $W(t)$ does not present any power-law regime. Instead, the lower bounds display a slow logarithmic decrease with time, saturating at a finite volume fraction. The saturation is a feature related to the finite partition used in the numerical calculation. This minimum decreases logarithmically with the number of cells in the phase space. The same behavior is observed for the whole family of generalized circle maps with an arbitrary inflexion z. The critical attractors within this family have all $d_f = 1$ although they exhibit a z-dependent multifractal singularity spectra. The z-independent scenario for $W(t)$ corroborates the conjecture that d_f is the relevant geometric exponent coupled to the equilibration dynamics.

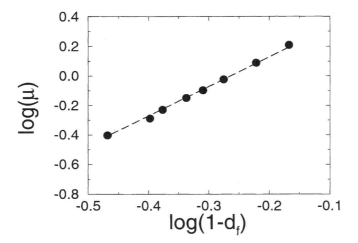

FIGURE 6 $\log_{10}(\mu)$ versus $\log_{10}(d_f)$. The parametric dependence of the dynamic exponent μ with the fractal dimension d_f of the critical attractor is very well fitted to the form $\mu \propto (1 - d_f)^2$. It indicates that d_f is the relevant geometric exponent coupled to the equilibration dynamics of the uniform ensemble.

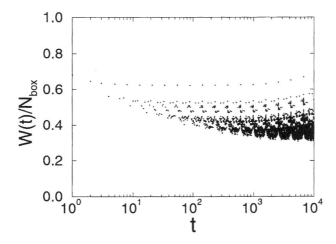

FIGURE 7 The volume occupied by the ensemble $W(t)$ as a function of discrete time in the standard critical circle map. The lower bounds display a slow logarithmic decay with time saturating at a finite volume fraction due to the finite partition of the phase space.

6   CONCLUDING REMARKS AND PERSPECTIVES

In this chapter it was shown that the slow dynamics of low-dimensional maps at the chaos threshold can be associated with the existence of a nonextensive Tsallis entropy that evolves in time at a constant rate. This feature generalizes the well-know description of the sensitivity to initial conditions at noncritical points at which it is an extensive Boltzmann-Gibbs-like entropy that evolves linearly in time. The entropic parameter q, which measures in some sense the degree of nonextensivity in phase space, was shown to be directly associated with geometric exponents characterizing the scaling properties of the critical dynamical attractor. For an initial ensemble concentrated in an infinitesimal region of the attractor, there is a nonextensive q-entropy that grows linearly in time, in a close relationship with the short-time mixing dynamics at critical points. In this case, the proper value of q_{mix} that determines this linearly growing entropy can be obtained from a simple scaling relation, connecting it with geometric exponents governing the scaling properties of the most rarefied and most concentrated regions of the attractor. On the other hand, for an ensemble initially spread over the entire phase space there is a proper q_{eq}-entropy that decreases linearly in time, resembling the equilibration dynamics at critical systems. In this case, the numerical results indicate that q_{eq} seems to be directly related to the fractal dimension of the attractor's support. Independent parameters q_{mix} and q_{eq} describe distinct dynamical regimes.

More recently, extensions of the above picture have been considered in order to describe the dynamics of low-dimensional conservative maps [5, 8]. In these works, an initial distribution very localized in a tiny region of the phase space is followed during the dynamic evolution. However, the volume of the distribution is conserved because of the Liouville's theorem and, therefore, any fine-grained quantity, for example an entropy defined as an integral over phase space, does not vary in time. With finite graining the entropy is observed to increase in time. For a fully chaotic map, such as the generalized cat map, the usual $q = 1$ entropy evolves linearly in time up to a saturation value determined by the fine graining of phase space. In partially chaotic maps, for example the standard map, a crossover was identified. There is a short-time dynamics, during which the linearly growing entropy is a nonextensive one [5]. After a characteristic crossover time, normal extensive behavior sets up. The crossover time is strongly dependent on the fraction of phase space corresponding to chaotic orbits. As the value of the fraction decreases, the period governed by the anomalous nonextensive behavior with slow dynamics increases.

Based on these findings a possible explanation for the emergence of nonextensive behavior in extended Hamiltonian systems with long-range interactions can be as follows: when nonlinear conservative maps with few degrees of freedom are coupled through short-range interactions, the volume in phase space with zero-Lyapunov exponent decreases exponentially with the number of coupled degrees of freedom [17]. This means that an extended Hamiltonian system coupled by short-range interactions becomes fully chaotic and, therefore, fast exponential mixing takes place, leading to ergodicity and extensivity. Coupling conservative systems with few degrees of freedom through slowly decaying long-range interactions preserves a finite volume of phase space with zero-Lyapunov exponent. Therefore an initial slow dynamical regime takes place and exponential mixing emerges only after a characteristic crossover time which depends on the range of the interactions and the number of particles. During this initial transient the system is nonergodic and one needs to perform a nonextensive statistical description of its global properties. Ergodicity and the consequent extensive behavior emerges only after a crossover time. This description is consistent with recent findings concerning the equilibrium properties of long-range interacting Hamiltonian systems [2, 24] which are also reported in this book.

Power-law sensitivity to initial conditions has also been reported in self-organized, driven dissipative systems. The difference between two initially close configurations is usually quantitatively measured as a Hamming distance and its time evolution shows a power-law increase when both configurations are exposed to the same external noise. The slow dynamics seems to be quite robust and emerges in quite distinct models such as biological evolution [34], rice piles [19], and coupled map lattices [30]. In all cases, there is a proper nonextensive entropy that evolves in time at a constant rate.

As a final remark, it is important to stress that, although the power-law sensitivity to initial conditions has provided some important clues about the emergence of nonextensive behavior in complex systems and its connections with the fractal structure of the dynamical attractor, some relevant questions are still open. Firstly, simple scaling relations between the parameter q of the linearly evolving nonextensive entropy and fractal exponents have only be established for one-dimensional dissipative maps. The extension of the existing relations to systems with further degrees of freedom is a challenge due to the intrinsic difficult to achieve reliable estimates of these exponents. Further, the studies of the slow dynamics of conservative and self-organized systems have focused only in the mixing regime. Under which conditions the equilibration dynamics emerges and becomes dominant is still unclear. These are questions that need to be addressed to achieve a more complete description relating these slow dynamical regimes and the nature of the statistical distributions of nonextensive systems.

7 ACKNOWLEDGMENTS

This work was based on several previous collaborations with C. Tsallis, U. M. S. Costa, A. R. Plastino, U. Tirnakli, F. A. B. F. de Moura, C. R. da Silva, and H. R. da Cruz. I would also like to thank the partial financial support from CNPq, CAPES, and FAPEAL.

REFERENCES

[1] Alemany, P. A. "Possible Connection of the Generalized Thermostatistics with a Scale-Invariant Statistical Thermodynamics." *Phys. Lett. A* **235** (1997): 452.

[2] Anteneodo C., and C. Tsallis. "Breakdown of Exponential Sensitivity to Initial Conditions: Role of the Range of Interactions." *Phys. Rev. Lett.* **24** (1998): 5313.

[3] Bai-lin, Hao. *Elementary Symbolic Dynamics.* Singapore: World Scientific, 1989.

[4] Bak, P., C. Tang, and K. Wiesenfeld. "Self-Organized Criticality—An Explanation of $1/f$ Noise." *Phys. Rev. Lett.* **59** (1987): 381.

[5] Baldovin, F. "Mixing and Approach to Equilibrium in the Standard Map." condmat/0109356.

[6] Baldovin, F., C. Tsallis, and B. Schulze. "Nonstandard Mixing in the Standard Map." condmat/0108501

[7] Ball, R. C., and R. Blumenfeld. "Universal Scaling of the Stress Field at the Vicinity of a Wedge Crack in 2 Dimensions and Oscillatory Self-Similar Corrections to Scaling." *Phys. Rev. Lett.* **65** (1990): 1784.

[8] Baranger, M., V. Latora, and A. Rapisarda. "Time Evolution of Thermodynamic Entropy for Conservative and Dissipative Chaotic Maps." *Chaos, Solitons and Fractals* **13** (2002): 471.

[9] Caceres, M. O. "Irreversible Thermodynamics in the Framework of Tsallis Entropy." *Physica A* **218** (1995): 471.

[10] Chame, A., and E. V. L. de Mello. "The Onsager Reciprocity Relations within Tsallis Statistics." *Phys. Lett. A* **228** (1997): 159.

[11] Costa, U. M. S., M. L. Lyra, A. R. Plastino, and C. Tsallis. "Power-Law Sensitivity to Initial Conditions within a Logisticlike Family of Maps: Fractality and Nonextensivity." *Phys. Rev. E* **56** (1997): 245.

[12] Curado, E. M. F., and C. Tsallis. "Generalized Statistical-Mechanics Connection with Thermodynamics." *J. Phys. A* **24** (1991): L69.

[13] Curado, E. M. F., and C. Tsallis. "Corrigenda: Generalized Statistical-Mechanics Connection with Thermodynamics." *J. Phys. A* **24** (1991): 3187.

[14] Curado, E. M. F., and C. Tsallis. "Corrigenda: Generalized Statistical-Mechanics Connection with Thermodynamics." *J. Phys. A* **25** (1992): 1019.

[15] de Moura, F. A. B. F., U. Tirnakli, and M. L. Lyra. "Convergence to the Critical Attractor of Dissipative Maps: Log-Periodic Oscillations, Fractality and Nonextensivity." *Phys. Rev. E* **62** (2000): 6361.

[16] Drozdz, S., F. Ruf, J. Speth, and M. Wojcik. "Imprints of Log-Periodic Self-Similarity in the Stock Market." *Eur. Phys. J. B* **10** (1999): 589.

[17] Falcioni, M., U. M. B. Marconi, and A. Vulpiani. "Ergodic Properties of High-Dimensional Symplectic Maps." *Phys. Rev. A* **44** (1991): 2263.

[18] Feigenbaum, M. J. "Quantitative Universality for a Class of Nonlinear Transformations." *J. Stat. Phys.* **19** (1978): 25.

[19] Gleiser, P. M. "Damage Spreading in a Rice Pile Model." *Physica A* **295** (2001): 311.

[20] Halsey, T. C., M. H. Jensen, L. P. Kadanoff, I. Procaccia, and B. I. Shraiman. "Fractal Measures and Their Singularities—The Characterization of Strange Sets." *Phys. Rev. A* **33** (1986): 1141.

[21] Hauser, P. R., C. Tsallis, and E. M. F. Curado. "Criticality of the Routes to Chaos of the $1 - A|X|^Z$ Map." *Phys. Rev. A* **30** (1984): 2074.

[22] Hilborn, R. C. *Chaos and Quantization.* New York: Cambridge University Press, 1994.

[23] Jensen, M. H., L.P. Kadanoff, and A. Libchaber. "Global Universality at the Onset of Chaos—Results of a Forced Rayleigh-Benard Experiment." *Phys. Rev. Lett.* **55** (1985): 2798.

[24] Latora, V., A. Rapisarda, and C. Tsallis. "Non-Gaussian Equilibrium in a Long-Range Hamiltonian System." *Phys. Rev. E* **64** (2001): 056134.

[25] Lyra, M. L. "Weak Chaos: Power-Law Sensitivity to Initial Conditions and Nonextensive Thermostatistics." *Ann. Rev. Comp. Phys.* **6** (1999): 31.

[26] Lyra, M. L., and C. Tsallis. "Nonextensivity and Multifractality in Low-Dimensional Dissipative Systems." *Phys. Rev. Lett.* **80** (1998): 53.

[27] Mariz, A. M. "On the Irreversible Nature of the Tsallis and Renyi Entropies." *Phys. Lett. A* **165** (1992): 409.

[28] Moon, F. C. *Chaotic and Fractal Dynamics: An Introduction for Applied Scientists and Engineers.* New York: Wiley, 1992.

[29] Newman, W. I., D. L. Turcotte, and A. M. Gabrielov. "Log-periodic Behavior of a Hierarchical Failure Model with Applications to Precursory Seismic Activation." *Phys. Rev. E* **52** (1995): 4827.

[30] Papa, A. R. R., and C. Tsallis. "Imitation Games: Power-Law Sensitivity to Initial Conditions and Nonextensivity." *Phys. Rev. E* **57** (1998): 3923.

[31] Rajagopal, A. K. "Dynamic Linear Response Theory for a Nonextensive System Based on the Tsallis Prescription." *Phys. Rev. Lett.* **76** (1996): 3469.

[32] Sornette, D. "Discrete-Scale Invariance and Complex Dimensions." *Phys. Rep.* **297** (1998): 239.

[33] Sornette D., A. johansen, and A. Arneodo. "Complex Fractal Dimensions Describe the Hierarchical Structure of Diffusion-Limited-Aggregate Clusters." *Phys. Rev. Lett.* **76** (1996): 251.

[34] Tamarit, F., S. A. Cannas, and C. Tsallis. "Sensitivity to Initial Conditions in the Bak-Sneppen Model of Biological Evolution." *Eur. Phys. J. B* **1** (1998): 545.

[35] Tirnakli, U., C. Tsallis, and M. L. Lyra. "Circular-like Maps: Sensitivity to the Initial Conditions, Multifractality and Nonextensivity." *Eur. Phys. J. B* **11** (1999): 309.

[36] Tirnakli, U., C. Tsallis, and M. L. Lyra. "Asymmetric Unimodal Maps at the Edge of Chaos." *Phys. Rev. E* **65** (2002): 036207.

[37] Tsallis, C. "Possible Generalization of Boltzmann-Gibbs Statistics." *J. Stat. Phys.* **52** (1988): 479.

[38] Tsallis, C. "Thermodynamic Stability Conditions for the Tsallis and Renyi Entropies—Comment." *Phys. Lett. A* **206** (1995): 389.

[39] Tsallis, C, L. R. da Silva, R. S. Mendes, R. O. Vallejos, and A. M. Mariz. "Specific Heat Anomalies Associated with Cantor-Set Energy Spectra." *Phys. Rev. E* **56** (1997): 4922.

[40] Tsallis, C., S. V. F. Levy, A. M. C. de Souza, and R. Maynard. "Statistical-Mechanical Foundation of the Ubiquity of the Levy Distributions in Nature." *Phys. Rev. Lett.* **77** (1996): 5422; Erratum **77** (1996): 5442.

[41] Tsallis, C., A. R. Plastino, and W.-M. Zheng "Power-Law Sensitivity to Initial Conditions—New Entropic Representation." *Chaos, Solitons & Fractals* **8** (1997): 885.

[42] Vallejos, R. O., R. S. Mendes, L. R. da Silva, and C. Tsallis. "Connection between Energy Spectrum, Self-Similarity, and Specific Heat Log-Periodicity." *Phys. Rev. E* **58** (1998): 1346.

Numerical Analysis of Conservative Maps: A Possible Foundation of Nonextensive Phenomena

Fulvio Baldovin

We discuss the sensitivity to initial conditions and the entropy production of low-dimensional conservative maps, focusing on situations where the phase space presents complex (fractal-like) structures. We analyze numerically the standard map as a specific example and we observe a scenario that presents appealing analogies with anomalies detected in long-range Hamiltonian systems. We see how the Tsallis nonextensive formalism handles this situation both from a dynamical and from a statistical mechanics point of view.

1 INTRODUCTION

In recent years, the Tsallis extension of the Boltzmann-Gibbs (BG) statistical mechanics [9, 26], usually referred to as nonextensive (NE) statistical mechanics, has become an intense and exciting research area (see, e.g., Tsallis [25]). The q-exponential distribution functions that emerge as a consequence of the NE formalism have been applied to an impressive variety of problems, ranging

Interdisciplinary Applications of Ideas from Nonextensive. . .
edited by Murray Gell-Mann and Constantino Tsallis, Oxford University Press.

from turbulence, to high-energy physics, epilepsy, protein folding, and financial analysis. Yet, the foundation of this formalism, as well as the definition of its area of applicability, is still not completely understood, and it stands as a present challenge in the affirmation of the whole proposal. An intensive effort is currently being made to investigate this point, precisely in trying to understand: (1) which mechanisms lead to a crisis of the BG formalism; and (2) in these cases, does the NE formalism provide a "way out" to some of the problems? A possible approach to these questions comes from the study of the underlying dynamics that gives the basis for a statistical mechanic treatment of the system. This idea is not new. Einstein, in his critical remark about the validity of the Boltzmann principle [10],[1] was one of the first to call attention to the relevance of a *dynamical foundation* of statistical mechanics. Another fundamental contribution is Krylov's seminal work [14] on the mixing properties of dynamical systems.

In one-dimensional (dissipative) systems, intensive effort has been made to analyze the properties of the systems at the *edge of chaos*, i.e., at the critical point that marks the transition between chaoticity and regularity [6, 8, 16, 19, 18, 23, 27].[2] It turns out that the NE formalism is able to describe the most prominent characteristics of all universality classes of unimodal maps at the edge of chaos. In particular, it has been found that the sensitivity to initial conditions has the form of a q-exponential with a value of q that can be deduced from one of the Feigenbaum's constants; the same value of q can be obtained as the result of a multifractal analysis and studying the properties of relaxation to equilibrium of the system. Moreover, a nice connection with the NE entropy production (once again characterized by the same value of q) has been observed.

In this chapter we will address the subject by considering Hamiltonian maps and focusing our attention on the paradigmatic case of the standard map. We will see how some of the analysis performed for dissipative systems at the edge of chaos can also describe the transition from chaoticity to regularity of conservative systems. Despite the simplicity of our example, some remarkable similarities can be found with anomalies that have been discovered in many-body long-range Hamiltonian systems [17, 21], so that the model, in addition to being important by itself, possibly catches fundamental phenomena of much more general Hamiltonian systems.

---

[1]Usually W is set equal to the number of complexions.... In order to calculate W, one needs a *complete* (molecular-mechanical) theory of the system under consideration. Therefore it is dubious whether the Boltzmann principle has any meaning without a complete molecular-mechanical theory or some other theory which describes the elementary processes. $S = \frac{R}{N} \log W + \text{const.}$ seems without content, from a phenomenological point of view, without giving, in addition, such an *Elementartheorie* [10].

[2]See also Grassberger and Scheunert [11]; Schneider et al. [24]; Anania and Politi [1]; and Hata et al. [12].

2 HAMILTONIAN MAPS

Time-independent Hamiltonian systems are dynamical systems characterized by an even number of dimensions $d = 2n$ (n is the number of degrees of freedom) and by a single function, the Hamiltonian H, that determines a complete set of differential equations for the d variables. If we are interested in studying the properties of an *isolated system* (say, in a microcanonical ensemble's perspective), we can exploit the fact that H is constant along a trajectory and limit our analysis to the constant-energy hypersurface of the phase space, thus reducing the order of the system to $2n - 1$. Moreover, in statistical mechanics we are mostly interested in the *recurrent trajectories*, i.e., in those trajectories that come back again and again, indefinitely, to any part of the phase space they have once visited. In this case we can obtain a great simplification of the problem (specifically for the numerical experiments) by taking a *Poincaré section* of the phase space. This is made by cutting the constant-energy hypersurface transversally and considering the successive intersections of each orbit with this transversal surface.[3] In this way we pass from a continuous-time system of $2n - 1$ differential equations to a discrete-time system of $d_M = 2n - 2$ iteration equations, called a *conservative* or *Hamiltonian map*. It is possible to show that the map **M** thus obtained is *symplectic*[4] (see, e.g., Ott [20])[5]; as a particular consequence of this property, we state that **M** is *volume preserving* $(\det(\partial M_i/\partial x_j) = 1)$. Typically, **M** will depend on one or more parameters that control the dynamical regimes of the system. Depending on the parameters and on the region of the phase space it is possible to observe *regular orbits* (periodic or quasi-periodic trajectories) and *chaotic orbits*. A typical phenomenon, characteristic of the separation between regular and chaotic regions, is the presence of fractal-like structures, for example, "islands around islands" (see, e.g., Zaslavsky and Niyazov [29]).

The lowest dimension where it is possible to reproduce such a scenario is a system with two degrees of freedom, so that its Poincaré section is two-dimensional. This is the case of the *standard map* (or *kicked rotator map*[6]), that is characterized by the equations:

$$x_{t+1} = y_t + \frac{a}{2\pi} \sin(2\pi x_t) + x_t \pmod 1,$$

$$(1)$$

---

[3]Obviously different orbits will have, in general, different recurrence time; nevertheless, statistically we are typically interested in characterizing behaviors with long times averages, so that the Poincaré section remains a significant description for these kinds of system properties.

[4]The same structure is obtained using a surface of section on $(2n+1)$-dimensional systems with a Hamiltonian that depends periodically on time. In this case the system is reduced to a $2n$-dimensional symplectic map.

[5]A nice introduction to the numerical exploration of Hamiltonian systems is also in Henon [13].

[6]The map can, in fact, be derived in a simple way from the periodic time-dependent Hamiltonian of the kicked rotator.

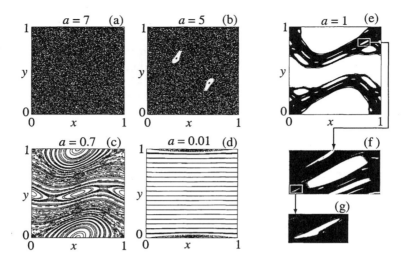

FIGURE 1   (a)–(d) Phase portrait of the standard map for typical values of a. $N = 20 \times 20$ orbits (black dots) were started with a uniform distribution in the unit square and traced for $0 \leq t \leq 200$. (e)–(g) Islands-around-islands. (e): $N = 100 \times 100$ initial data were started inside a small square of side $\sim 10^{-2}$ and traced for $0 \leq t \leq 5000$. (f) is a magnification of the island inside the rectangle in (e). (g) is a magnification of the island inside the rectangle in (f).

$$y_{t+1} = y_t + \frac{a}{2\pi} \sin(2\pi x_t) \qquad (\mathrm{mod}\ 1),$$

where $a \in \mathbb{R}$ ($t = 0, 1, ...$). The system is integrable[7] when $a = 0$, while, roughly speaking, chaoticity increases when $|a|$ increases; in figure 1 we display the phase portrait of the map (2) for different values of a and we illustrate the islands-around-islands structure.

3   SENSITIVITY TO INITIAL CONDITIONS

An important concept to characterize and quantify the chaotic behavior of dynamical systems is the sensitivity to initial conditions. Consider, in the phase

---

[7] A time-independent Hamiltonian system is said to be *integrable* if it has n independent integrals (global constants of motion). In this case, via a canonical transformation, it can be reduced to *normal form*. In the new coordinates, the *action-angle variables*, the equation of motions are those of n coupled harmonic oscillators. The fundamental theorem that explains the passage from an integrable to a chaotic system is the KAM theorem (see, e.g., Ott [20]).

space, an initial condition x_0 and a nearby initial data x_0' along a direction u_0. We can define $|\Delta x(0)|$ as the distance between x_0 and x_0'. Each initial condition will generate a different orbit, so that we can follow the evolution of the distance $|\Delta x(t)|$. The sensitivity to initial condition $\xi(t, x_0, u_0)$ is defined as the limit $\xi(t, x_0, u_0) \equiv \lim_{|\Delta x(0)| \to 0} |\Delta x(t)| / |\Delta x(0)|$. Inside a chaotic region, for typical orientations of the vector u_0,[8] the sensitivity to initial conditions grows exponentially. As a consequence of the symplectic structure, chaotic regions are, in fact, characterized by d_M Lyapunov exponents coupled in pairs, where each member of the pair is the opposite of the other. A typical vector u_0 has a nonzero component along the direction where the Lyapunov exponent is largest; hence, at least after some iteration steps, the maximum Lyapunov exponent will drive the exponential separation of the orbits. On the other hand, inside a regular region, the growth of ξ is linear with respect to the iteration time. Finally, at the separation between chaotic and regular regions, the presence of fractal-like structures imposes constraints on the path followed by the trajectories, so that anomalous behaviors may arise when the maximum Lyapunov exponent goes to zero. In connection with these ideas and with the NE formalism, it has been conjectured [27] that in many cases, when the maximum Lyapunov exponent goes to zero, the sensitivity to initial condition could be described, instead of by an exponential behavior, by a q-exponential:

$$\xi(t) = \exp_q(\lambda_q t) \equiv [1 + (1 - q)\lambda_q t]^{\frac{1}{1-q}} \quad (q \in \mathbb{R}), \quad (2)$$

that is characterized by the index q and by the *generalized Lyapunov coefficient* λ_q (with this notation, the usual expression for the Lyapunov exponent λ_1 is recovered in the limit $q \to 1$). We notice for future reference that eq. (2) is a solution of the differential equation $\dot{\xi} = \lambda_q \xi^q$, and that the inverse function of the q-exponential is the q-logarithm defined as $\ln_q(x) \equiv (x^{1-q} - 1)/(1 - q)$. Recently, this conjecture has been proved correct for the pitchfork and tangent bifurcations [5, 22], and for the edge of chaos [6] of one-dimensional unimodal maps (see Robledo [23]).

In ensemble theory, in correspondence with a macroscopic state of the system, there exist many possible microscopic states, distributed, for example, over a constant-energy surface. If we are trying to investigate the dynamical basis of the ensemble approach, we are then mostly interested in studying *global behaviors* over this surface. Taking inspiration in the microcanonical ensemble, we can try to characterize the "global chaoticity" of a map by sampling uniformly all the regions of the phase space and then averaging the sensitivity to initial conditions of all different regions. In doing so we expect that if a chaotic region is present, at least after enough time, the rapidity of the exponential increment will lead to an exponential growth of the global sensitivity to initial conditions. Nevertheless, if enough of the phase space is occupied by regular or fractal-like structures, for small or intermediate time, linear or even *power-law* behaviors

---

[8]This refers that all the vectors except a Lebesgue set measure zero.

may appear. In other words, it is possible to observe *crossovers* between different dynamical regimes. A way to model this more complex situation is to admit that ξ obeys a more general differential equation, i.e.: $\dot{\xi} = \lambda_1 \xi + (\lambda_q - \lambda_1)\xi^q$ ($q < 1$, $\lambda_1, \lambda_q > 0$). The solution [28]

$$\xi(t) = \left[1 - \frac{\lambda_q}{\lambda_1} + \frac{\lambda_q}{\lambda_1} e^{(1-q)\lambda_1 t}\right]^{\frac{1}{1-q}}, \tag{3}$$

presents, in the case $0 < \lambda_1 \ll \lambda_q$, three asymptotical behaviors, namely

1. linear: $\xi \sim 1 + \lambda_q t$, for $0 \leq t \ll t_{\text{cross1}} \equiv \frac{1}{(1-q)\lambda_q}$;

2. power law: $\xi \sim [(1-q)\lambda_q]^{\frac{1}{1-q}} t^{\frac{1}{1-q}}$, for $t_{\text{cross1}} \ll t \ll t_{\text{cross2}} \equiv \frac{1}{(1-q)\lambda_1}$; and

3. exponential: $\xi \sim (\frac{\lambda_q}{\lambda_1})^{\frac{1}{1-q}} e^{\lambda_1 t}$, for $t \gg t_{\text{cross2}}$.

To see if the above-stated conjectures find numerical confirmation, let us test, for some different values of the parameter a, the global sensitivity to initial conditions of the standard map. The experiment is performed as follows. For a fixed a, we consider a partition of the phase space made of small squares of equal size. A square of the partition is chosen randomly and two initial data are set inside it, at random positions. Using the dynamic eq. (2), the function $\xi(t)$ is then calculated numerically. The same calculation is repeated many times, changing the square when the initial data are set, and finally averaging over the different results. When the experiment is performed for rather large values of $|a|$ (say, larger than 2), the logarithm of ξ displays a linear increase with the iteration time ($\xi(t)$ is exponential), and then a saturation occurs because a separation of the order of the total size of the phase space is reached. In figure 2 we present some preliminary results that show how, reducing $|a|$, a crossover between a power-law and an exponential behavior happens (an accurate analysis of this experiment is currently being made [2]). In fact, the linear increase of the logarithm of ξ is pushed forward in time (fig. 2(a)) and correspondingly the q-logarithm of ξ, for a special value of q close to 0.3, displays a linear increase during the initial steps (fig. 2(b)). The crossover time increases when a decreases.

4   ENTROPY PRODUCTION

In the previous section we have seen that anomalous global dynamical behaviors that arise in the presence of a complex structure of the phase space may be well-modeled by the NE formalism. The next question is how to connect these dynamical behaviors with statistical mechanics. An interesting connection can be made by studying the average *entropy production* of the system using the following experiment.

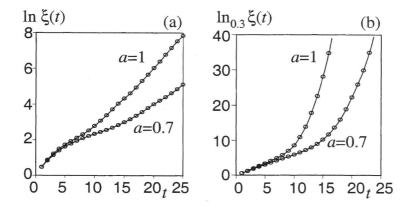

FIGURE 2 Sensitivity to initial conditions of the standard map for small values of a. $|x_0 - x_0'| \sim 10^{-9}$ and results are averages over 5000 histories. In (a) $\ln(\xi(t))$ displays a linear increase only after a certain time. Correspondingly, in (b) $\ln_{0.3}(\xi(t))$ exhibits a linear increase for the initial steps. The lines are guides to the eye.

First, we introduce a coarse-graining partition of the phase space by dividing it into W cells of equal size. A statistical ensemble can then be introduced considering many copies of the system (N points) distributed in the phase space. The occupation number N_i of each cell i ($\sum_{i=1}^{W} N_i = N$) provides a probability distribution $p_i \equiv N_i/N$, and it is then possible to define the Tsallis entropy:

$$S_q = \frac{1 - \sum_{i=1}^{W} p_i^q}{q - 1} \quad (q \in \mathbb{R}), \tag{4}$$

where, in the limit $q \to 1$, we recover the usual BG definition $S_1 = -\sum_{i=1}^{W} p_i \ln p_i$. Then, we start at $t = 0$ with a far-from-equilibrium initial condition putting all the N points randomly or uniformly distributed inside a *single* cell. Using the dynamic equations of the map, at each iteration step the points spread in the phase space; hence, the q-entropy (4), that is zero at $t = 0$ $\forall q$, typically begins to increase. After a certain number of iterations, the entropy saturates and equilibrium is attained because of the finiteness of W. The intermediate stage, before saturation and after a possible initial transient, describes the entropy production characteristic of the cell where the initial data were set. Once again, a *global* description of the entropy production is obtained averaging many different histories, with the initial cells randomly chosen in the whole phase space.

The connection between the entropy production and the sensitivity to initial conditions, in the case of "usual" chaotic situations in the phase space of conservative maps (here and in the following we refer to "usual" chaoticity as a situation where the chaotic region is regular, without structures), was made in Latora and Baranger [15]; they observed that the intermediate stage of the entropy growth is *linear* for the BG entropy S_1, with a slope that is equal to the sum of the positive Lyapunov exponents.[9] In Baldovin et al. [7] and Baldovin [3], the same experiment was performed for the standard map, but focusing on small values of the parameter a, in order to study the effects on the entropy produc-tion due to the presence of the fractal-like structures in the phase space. Figure 3 shows how, reducing a, the stage where S_1 increases linearly (before saturation) is pushed forward in time. Correspondingly, in figure 4 we can see that S_q, with $q \simeq 0.3$, displays a linear stage for the first iteration steps. This means that the dynamical crossover observed for the sensitivity to initial conditions is also detectable when one analyzes the entropy production. Moreover, the "anoma-lous" initial stage is characterized by a linear increase of a q-entropy with a value of q that is comparable with the one that characterizes the q-exponential initial sensitivity to initial conditions. Notice also that the crossover time di-verges when $a \to 0$ (see inset of fig. 4). Our results suggest that in the limit $\lim_{t \to \infty} \lim_{a \to 0} \lim_{W \to \infty}$, only for $q = q^* \simeq 0.3$ we have a finite, non-vanishing entropy production rate $S_q(t)/t$, whereas the entropy production rate vanishes (diverges) for $q > q^*$ ($q < q^*$). We stress that as in other situations where the NE formalism applies, the order of the limit is not permutable [25]. indexsensitivity to initial conditions, and entropy production

5   TOWARD THE GENERALIZATION OF PESIN EQUALITY

Let us try to analyze qualitatively the motivations that are at the basis of the connection between the sensitivity to initial conditions and the entropy produc-tion, taking as an example the case of a symplectic two-dimensional map.

Because of the symplectic structure, in the "usual" chaotic situation, there are two opposite Lyapunov exponents. When we set many initial data inside a small region of the phase space, after few iterations the points will tend to align along a line that is tangent to the directions of the local maximum Lyapunov exponents. For simplicity, let us first think that initially the N points are uni-formly distributed along the direction of the local maximum Lyapunov exponent, with a small distance of separation $|\Delta \mathbf{x}(0)|$. If the average positive Lyapunov ex-ponent of the map is λ_1 and if the dependence on the local position is smooth enough so that Lyapunov exponents do not vary much, after t iterations the points will be almost uniformly distributed, and the average separation will be $|\Delta \mathbf{x}(t)| = |\Delta \mathbf{x}(0)| \exp(\lambda_1 t)$. Now, a sufficiently good coarse graining (W large) resolves possible sharp bending of the line where the points lie. So that, after the t iterations, we will measure on the average an almost uniform probability

---

[9]The sum of the positive Lyapunov exponents coincides, via Pesin equality, with the Kolmogorov-Sinai entropy rate.

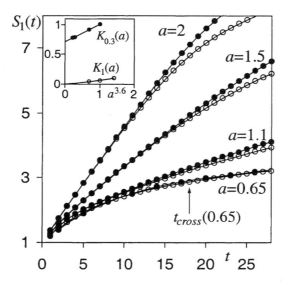

FIGURE 3 $S_1(t)$ for different values of $a \leq 2$. For full circles, the number of cell W is five times larger than the number of cells for empty circles. About 5000 different histories were averaged. Inset: Slopes of $S_1(t)$ $(K_1(a))$ and of $S_{0.3}(t)$ $(K_{0.3} \simeq 0.71 + 0.30\ a^{3.6})$ in their linear regimes (see also fig. 4).

distribution that occupies a number of cells equal to

$$W_{\text{eff}} \sim \frac{N|\Delta\mathbf{x}(t)|}{N|\Delta\mathbf{x}(0)|} = \exp(\lambda_1 t)\,. \tag{5}$$

But the BG entropy of a uniform distribution of W_{eff} different possibilities is equal to

$$S_1 = \ln(W_{\text{eff}}) \sim \lambda_1 t\,, \tag{6}$$

that is precisely the connection observed numerically in Latora and Baranger [15].[10] In practice, the initial distribution occupies a small surface rather than lying along the direction of the local maximum Lyapunov exponent. Nevertheless, as a consequence of the contraction produced by the negative Lyapunov exponent, what is relevant is the projection of the initial positions along the direction of the of the local maximum Lyapunov exponent. If we start with a uniform distribution inside the initial region, this projection will produce a uniform distribution along that direction.

---

[10]If the dimension of the phase space is larger than 2, there is more than one positive Lyapunov exponent. In this case $W_{\text{eff}} \sim \exp(\sum_i \lambda_1^i t)$, where the sum is performed over the positive Lyapunov exponents λ_1^i, and $S_1 \sim \sum_i \lambda_1^i t$.

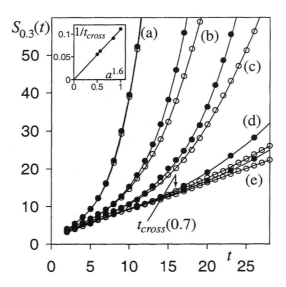

FIGURE 4 $S_{0.3}(t)$ for: $a = 1.5$ (a); $a = 1.1$ (b); $a = 0.9$ (c); $a = 0.7$ (d); $a = 0.65$ (e). For full circles, the number of cell W is five times larger than the number of cells for empty circles. About 5000 different histories were averaged. Inset: $t_{cross}(a)$. Notice that $t_{cross}(a)$ diverges for $a \to 0$.

As we have seen, the presence of structures in the phase space may slow-down the sensitivity to initial conditions to a power-law behavior. In this case, the previous hypotheses must be checked even more carefully.[11] However, if we repeat the same reasoning, just substituting the exponential separation with a q-exponential separation, we have that $W_{\text{eff}} \sim \exp_q(\lambda_q t)$ ($q < 1$). And the q-entropy of a uniform probability distribution with W_{eff} possibilities is given by

$$S_q = \ln_q(W_{\text{eff}}) \sim \lambda_q t. \tag{7}$$

Equation (7) represents a sort of generalization of the Pesin equality (see, for example, Ott [20]), linking dynamical quantities to the NE statistical entropy.[12]

Note that the above qualitative scheme can also apply for higher-dimensional phase space, in the case that only a single (local) direction drives the separation of the points.

---

[11]For example, we observe that the presence of structure is likely to produce inhomogeneities in the distribution of the points, so that, $\forall q \in \mathbb{R}$, the entropy S_q may be smaller than its maximum $\ln_q(W_{\text{eff}})$, obtained at equiprobability.

[12]Along these lines of reasoning, very recently the validity of eq. (7) has been analytically proved for th edge of chaos of unimodal maps [4].

6 CONCLUSIONS

The main feature presented in the previous sections is the observation, during the transition from chaoticity to regularity of the standard map, of a crossover between an "anomalous" dynamical regime and the "usual" one. The duration of the "anomalous" regime increases with the complexity of the structure of the phase space. We have also seen that in these conditions the average sensitivity to initial conditions can be well described inside the NE formalism and that a previously known connection between the sensitivity to initial condition and the BG entropy, valid for the "usual" regime, seems to naturally generalize inside this formalism. These observations present a striking similitude to anomalies observed in the many-body Hamiltonian system of the long-range coupled rotators [17, 21]. Under certain conditions, the maximum Lyapunov exponent of these systems goes to zero as the number of rotators is increased, and a longstanding non-Boltzmannian meta-equilibrium state has been observed, whose duration diverges when the number of rotators goes to infinity. For a finite number of rotators, after the meta-equilibrium state, there is a crossover to the BG equilibrium. The presence of the meta-equilibrium state may be associated with an "anomalous" dynamical behavior of the kind we have illustrated in this paper, and the effect produced by an *increase* in the number of rotators in the structure of the phase space may be analogous to that originated by a *decrease* of a in the case of the standard map.

In summary, the discussion of the previous sections, with the coherent description of the dynamical crossover both in terms of the sensitivity to initial conditions and of the entropy production inside the Tsallis formalism, may represent a step towards a dynamical foundation of nonextensive phenomena that appear to be so widespread in nature.

ACKNOWLEDGMENTS

We would like to thank C. Tsallis, A. Robledo, and L. G. Moyano for useful discussions and comments, and M. Zamberlan for encouragement. We have benefited from partial support by CAPES, PRONEX, CNPq, and FAPERJ (Brazilian agencies).

REFERENCES

[1] Anania, G., and A. Politi. "Dynamical Behavior at the Onset of Chaos." *Europhys. Lett.* **7** (1988): 119.

[2] Añaños, G. F. J., F. Baldovin, and C. Tsallis. "Nonstandard Sensitivity to Initial Conditions in Conservative Maps." (2003): in progress.

[3] Baldovin, F. "Mixing and approach to Equilibrium in the Standard Map." *Physica A* **305** (2002): 124.

[4] Baldovin, F., and A. Robledo. "Exact Connection between Edge of Chaos Dynamics and Tsallis' Entropy." (2003): in progress.

[5] Baldovin, F., and A. Robledo. "Sensitivity to Initial Conditions at Bifurcations in One-Dimensional Nonlinear Maps: Rigorous Nonextensive Solutions." *Europhys. Lett.* **60** (2002): 518.

[6] Baldovin, F., and A. Robledo. "Universal Renormalization-Group Dynamics at the Onset of Chaos in Logistics Maps and Nonextensive Statistical Mechanics." *Phys. Rev. E* **66** (2002): 045104(R).

[7] Baldovin, F., C. Tsallis, and B. Schulze. "Nonstandard Entropy Production in the Standard Map." *Physica A* **320** (2003): 184.

[8] Borges, E. P., C. Tsallis, G. F. J. Añaños, and P. M. C. de Oliveira. "Nonequilibrium Probabilistic Dynamics of the Logistic Map at the Edge of Chaos." *Phys. Rev. Lett.* **89** (2002): 254103.

[9] Curado, E. M. F., and C. Tsallis. "Generalized Statistical Mechanics; Connection with Thermodynamics." *J. Phys. A* **24** (1991): L69. [Corrigenda: **24** (1991): 3187 and **25** (1992): 1019].

[10] Einstein, A. "Allgemeines über das Boltzmannsche Prinzip." *Annal. Physik* **33** (1910): 1275. Translation: Abraham Pais. *Subtle is the Lord: Science and Life of Albert Einstein.* New York: Oxford University Press, 1982.

[11] Grassberger, P., and M. Scheunert. "Some More Universal Scaling Laws for Critical Mappings." *J. Stat. Phys.* **26** (1981): 697.

[12] Hata, H., T. Horita, and H. Mori. "Dynamic Description of the Critical 2^∞ Atrractor and 2^m-Band Chaos." *Progr. Theor. Phys.* **82** (1989): 897.

[13] Henon, M. "Numerical Exploration of Hamiltonian Systems." In *Chaotic Behavior of Deterministic Systems*, edited by G. Ioos, R. H. G. Helleman, and R. Stora. Amsterdam: North-Holland, 1983.

[14] Krylov, N. "Relaxation Processes in Statistical Systems." *Nature* **153** (1944): 709.

[15] Latora, V., and M. Baranger. "Kolmogorov-Sinai Entropy Rate versus Physical Entropy." *Phys. Rev. Lett.* **82** (1999): 520.

[16] Latora, V., M. Baranger, A. Rapisarda, and C. Tsallis. "The Rate of Entropy Increase at the Edge of Chaos." *Phys. Lett. A* **273** (2000): 97.

[17] Latora, V, A. Rapisarda, and C. Tsallis. "Non-Gaussian Equilibrium in a Long-Range Hamiltonian System." *Phys. Rev. E* **64** (2001): 056134.

[18] Lyra, M. L. "Nonextensive Entropies and Sensitivity to Initial Conditions of Complex Systems." This volume.

[19] Lyra, M. L., and C. Tsallis. "Nonextensivity and Multifractality in Low-Dimensional Dissipative Systems." *Phys. Rev. Lett.* **80** (1998): 53.

[20] Ott, E. *Chaos in Dynamical Systems.* Cambridge: Cambridge University Press, 1993.

[21] Rapisarda, A., and V. Latora. "Nonextensive Effects in Hamiltonian Systems." This volume.

[22] Robledo, A. "Criticality in Nonlinear One-Dimensional Maps: RG Universal Map and Nonextensive Entropy." February 2002. arXiv.org e-Print Archive, Quantum Physics, Cornell University. September 2002. ⟨http://xxx.lanl.gov/abs/cond-mat/0202095⟩.

[23] Robledo, A. "Unifying Laws in Multidisciplinary Power-Law Phenomena: Fixed-Point Universality and Nonextensive Entropy." This volume.

[24] Schneider, T., A. Politi, and D. Wurtz. "Resistance and Eigenstates in a Tight-Binding Model with Quasiperiodic Potential." *Z. Phys. B* **66** (1987): 469.

[25] Tsallis, C. "Nonextensive Statistical Mechanics: Construction and Physical Interpretation." This volume.

[26] Tsallis, C. "Possible Generalization of B oltzmann-Gibbs Statistics." *J. Stat. Phys.* **52** (1988): 479.

[27] Tsallis, C., A. R. Plastino, and W.-M. Zheng. "Power-Law Sensitivity to Initial Conditions—New Entropic Representation." *Chaos, Solitons and Fractals* **8** (1997): 885.

[28] Tsallis, C., G. Bemski, and R. S. Mendes. "Is Reassociation in Folded Proteins a Case of Nonextensivity?." *Phys. Lett. A* **257** (1999): 93.

[29] Zaslavsky, G. M., and B. A. Niyazov. "Fractional Kinetics and Accelerator Modes." *Phys. Rep.* **283** (1997): 73.

Nonextensive Effects in Hamiltonian Systems

Andrea Rapisarda
Vito Latora

1 INTRODUCTION

The Boltzmann-Gibbs formulation of equilibrium statistical mechanics depends crucially on the nature of the Hamiltonian of the N-body system under study, but this fact is clearly stated only in the introductions of textbooks and, in general, it is very soon neglected. In particular, the very same basic postulate of equilibrium statistical mechanics, the famous *Boltzmann principle* $S = k \log W$ of the *microcanonical ensemble*, assumes that dynamics can be automatically and easily taken into account, although this is not always justified, as Einstein himself realized [20]. On the other hand, the Boltzmann-Gibbs *canonical ensemble* is valid only for sufficiently short-range interactions and does not necessarily apply, for example, to gravitational or unscreened Colombian fields for which the usually assumed entropy extensivity postulate is not valid [5]. In 1988, Constantino Tsallis proposed a generalized thermostatistics formalism based on a nonextensive entropic form [24]. Since then, this new theory has been encountering an increasing number of successful applications in different fields (for some recent examples see Abe and Suzuki [1], Baldovin and Robledo [4], Beck et al. [8], Kaniadakis et al. [12], Latora et al. [16], and Tsallis et al. [25]) and seems to be the best candidate for a generalized thermodynamic formalism which should be valid when nonextensivity, long-range correlations, and fractal structures in phase space cannot be neglected: in other words, when *the dynamics play a nontrivial role* [11] and fluctuations are quite large and non-Gaussian [6, 7, 8, 24, 26].

In this contribution we consider a nonextensive N-body classical Hamiltonian system, with infinite range interaction, the so-called *Hamiltonian mean field* (HMF) model, which has been intensively studied in the last several years [3, 13, 14, 15, 17, 18, 19]. The out-of-equilibrium dynamics of the model exhibits a series of anomalies like negative specific heat, metastable states, vanishing Lyapunov exponents, and non-Gaussian velocity distributions. After a brief overview of these anomalies, we show how they can be interpreted in terms of nonextensive thermodynamics according to the present understanding.

2   METASTABILITY AND OTHER ANOMALIES

The Hamiltonian mean field (HMF) model describes a system of N planar classical spins interacting through an infinite-range potential [3]. The Hamiltonian of the model can be written as

$$H = \sum_{i=1}^{N} \frac{p_i^2}{2} + \frac{1}{2N} \sum_{i,j=1}^{N} [1 - \cos(\theta_i - \theta_j)] \ , \tag{1}$$

where θ_i is the ith angle and p_i the conjugate variable representing the angular momentum or the rotational velocity, since unit mass is assumed. Notice that the summation in the potential is extended to all couples of spins and not restricted to first neighbors. The order parameter is the magnetization $\mathbf{M} = 1/N \sum_{i=1}^{N} \mathbf{m}_i$, where $\mathbf{m}_i = [\cos(\theta_i), \sin(\theta_i)]$. The canonical analytical solution of the model predicts a second-order phase transition from a low-energy ferromagnetic phase with magnetization $M \sim 1$ to a high-energy one, where the spins are homogeneously oriented on the unit circle and $M \sim 0$. The *caloric curve*, i.e., the dependence of the energy density $U = E/N$ on the temperature T, is given by $U = T/2 + 1/2(1 - M^2)$ and shown in the inset (b) of figure 1 as a full curve. The critical point is at $U_c = 0.75$ (reported as a dashed vertical line) corresponding to a critical temperature $T_c = 0.5$ [3]. The dynamical behavior of the model can be investigated in the microcanonical ensemble by starting the system out of equilibrium and numerically integrating the equations of motion [17, 14]. For example, one can use *water-bag initial conditions*, i.e., $\theta_i = 0$ for all i ($M = 1$) and uniformly distributed velocities. As shown in the inset (b) of figure 1, microcanonical simulations (open circles) are, in general, in good agreement with the equilibrium prediction, except for a region below U_c, where dynamics characterized by Lévy walks, anomalous diffusion [19], and a negative specific heat

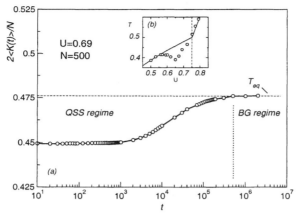

FIGURE 1   The temporal evolution of twice the average kinetic energy per particle is plotted for $N = 500$ and energy density $U = E/N = 0.69$ (a). Two different regimes are clearly visible. The first plateau corresponds to a metastable quasi-stationary state (QSS) regime, while the second one refers to the Boltzmann-Gibbs (BG) equilibrium regime. We report as a dashed line the final equilibrium temperature which for $U = 0.69$ is equal to $T_{eq} = 0.4756$.' We plot, in the inset (b), the equilibrium *caloric curve* (T vs. U) as a full line in comparison with the numerical simulations relative to the QSS regime (open circles). Also, in this case, $N = 500$.

have also been found [13]. In figure 1(a) we report the time evolution of twice the average kinetic energy per particle $2\langle K \rangle / N$, a quantity that coincides with the temperature T. Brackets denote time averages. The system, started with out-of-equilibrium initial conditions, rapidly reaches a metastable quasi-stationary state (QSS), which does not coincide with the equilibrium prediction. In fact, after a short transient time [15] not reported here, the quantity $2\langle K \rangle / N$ assumes a fixed value (the plateau in fig. 1) corresponding to an N-dependent temperature $T_{QSS}(N)$ that is lower than the equilibrium prediction, which is also reported as a dashed line. In correspondence with this plateau, one gets a value of $M_{QSS} \sim 0$ for the magnetization. If we want to observe relaxation to the equilibrium state with temperature $T_{eq} = 0.4756$ and magnetization $M_{eq} = 0.307$, we have to wait for a time longer than 10^5 for a system with $N = 500$ as shown in figure 1. However, this time diverges with the size of the system. In figure 2 we illustrate the validity of three scaling laws [18]:

1. $\tau_{QSS} \propto N$, τ_{QSS} being the lifetime of the metastable state;
2. $T_{QSS}(N) - T_{N=\infty} \propto N^{-1/3}$, where $T_{QSS}(N)$ is the temperature of the plateau and $T_{N=\infty} = 0.38$ is the asymptotic temperature, a value obtained analytically as the metastable prolongation, at energies below U_c, of the high-energy

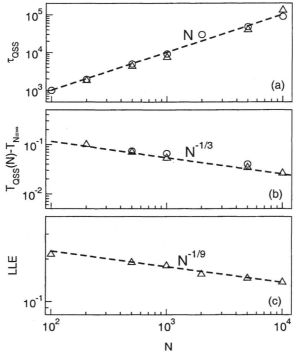

FIGURE 2 Scaling as a function of the size N of the system. In panel (a) we plot the QSS lifetime τ_{QSS}. In panel (b) we show the difference of the QSS temperature with the asymptotic one $T_{N=\infty} = 0.38$. In panel (c) we report the largest Lyapunov exponent (LLE) corresponding to the QSS temporal regime. We also plot the corresponding power laws as a dashed lines in all the three panels. Circles and triangles correspond to different out-of-equilibrium initial conditions.

solution ($M = 0$); see figure (b). This also implies that $M_{QSS} \propto N^{-1/6}$ [18]; and

3. $LLE \propto N^{-1/9}$, where LLE indicates the largest Lyapunov exponent [15].

These numerical results clearly indicate that the infinite-time limit, $t \to \infty$, and the thermodynamic limit, $N \to \infty$, *do not commute*. Thus, if the latter is performed before the former, the system does not relax to the Boltzmann-Gibbs equilibrium and lives forever in the QSS regime, where the Lyapunov exponents are zero and, therefore, anomalous mixing features are expected.

Indeed, similar anomalies have also been observed recently in other nonextensive Hamiltonian systems [9, 22] and, in particular, in the generalized version of the HMF model, the so-called $\alpha - XY$ [2, 10, 23], where α is the exponent that controls the range of the interaction which goes like $r^{-\alpha}$. The two extreme cases are: (i) $\alpha = 0$, which corresponds to the fully coupled HMF model; and (ii) $\alpha = \infty$, which means nearest-neighbors interaction. In general, for $\alpha < d$, d being the dimensionality of the system, both the dynamics and the thermodynamics are very similar to the case of the HMF model (see Campa et al. [10] and Tamarit and Anteneodo [23] and references therein).

3 APPLICATION OF NONEXTENSIVE THERMODYNAMICS

In the very same QSS regime in which the thermodynamic anomalies discussed above are observed, one has probability distribution functions and relaxation phenomena that can be interpreted by means of the generalized q-exponential curves predicted by nonextensive thermodynamic formalism [24]. We discuss two examples in the following.

3.1 VELOCITY PROBABILITY DISTRIBUTION FUNCTIONS

During the dynamical evolution, the velocity probability distribution functions (Pdfs), initialized with a uniform distribution, quickly acquire and maintain, for the entire duration of the metastable state, a *non-Gaussian shape*. The QSS velocity Pdfs for three different N values are reported in figure 3(a). In the QSS regime, at small velocities, the Pdfs are wider than the Gaussian equilibrium curve (also reported for comparison), but show a faster decrease for $p > 1.2$. The enhancement for velocities around $p \sim 1$ is consistent with the anomalous diffusion and the Lévy walks observed in the QSS regime [14].

From a dynamical point of view, the stability of the QSS velocity Pdf can be explained by the fact that, for $N \to \infty$ we have $M_{QSS} \to 0$. Therefore, the force on each single spin also vanishes, i.e., $F_i = (-M_x \sin\theta_i + M_y \cos\theta_i) \to 0$. On the other hand, when N is finite, the magnetization is not exactly zero and this generates a small random force, whose strength depends on N, which makes the system eventually evolve into the usual Maxwell-Boltzmann distribution after some time. When this happens, Lévy walks disappear and anomalous diffusion gives way to Brownian diffusion [19]. We fit the non-Gaussian Pdf in figure 3(a) by using the one-particle prescription of the generalized thermodynamics [18]

$$P(p) = \left[1 - (1-q)\frac{p^2}{2T} \right]^{1/(1-q)}. \tag{2}$$

This formula recovers the Maxwell-Boltzmann distribution for $q = 1$. The best fit, shown in figure 4 as a full curve, is obtained by a curve with the entropic index $q = 7$ and the asymptotic temperature $T = 0.38$. The agreement between numerical results and the theoretical curve improves with the size of the system. A finite-size scaling confirming the validity of the fit is reported in figure 3(b).

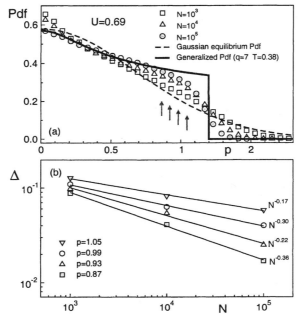

FIGURE 3 In panel (a) we show the velocity Pdfs in the QSS regime for $N = 10^3, 10^4, 10^5$. For comparison, a Gaussian Pdf with the equilibrium value of the temperature $T = 0.4756$ (dashed curve) and a truncated generalized Pdf (eq. (2)) with $q = 7$ and $T = 0.38$ (full curve) are also shown, see text. In panel (b) we show the scaling of the difference between the generalized Pdf and the numerical one $\Delta = |Pdf_{th}(p) - Pdf_{num}(p)|$ vs. N. The points indicated by the arrows in panel (a) are considered.

There, the quantity $\Delta = |Pdf_{th}(p) - Pdf_{num}(p)|$—that is, the difference between the theoretical points and the numerical ones—is shown to go to zero as a power of N, for four values of p. Since $q > 3$, the theoretical curve does not have a finite integral and, therefore, it needs to be truncated with a sharp cutoff to make the total probability equal to one. It is, however, clear that the fitting value $q = 7$ is only an effective nonextensive entropic index [18]. Generalized nonextensive Pdfs have been successfully used to describe turbulent dissipative fluids [8, 12]. Our results indicate Tsallis statistics can also be applied to nonextensive Hamiltonian many-body systems.

3.2 TEMPERATURE RELAXATION

Another interesting application of the generalized nonextensive formalism has been observed recently in this model [21]. Investigating with more accuracy the process of relaxation from the QSS regime to the final equilibrium one, for sys-

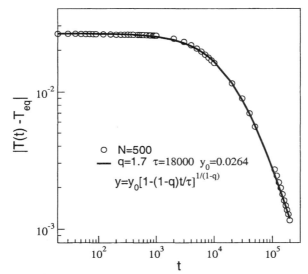

FIGURE 4 We show (open circles) the temporal evolution of the difference between the temperature $T(t) = 2\langle K(t)\rangle/N$ extracted from the numerical simulation with $N = 500$ and the equilibrium value $T_{eq} = 0.4756$. The same data used in figure 1 are plotted. The relaxation is very well reproduced with a q-exponential formula; see eq. (3) in the text. We show the best fit obtained with the entropic index $q = 1.7$, a saturation value $y_0 = 0.0264$, and charateristic time $\tau = 1.8 \cdot 10^3$ as a solid line.

tems of different sizes, it has been found that the temporal evolution of the difference between the temperature $T(t) = 2\langle K(t)\rangle/N$, calculated from the simulation, and the equilibrium value T_{eq} follows exactly a q-exponential curve. That is, the system relaxes to equilibrium according to the formula

$$y(t) = y_0 \left[1 - (1-q)\frac{t}{\tau}\right]^{1/(1-q)} , \qquad (3)$$

where q is the entropic index, τ a characteristic time, and y_0 a saturation value [24]. This is shown in figure 4, where we plot the numerical simulations for $U = 0.69$ and $N = 500$. The same data shown in figure 1 are used. One finds a value of the entropic index equal to $q = 1.7$ and a characteristic relaxation time $\tau = 18000$. A complete study of this relaxation is also in progress for the $\alpha - XY$ model as a function of the range of the interaction and will be published elsewhere [21]. From this investigation, it seems that the general validity of the q-exponential relaxation is confirmed and there are also preliminary indications of a universal law for the index q as a function of the range of the interaction.

Before closing this subsection, we note that a very similar relaxation process has also been found recently for an interesting echo experiment on the Internet [1].

4 CONCLUSIONS

We have discussed the application of Tsallis statistics to nonextensive Hamiltonian systems. In particular, the HMF model has been discussed. We have shown the existence of several anomalies and their connection to the generalized thermodynamic Tsallis formalism. Similar evidence has been confirmed for other classical long-range many-body models [9, 10, 22]. We are convinced that in the near future a more complete and detailed series of studies will clarify the role of nonextensive thermodynamics in Hamiltonian systems and will help us reach the same level of understanding that we have today for dissipative ones [4, 8, 12].

ACKNOWLEDGMENTS

The results discussed here are part of a work in progress in collaboration with C. Tsallis and A. Giansanti.

REFERENCES

[1] Abe, S., and N. Suzuki. "Itineration of the Internet over Nonequilibrium Stationary States in Tsallis Statistics." *Phys. Rev. E* **67** (2003): 016106.

[2] Anteneodo, C., and C. Tsallis. "Breakdown of Exponential Sensitivity to Initial Conditions: Role of the Range of Interactions." *Phys. Rev. Lett.* **80** (1998): 5313–5316.

[3] Antoni, M., and S. Ruffo. "Clustering and Relaxation in Hamiltonian Long-Range Dynamics." *Phys. Rev. E* **52** (1995): 2361–2374.

[4] Baldovin, F., and A. Robledo. "RG Universal Dynamics at the Onset of Chaos in Unimodal Maps and Nonextensive Statistical Mechanics." *Phys. Rev. E* **66** (2002): 045104.

[5] Balescu, R. *Equilibrium and Nonequilibrium Statistical Mechanics.* Wiley, 1974.

[6] Beck, C. "Dynamical Foundation of Nonextensive Statistical Mechanics." *Phys. Rev. Lett.* **87** (2001): 180601.

[7] Beck, C., and E. G. D. Cohen. "Superstatistics." [cond-mat/0205097].

[8] Beck, C., G. S. Lewis, and H. L. Swinney. "Measuring Nonextensitivity Parameters in a Turbulent Couette-Taylor Flow." *Phys. Rev. E* **63** (2001): 035303(R).

[9] Borges, E. P., and C. Tsallis. "Negative Specific Heat in a Lennard-Jones-like Gas with Long-Range Interactions." *Physica A* **305** (2002): 148–151.

[10] Campa, A., A. Giansanti, and D. Moroni. "Metastable States in a Class of Long-Range Hamiltonian Systems." *Physica A* **305** (2002): 137–143.

[11] Cohen, E. G. D. "Statistics and Dynamics." *Physica A* **305** (2002): 19–26.

[12] Kaniadakis, G., M. Lissia, and A. Rapisarda, eds. Proceedings of the International Conference "Nonextensive Thermodynamics and Physical Applications." *Physica A* **305** (2002): 1–350.

[13] Latora, V., and A. Rapisarda. "Microscopic Dynamics of a Phase Transition: Equilibrium vs. Out-of-Equilibrium Regime." *Nucl. Phys. A* **681** (2001): 406c–413c.

[14] Latora, V., A. Rapisarda, and S. Ruffo. "Chaos and Statistical Mechanics in the Hamiltonian Mean Field Model." *Physica D* **131** (1999): 38–54.

[15] Latora, V., A. Rapisarda, and C. Tsallis. "Fingerprints of Nonextensive Thermodynamics in a Long-Range Hamiltonian System." *Physica A* **305** (2002): 129–136.

[16] Latora, V., M. Baranger, A. Rapisarda, and C. Tsallis. "Generalization to Nonextensive Systems of the Rate of Entropy Increase: The Case of the Logistic Map." *Phys. Lett. A* **273** (2000): 97–103.

[17] Latora, V., A. Rapisarda, and S. Ruffo. "Lyapunov Instability and Finite Size Effects in a System with Long-Range Forces." *Phys. Rev. Lett.* **80** (1998): 692–695.

[18] Latora, V., A. Rapisarda, and C. Tsallis. "Non-Gaussian Equilibrium in a Long-Range Hamiltonian System." *Phys. Rev. E* **64** (2001) : 056134–1.

[19] Latora,V., A. Rapisarda, and S. Ruffo. "Superdiffusion and Out-of-Equilibrium Chaotic Dynamics with Many Degrees of Freedoms." *Phys. Rev. Lett.* **83** (1999): 2104–2107 .

[20] Pais, A. *Subtle is the Lord.* New York: Oxford University Press, 1982.

[21] Rapisarda, A., C. Tsallis, and A. Giansanti. "Relaxation to Equilibrium in the d–xy Model." Unpublished manuscript (in preparation), 2002.

[22] Sota, Y., O. Iguchi, M. Morikawa, T. Tatekawa, and K. Maeda. "Origin of Scaling Structure and Non-Gaussian Velocity Distribution in a Self-Gravitating Ring Model." *Phys. Rev. E* **64** (2001): 056133.

[23] Tamarit, F., and C. Anteneodo. "Rotators with Long-Range Interactions: Connection with the Mean-Field Approximation." *Phys. Rev. Lett.* **84** (2000): 208–211.

[24] Tsallis, C. "Possible Generalization of Boltzmann-Gibbs Statistics." *J. Stat. Phys.* **52** (1988): 479–487.

[25] Tsallis, C., J. C. Anjos, and E. P. Borges. "Fluxes of Cosmic Rays: A Delicately Balanced Anomalous Thermal Equilibrium." [astro-ph/0203258].

[26] Wilk, G., and Z. Wlodarczyk. "Interpretation of the Nonextensivity Parameter q in Some Application of Tsallis Statistics and Lévy Distributions." *Phys. Rev. Lett.* **84** (2000): 2770–2773.

A Hamiltonian Approach for Tsallis Thermostatistics

J. S. Andrade, Jr.
M. P. Almeida
A. A. Moreira
A. B. Adib
G. A. Farias

1 INTRODUCTION

Since the pioneering work of Tsallis in 1988 [15] in which a nonextensive generalization of the Boltzmann-Gibbs (BG) formalism for statistical mechanics was proposed, intensive research has been dedicated to the development of the conceptual framework behind this new thermodynamical approach and to its application to realistic physical systems. In order to justify the Tsallis generalization, it has been frequently argued that the BG statistical mechanics has a domain of applicability restricted to systems with short-range interactions and non-(multi)fractal boundary conditions [14]. Moreover, it has been recalled that anomalies displayed by mesoscopic dissipative systems and strongly non-Markovian processes represent clear evidence of the departure from BG thermostatistics. These types of arguments have been duly reinforced by recent convincing examples of physical systems that are far better described in terms of the generalized formalism than in the usual context of the BG thermodynamics (see Tsallis [14] and references therein). It thus became evident that the intrinsic

Nonextensive Entropy—Interdisciplinary Applications
edited by Murray Gell-Mann and Constantino Tsallis, Oxford University Press   123

nonlinear features present in the Tsallis formalism that lead naturally to power laws represent powerful ingredients for the description of complex systems.

In the majority of studies dealing with the Tsallis thermostatistics, the starting point is the expression for the generalized entropy S_q,

$$S_q = \frac{k}{q-1} \left[1 - \int [f(x)]^q dx \right] \,, \tag{1}$$

where k is a positive constant, q a parameter, and f is the probability distribution. Under a different framework, some interesting studies [8] have shown that the parameter q can be somehow linked to the system sensibility on initial conditions. Few works have been committed to substantiate the form of entropy (1) in physical systems based entirely on first principles [1, 13]. For example, it has been demonstrated that it is possible to develop dynamical thermostat schemes that are compatible with the generalized canonical ensemble [12].

In a recent study by Almeida [5], a derivation of the generalized canonical distribution is presented from first principle statistical mechanics. As a consequence, it is shown that the particular features of a macroscopic subunit of the canonical system, namely, the heat bath, determines the nonextensive signature of its thermostatistics and therefore its power-law behavior. More precisely, it is exactly demonstrated in Almeida [5] that if one specifies

$$\frac{d}{dE} \left(\frac{1}{\beta} \right) \propto q - 1 \,, \tag{2}$$

where q is a constant and $\beta = d \ln \Omega(E)/dE$, with $\Omega(E)$ being the density of states (structure function) of the heat bath, the generalized canonical distribution maximizing (1) for the Hamiltonian H of the system is recovered,

$$f(H) \propto [1 + \beta(q-1)(E-H)]^{\frac{1}{q-1}} \,. \tag{3}$$

Here, E is a conserved quantity and denotes the energy of the extended system (system+heat bath). Equation (2) provides a very simple but meaningful connection between the generalized q-statistics and the thermodynamics of nonextensive systems. It is analogous to state that, if the condition of an infinite heat bath capacity is violated, the resulting canonical distribution can no longer be of the exponential form and therefore should not follow the traditional BG thermostatistics.

Inspired by the conjecture proposed in Almeida [5], we first review recent results [2] showing that finite systems whose Hamiltonians obey a generalized homogeneity relation follow precisely the nonextensive Tsallis thermostatistics. In the thermodynamical limit, these results indicate that the Boltzmann-Gibbs statistics are always recovered, regardless of the type of potential among interacting particles. In the second part, we follow the same type of conceptual arguments to develop a variant of the Nosé thermostat and derive the Hamiltonian of a nonextensive system that is compatible with the canonical ensemble of the generalized Tsallis thermostatistics [6]. For the case of a simple one-dimensional harmonic oscillator, we confirm by numerical simulation of the dynamics that the distribution of energy H follows precisely the canonical q-statistics for different values of the parameter q. The approach is further tested for classical many-particle systems by means of molecular dynamics simulations.

2   TSALLIS THERMOSTATISTICS FOR FINITE SYSTEMS

The theoretical approach presented here for the Tsallis thermostatistics is entirely based on standard methods of statistical mechanics [2, 5]. In a shell of constant energy, we consider a system whose Hamiltonian can be written as a sum of two parts,

$$H(\mathbf{r}) = H_1(\mathbf{r}_1) + H_2(\mathbf{r}_2, \mathbf{r}_3, \dots) \,, \tag{4}$$

where,

$$\mathbf{r} = (\mathbf{r}_1, \mathbf{r}_2, \mathbf{r}_3, \dots) = (r_1, \dots, r_{2N}) \,, \qquad \text{with}$$
$$\mathbf{r}_1 = (r_1, \dots, r_{n_1}),$$
$$\mathbf{r}_2 = (r_{n_1+1}, \dots, r_{n_1+n_2}),$$
$$\mathbf{r}_3 = (r_{n_1+n_2+1}, \dots, r_{n_1+n_2+n_3}), \dots \,.$$

The aim is to show that, if H_2 satisfies a generalized homogeneity relation of the type

$$\lambda H_2(\mathbf{r}_2, \mathbf{r}_3, \dots) = H_2(\lambda^{1/a_2}\mathbf{r}_2, \lambda^{1/a_3}\mathbf{r}_3, \dots) \,, \tag{5}$$

where $\lambda, a_2, a_3, \dots$ are non-null real constants, then the correct statistics for $H_1(\mathbf{r}_1)$ is the one proposed by Tsallis. The foregoing derivation is based on a simple scaling argument, but we shall draw parallels to Almeida [5] whenever appropriate.

The structure function (density of states) for H_2 at the energy level E is given by

$$\Omega_2(E) = \int \delta(H_2(\mathbf{r}_2, \mathbf{r}_3, \dots) - E)\mathrm{d}V_2 \,, \tag{6}$$

where $\mathrm{d}V_2 = \mathrm{d}^{n_2}\mathbf{r}_2\mathrm{d}^{n_3}\mathbf{r}_3 \dots$ is the volume element in the subspace spanned by $(\mathbf{r}_2, \mathbf{r}_3, \dots)$. For systems satisfying eq. (5), this function can be evaluated by taking $\lambda > 0$ and computing

$$
\begin{aligned}
\Omega_2(\lambda E_0) &= \int \delta(H_2(\mathbf{r}_2, \mathbf{r}_3, \dots) - \lambda E_0) dV_2 \\
&= \frac{1}{\lambda} \int \delta(\lambda^{-1} H_2(\mathbf{r}_2, \mathbf{r}_3, \dots) - E_0) dV_2 \\
&= \frac{1}{\lambda} \int \delta(H_2(\lambda^{-\frac{1}{a_2}} \mathbf{r}_2, \lambda^{-\frac{1}{a_3}} \mathbf{r}_3, \dots) - E_0) dV_2 \\
&= \frac{1}{\lambda} \int \delta(H_2(\mathbf{r}_2', \mathbf{r}_3', \dots) - E_0) \lambda^{\frac{n_2}{a_2} + \frac{n_3}{a_3} + \cdots} dV_2' \\
&= \lambda^{\frac{1}{(q-1)}} \Omega_2(E_0) \,,
\end{aligned}
\tag{7}
$$

where we define

$$
\frac{1}{q-1} \equiv \sum_{i=2,3,\dots} \frac{n_i}{a_i} - 1,
\tag{8}
$$

and denote $\mathbf{r}_2' = \lambda^{-1/a_2} \mathbf{r}_2$, $\mathbf{r}_3' = \lambda^{-1/a_3} \mathbf{r}_3$, etc., and $dV_2' = d^{n_2} \mathbf{r}_2' d^{n_3} \mathbf{r}_3' \dots$. Hence, if Ω_2 is defined at a value E_0, it is also defined at every $E = \lambda E_0$, with $\lambda > 0$. We can then write

$$
\Omega_2(E) = \left(\frac{E}{E_0} \right)^{\frac{1}{q-1}} \Omega_2(E_0) \,,
\tag{9}
$$

and express the canonical distribution law over the phase space of H_1 as [10]

$$
\begin{aligned}
f(\mathbf{r}_1) &= \frac{\Omega_2(H - H_1(\mathbf{r}_1))}{\Omega(H)} \\
&= \frac{\Omega_2(H)}{\Omega(H)} \left(1 - \frac{H_1(\mathbf{r}_1)}{H} \right)^{\frac{1}{q-1}} \,,
\end{aligned}
\tag{10}
$$

where H is the total energy of the joint system composed by H_1 and H_2, and $\Omega(H)$ is its structure function,

$$
\Omega(H) = \int \delta(H_1(\mathbf{r}_1) + H_2(\mathbf{r}_2, \mathbf{r}_3, \dots) - H) dV_1 dV_2 \,,
\tag{11}
$$

where dV_1 and dV_2 are the infinitesimal volume elements of the phase spaces of H_1 and H_2, respectively. Comparing eq. (10) with the distribution in the form of eq. (3), we get the following relation between q, β, and H:

$$
\beta = \frac{1}{(q-1)H} \,.
\tag{12}
$$

As already mentioned, previous studies have shown that the Tsallis distribution (eq. (3)) is compatible with some anomalous "canonical" configurations where the heat bath is finite [13] or composes a peculiar type of extended phase-space dynamics [6]. As demonstrated in Almeida [5] and Adib [2], the observation of the Tsallis distribution may simply reflect the finite size of a thermal environment with the property (5), the thermodynamical limit corresponding to $q \to 1$ in eq. (8). We emphasize that, although similar conclusions could be drawn from Plastino [13] and Andrade [6], the theoretical framework introduced in Almeida [5] and Adib [2] permits us to put forward a rigorous realization of the q-thermostatistics: it stems from the *weak* coupling of a system to a "heat bath" whose Hamiltonian is a homogeneous function of its coordinates, the value of q being completely determined by its degree of homogeneity (eq. (8)).

2.1   APPLICATION TO A CLASSICAL SYSTEM OF N PARTICLES

As a specific application of the previous results, we investigate the form of the momenta distribution law for a classical N-body problem in d dimensions. The Hamiltonian of such a system can be written as

$$
\begin{aligned}
H(\mathbf{p}, \mathbf{q}) &= \frac{1}{2} \sum_{i=1}^{N} \mathbf{p}_i^2 + V(\mathbf{q}_1, \dots, \mathbf{q}_N) \\
&= H_1(\mathbf{p}_1, \dots, \mathbf{p}_N) + H_2(\mathbf{q}_1, \dots, \mathbf{q}_N),
\end{aligned}
\tag{13}
$$

where we define $H_1 \equiv \frac{1}{2} \sum_{i=1}^{N} \mathbf{p}_i^2$, $\mathbf{p}_i = (p_{i1}, \dots, p_{id})$ is the linear momentum vector of an arbitrary particle i (hence the number of degrees of freedom of system 1 is $n_1 = Nd$), and H_2 (the "bath") is due to a homogeneous potential $V(\mathbf{q})$ of degree α; i.e., $\lambda V(\mathbf{q}) = V(\lambda^{1/\alpha}\mathbf{q})$ with $\mathbf{q} = (\mathbf{q}_1, \dots, \mathbf{q}_N)$. At this point, we emphasize that the distinction between "system" and "bath" is merely formal and does not necessarily involve a physical boundary. It relies solely on the fact that we can decompose the total Hamiltonian into two parts [10]. By making the correspondences $\mathbf{r}_1 = (\mathbf{p}_1, \dots, \mathbf{p}_N)$, $\mathbf{r}_2 = (\mathbf{q}_1, \dots, \mathbf{q}_N)$, $n_1 = Nd$, $n_2 = Nd$, $a_1 = 2$, and $a_2 = \alpha$, the homogeneity relation (5) is satisfied. From eq. (10) it then follows that

$$
f(\mathbf{p}_1, \dots, \mathbf{p}_N) \propto \left[H - \frac{1}{2} \sum_{i=1}^{N} \mathbf{p}_i^2 \right]^{\frac{1}{q-1}},
\tag{14}
$$

where the nonextensivity measure q is given by

$$
\frac{1}{q-1} = \frac{Nd}{\alpha} - 1 \quad \Rightarrow \quad q = \frac{Nd}{Nd - \alpha}.
\tag{15}
$$

It is often argued that the range of the forces should play a fundamental role in deciding between the BG or Tsallis formalisms to describe the thermostatistics of an N-body system [14]. Recall that, for a d-dimensional system, an interaction is said to be long ranged if $\alpha \geq -d$. Within this regime, the Tsallis thermostatistics is expected to apply, while for $\alpha < -d$ the system should follow the standard BG behavior [14]. This conjecture is not confirmed by the results of the problem at hand. Indeed, eq. (14) is consistent with the generalized q-distribution (eq. (3)) no matter what the value of α is, as long as it is non-null and N is finite. In the limit $N \to \infty$, however, we always get $q \to 1$, with the value of α determining the shape of the curve $q = q(N)$. If $\alpha > 0$, q approaches the value 1 from above, while for $\alpha < 0$ the value of q is always less than 1. Therefore, for (ergodic) classical systems with N particles interacting through a homogeneous potential at constant β, the *equilibrium* distribution of momenta always goes to the Boltzmann distribution, $f(\mathbf{r}_1) \propto \exp[-\beta H_1(\mathbf{r}_1)]$, when $N \to \infty$.

In order to corroborate this method, numerical simulations have been performed to investigate the statistical properties of a linear chain of anharmonic oscillators [2]. Besides the kinetic term, the Hamiltonian includes both on-site and nearest-neighbors quartic potentials, i.e.,

$$H = \sum_{i=1}^{N} \frac{p_i^2}{2} + \sum_{i=1}^{N} \frac{q_i^4}{4} + \sum_{i=1}^{N} \frac{(q_{i+1} - q_i)^4}{4}. \tag{16}$$

The choice of this system is inspired by the so-called Fermi-Pasta-Ulam (FPU) problem, originally devised to test whether statistical mechanics is capable or not of describing dynamical systems with a small number of particles [9]. The equations of motion are obtained from eq. (16) and integrated numerically. A detailed analysis concerning the ergodicity of this dynamical system is presented in Adib [2]. In figure 1 we show the logarithmic plot of the resulting distribution $f(\mathbf{p}_1, \mathbf{p}_2, \dots)$ against the transformed variable $H - H_1$ for systems with $N = 8$, 16, 32, and 64 oscillators. Assuming ergodicity, the distribution of momenta from the fluctuations in time of H_1 may be computed as

$$f(H_1) \propto H_1^{(N/2)-1} f(\mathbf{p}_1, \mathbf{p}_2, \dots), \tag{17}$$

where the $H_1^{(N/2)-1}$ factor accounts for the degeneracy of the momenta consistent with the magnitude of H_1 (cf. Khinchin [10]). Indeed, we observe in all cases that the fluctuations in H_1 follow very closely the prescribed power-law behavior (eq. (14)), with exponents given by eq. (15). These results, therefore, provide clear evidence for the validity of the dynamical approach proposed in Adib [2] to the generalized thermostatistics of finite systems.

3 MOLECULAR DYNAMICS FOR THE GENERALIZED THERMOSTATISTICS: THE THERMOSTAT OF TSALLIS

As already mentioned in the introduction, eq. (2) provides a very simple but meaningful connection between the generalized q-statistics and the thermodynamics of nonextensive systems. It is analogous to state that, if the condition

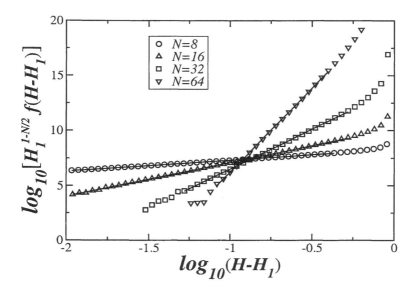

FIGURE 1   Logarithmic plot of the rescaled distribution $H_1^{1-N/2} f(H - H_1)$ as a function of the transformed variable $H - H_1$ for $N = 8$ (circles), 16 (triangles up), 32 (squares), and 64 (triangles down) anharmonic oscillators. The solid straight lines are the best fit to the simulation data of the expected power-law behavior eq. (14). The slopes are 1.0068 (1.0), 3.07 (3.0), 7.21 (7.0), and 15.27 (15.0) for $N = 8$, 16, 32, and 64, respectively, and the numbers in parentheses indicate the expected values obtained from eq. (15).

of an infinite heat bath capacity is violated, the resulting canonical distribution can no longer be of the exponential form and therefore should not follow the traditional BG thermostatistics. In this section, we review how the conjecture proposed in Almeida [5] has been used to develop a variant of the original Nosé thermostat [11] that is consistent with the q-thermostatistics [6].

We consider a system of N particles having coordinates x_i', masses m_i, and potential energy $\Phi(x')$. As in the extended system method originally proposed by Nosé [11], here we also introduce an additional degree of freedom through a variable s, which will play the role of an external heat bath and which will keep the average of the kinetic energy at a constant value. In practice, this is achieved by simply rescaling the *real* variables in terms of a new set of *virtual* variables

$$\mathbf{x}_i' = \mathbf{x}_i \ , \quad \mathbf{p}_i' = \frac{\mathbf{p}_i}{s^\lambda} \ , \quad p_s' = \frac{p_s}{s^\lambda} \ , \quad t' = \int \frac{dt}{s} \ , \tag{18}$$

where λ is a rescaling exponent, and $(\mathbf{x}_i', \mathbf{p}_i', t')$ and $(\mathbf{x}_i, \mathbf{p}_i, t)$ are the real and virtual coordinates, momenta and time, respectively. At this point, we postulate that a generalized Hamiltonian for the extended system can be written as

$$H_q(\mathbf{x}, \mathbf{p}, p_s, s) = \sum_{i=1} \frac{\mathbf{p}_i^2}{2m_i s^{2\lambda}} + \Phi(\mathbf{x}) + \frac{p_s^2}{2Q} + \frac{1}{\alpha} \frac{s^\gamma - 1}{\gamma} , \qquad (19)$$

where the first two terms on the right side represents the energy of the physical system that is free to fluctuate. The virtual variable p_s also has a real counterpart, $p_s' = p_s/s^\lambda$, and has been introduced to allow for a dynamical description of the variable s. More precisely, the third term $p_s^2/2Q$ corresponds to the kinetic energy of the heat bath and the parameter Q is an inertial factor associated with the motion of the variable s. The last term of the Hamiltonian (19) is a power-law potential in s with α and γ as parameters. As we show next, it provides the essential link between the concept of extended phase-space dynamics and the generalized canonical ensemble. A similar Hamiltonian expression has been previously used in Nosé [11] as a counterexample to demonstrate that $\ln(s)$ is the only possible potential form leading to the BG canonical distribution.

We start by considering the quasi-ergodic hypothesis and writing the time average of a given quantity (in the virtual time scale) $A(\mathbf{x}, \mathbf{p})$ as

$$\overline{A} = \frac{1}{Z} \int \int \int \int d\mathbf{x} \, d\mathbf{p} \, dp_s \, ds \, A\delta(H_q - E) \quad \text{with} \qquad (20)$$

$$Z = \int \int \int \int d\mathbf{x} \, d\mathbf{p} \, dp_s \, ds \, \delta(H_q - E) ,$$

where Z is analogous to a microcanonical normalization factor for the generalized Hamiltonian (19). Transforming the virtual momenta $\mathbf{p}$ and coordinates $\mathbf{x}$ back to real variables, changing the order of integration and rewriting the volume element as $d\mathbf{x} \, d\mathbf{p} = s^{g\lambda} d\mathbf{x}' d\mathbf{p}'$, where g is the number of degrees of freedom, we obtain

$$\overline{A} = \frac{1}{Z} \int \int d\mathbf{x}' d\mathbf{p}' A \int \int dp_s \, ds \, s^{g\lambda} \delta(H_q - E) . \qquad (21)$$

If we now make use of the property of the δ function, $\delta[h(s)] = \delta(s - s_0)/h'(s_0)$, where s_0 is the zero of h, it follows that

$$\overline{A} = \frac{1}{Z} \int \int d\mathbf{x}' \, d\mathbf{p}' \, A$$

$$\times \int dp_s \, \alpha \left[1 + \alpha\gamma \left(E - H - \frac{p_s^2}{2Q} \right) \right]^{\frac{g\lambda+1}{\gamma} - 1} , \qquad (22)$$

where

$$H = H(\mathbf{x}', \mathbf{p}') = \sum_{i=1} \frac{\mathbf{p}_i'^2}{2m_i} + \Phi(\mathbf{x}') . \qquad (23)$$

By integration with respect to p_s we get

$$\overline{A} = \frac{1}{Z}\left(\frac{\alpha Q}{2\gamma}\right)^{1/2} B\left(\frac{1}{2}, \frac{g\lambda+1}{\gamma}\right)$$
$$\times \int\int d\mathbf{x}'\,d\mathbf{p}'\,A[1 + \alpha\gamma(E-H)]^{\frac{g\lambda+1}{\gamma}-\frac{1}{2}}, \qquad (24)$$

where B is the beta function, $B(z, w) = \int_0^1 t^{z-1}(1-t)^{w-1}dt$. Finally, if we define

$$\alpha \equiv \frac{\beta(q+1)}{2(g\lambda+1)} \quad \text{and} \quad \gamma \equiv 2(g\lambda+1)\frac{(q-1)}{(q+1)}, \qquad (25)$$

the generalized canonical average is recovered

$$\overline{A} = \frac{1}{Z'}\int\int d\mathbf{x}'d\mathbf{p}'\,A\,[1+\beta(q-1)(E-H)]^{\frac{1}{(q-1)}} \quad \text{with} \qquad (26)$$
$$Z' = \int\int d\mathbf{x}'d\mathbf{p}'\,[1+\beta(q-1)(E-H)]^{\frac{1}{(q-1)}},$$

and we have thus proved that, under conservation of the extended Hamiltonian (eq. (19)), the fluctuations in the energy $H(\mathbf{x}', \mathbf{p}')$ of the physical system should be consistent with the canonical formulation of the nonextensive q-thermostatistics [7]. To obtain the time average in the real time scale, it is necessary to replace everywhere $g\lambda$ by $(g\lambda - 1)$ [11]. In addition, if the total momentum of the system is conserved, $\sum_{i=1}\mathbf{p}'_i = 0$, it is necessary to replace g by $g - 3$ in the expressions of α and γ, and insert the δ-function $\delta(\sum_{i=1}\mathbf{p}'_i)$ in the integrals of eq. (26) [11].

3.1 NUMERICAL APPLICATIONS OF THE GENERALIZED THERMOSTAT

As a simple realization of the generalized thermostat scheme, we consider an extended system composed of a single one-dimensional harmonic oscillator coupled to a heat bath whose thermal capacity obeys essentially eq. (2). From eq. (19), such a system can be described by the following extended Hamiltonian [6]:

$$H_q(x, p, p_s, s) = \frac{p^2}{2s^{2\lambda}} + \frac{x^2}{2} + \frac{p_s^2}{2Q} + \frac{1}{\alpha}\frac{s^\gamma - 1}{\gamma}. \qquad (27)$$

Here $m = 1$ for simplicity and we choose to set $\lambda = 2$ because the nonlinear dynamics for this case when $q = 1$ (i.e., for the BG thermostatistics) has been shown to be sufficiently chaotic to generate average properties of the canonical ensemble [16]. From eq. (27) and the scaling relations (eq. (18)), we obtain the equations of motion for the extended system in the *real* phase space,

$$\frac{dx'}{dt'} = \frac{p'}{s},$$

$$\frac{dp'}{dt'} = -\frac{x'}{s} - \frac{2s^2 p'_s p'}{Q},$$

$$\frac{ds'}{dt'} = \frac{s^3 p'_s}{Q},$$

$$\frac{dp'_s}{dt'} = \frac{1}{s^2}\left(2p'^2 - \frac{1}{\alpha}s^\gamma\right) - \frac{2s^2 p'^2_s}{Q}. \tag{28}$$

A fifth-order Runge-Kutta subroutine is then used to numerically solve this set of nonlinear differential equations. To ensure the conservation of energy H_q and the stability of integration, all runs have been performed with 10^8 time steps of $\Delta t' = 10^{-4}$ each. The density maps shown in figures 2(a) and (b) for $q = 0.8$ and 1.2, respectively, provide clear evidence that the dynamics of both systems fills space. For $q > 1$ [see fig. 2(b)], the accessible phase space lies in a compact set, whereas the phase-space support for $q < 1$ [see fig. 2(a)] is infinite. The former situation is compatible with the necessary cut-off condition on energy for $q > 1$ [14]. In figure 3 we show the logarithmic plot of the distributions of the transformed variable $\chi = 1 + \beta(q-1)(E - H)$, where $H = p'^2/2 + x'^2/2$, for three different values of the parameter $\gamma = 4(q - 1)/(q + 1)$ corresponding to $q = 0.7$, 0.8, and 0.9. Indeed, we observe in all cases that the fluctuations in χ follow very closely the prescribed power-law behavior, $\rho(\chi) \propto \chi^{1/(q-1)}$, and therefore confirm the validity of our dynamical approach to the generalized canonical ensemble. As shown in figure 4, the simulations performed for $q > 1$ are also compatible with the expected scaling behavior. However, instead of the long-range tail obtained for the case $q < 1$, a rather unusual power law with positive exponent is observed.

Now we focus on a more complex application of the thermostat scheme introduced in Andrade [6]. The basic idea is to simulate, through molecular dynamics (MD), the nonextensive behavior of a classical many-particle system. For completeness, we start by rewriting the expression for the extended Hamiltonian (eq. (19)) in terms of the usual q-thermostatistics parameters (i.e., q and β)

$$H_q(\mathbf{x}, \mathbf{p}, p_s, s) = \sum_{i=1} \frac{\mathbf{p}_i^2}{2m_i s^{2\lambda}} + \Phi(\mathbf{x}) + \frac{p_s^2}{2Q} + \frac{1}{\beta}\ln_q\left(s^{\frac{2(g\lambda+1)}{q+1}}\right), \tag{29}$$

where $\ln_q(s) \equiv (s^{q-1} - 1)/(q - 1)$ [14]. From eq. (29), it is then possible to derive the equations of motion for any value of q and any type of effective potential of interaction. We consider a cell containing 108 identical particles that interact through the Lennard–Jones potential, $\Phi(\Delta x_{ij}) = 4\epsilon[(\sigma/\Delta x_{ij})^{12} - (\sigma/\Delta x_{ij})^6]$, where Δx_{ij} is the distance between particles i and j, ϵ is the minimum energy, and σ the zero of the potential. Distance, energy, and time are measured in units of σ, ϵ, and $(m\sigma)^2/\epsilon$, respectively, and the equations of motion are numerically

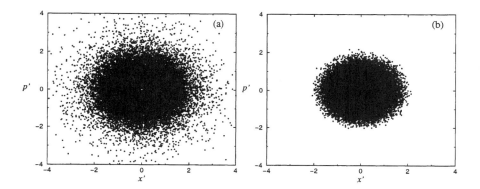

FIGURE 2 (a) Density plot of the harmonic oscillator dynamics subjected to the generalized thermostat scheme for $q = 0.8$. The initial conditions are $[x'(0) = 0.5, p'(0) = 0.5, s(0) = 1.0, p'_s(0) = 0.0]$ and the thermostat parameters have been set to $\alpha = 1.0$ and $Q = 1.0$. (b) Same as (a) but for $q = 1.2$.

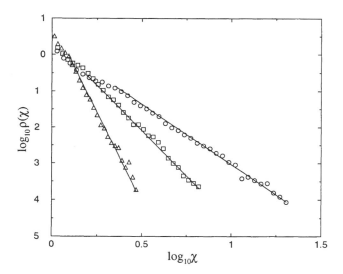

FIGURE 3 Logarithmic plot of the distributions of the transformed variable χ for $q = 0.7$ (circles), 0.8 (squares), and 0.9 (triangles). From right to left, the three straight lines with slopes -3.33, -5.0, and -10.0 correspond to the expected power-law behavior $\rho(\chi) \propto \chi^{1/(q-1)}$.

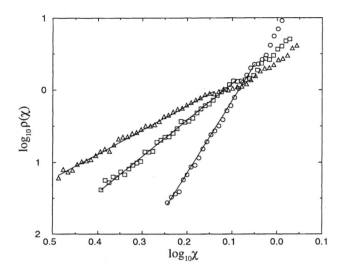

FIGURE 4 Logarithmic plot of the distributions of the transformed variable χ for $q = 1.1$ (circles), 1.2 (squares), and 1.3 (triangles). From right to left, the three straight lines with slopes 10.0, 5.0, and 3.33 correspond to the expected power-law behavior $\rho(\chi) \propto \chi^{1/(q-1)}$.

integrated using a predictor-corrector algorithm [4]. In all simulations performed, the relative fluctuation around the average of the total energy of the system has always been smaller than 10^{-6}.

Comparedindexmolelcular dynamics, for generalized thermostatistics to the previous example of a single harmonic oscillator, the complexity of the many-particle system hinders a quantitative prediction of the statistical behavior of its energy fluctuations. Because the exact form or even a plausible approximation of the density of states $\Omega(H)$ is difficult to obtain in this case, we restrict ourselves to the qualitative analysis of the resulting energy distribution $\rho(H) \propto \Omega(H)f(H)$. Furthermore, additional simulation tests have been performed with a different number of particles and physical conditions to confirm that the MD system always leads to unstable trajectories in phase space whenever the value of q is set to be smaller than a given threshold q_{min}. In spite of these limitations, however, the results shown in figure 5 clearly indicate the tendency for a broader distribution of energy when $q < 1$ (q has been set to be slightly larger than $q_{min} \approx 0.9940$ in this case). For $q > 1$, on the other hand, the resulting distribution of energy is notably more confined than the Gaussian-like distribution obtained for $q = 1$ (see fig. 5).

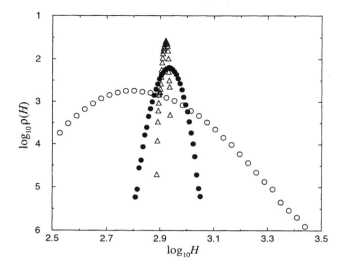

FIGURE 5 Logarithmic plot of the energy distributions for $q = 0.9941$ (circles), 1.0 (full circles) and 1.1 (triangles). In all three cases, the MD simulations have been performed with 108 particles, $\beta = 0.2$, and a density of 0.1 particles/σ^3.

4 CONCLUSIONS

In conclusion, we review here the results of some recent studies [2, 5, 6] on the generalized Tsallis thermostatistics that are entirely based on standard methods of statistical mechanics. They provide a deterministic connection between the generalized entropy and microdynamics and reveal a possible origin for the intrinsic nonlinear features present in the Tsallis formalism that lead naturally to power-law behavior. In particular, it has been shown that an adequate physical setting for the Tsallis formalism should be found in the physics of finite systems [2]. Therefore, the generalized Tsallis formalism can be applied to homogeneous Hamiltonian systems to engender an adequate theoretical framework for the statistical mechanics of finite systems. These results clearly indicate that the range of the interacting potential should play no role in the equilibrium statistical properties of a system in the thermodynamic limit. Under these conditions, the conventional BG thermostatistics remains valid and general. In a similar context, we have also reviewed the methodology proposed in Andrade [6] where it has been demonstrated that the essential features of the generalized canonical distribution can be captured with a proper extension of the standard Nosé thermostat [11]. More precisely, a Hamiltonian approach has been developed to the nonextensive q-thermostatistics that leads explicitly to the observation of power-law behavior.

ACKNOWLEDGMENTS

This work has been supported by the Brazilian Agencies CNPq, CAPES, and FUNCAP.

REFERENCES

[1] Abe, S., and A. K. Rajagopal. "Nonuniqueness of Canonical Ensemble Theory Arising from Microcanonical Basis." *Phys. Lett. A* **272** (2000): 341–345.

[2] Adib, A. B., A. A. Moreira, J. S. Andrade, Jr., and M. P. Almeida. "Tsallis Thermostatistics for Finite Systems: A Hamiltonian Approach." April 2002. arXiv.org e-Print Archive, Quantum Physics, Cornell University. September 2002. ⟨http://xxx.lanl.gov/abs/cond-mat/0204034⟩.

[3] Alemany, P. A., and D. H. Zanette "Fractal Random Walks from a Variational Formalism for Tsallis Entropies." *Phys. Rev. E* **49** (1994): R956–R958.

[4] Allen, M. P., and D. J. Tildesley. *Computer Simulation of Liquids*. Oxford: Pergamon, 1987.

[5] Almeida, M. P. "Generalized Entropies from First Principles." *Physica A* **300** (2001): 424–432.

[6] Andrade, Jr., J. S., M. P. Almeida, A. A. Moreira, and G. A. Farias. "Extended Phase-Space Dynamics for the Generalized Nonextensive Thermostatistics." *Phys. Rev. E* **65** (2002): 036121.

[7] Cho, K., J. D. Joannopoulos, and L. Kleinman. "Constant-Temperature Molecular Dynamics with Momentum Conservation." *Phys. Rev. E* **47** (1993): 3145–3151.

[8] Costa, U. M. S., M. L. Lyra, A. R. Plastino, and C. Tsallis. "Power-Law Sensitivity to Initial Conditions within a Logisticlike Family of Maps: Fractality and Nonextensivity." *Phys. Rev. E* **56** (1997): 245–250.

[9] Fermi, E., J. Pasta, S. Ulam, and M. Tsingou. "Studies in Nonlinear Problems I." Preprint LA-1940, Los Alamos National Lab, Los Alamos, NM, November 7, 1955; reprinted in E. Fermi, *Collected Papers Vol. II,* 978 (Chicago: University of Chicago Press, 1965).

[10] Khinchin, A. I. *Mathematical Foundations of Statistical Mechanics*. New York: Dover 1949.

[11] Nosé, S. "A Unified Formulation of the Constant Temperature Molecular-Dynamics Methods." *J. Chem. Phys.* **81** (1984): 511–519.

[12] Plastino, A. R., and C. Anteneodo. "A Dynamical Thermostating Approach to Nonextensive Canonical Ensembles." *Ann. Phys.* **255** (1997): 250–269.

[13] Plastino, A. R., and A. Plastino. "From Gibbs Microcanonical Ensemble to Tsallis Generalized Canonical Distribution." *Phys. Lett. A* **193** (1994): 140–143.

[14] Tsallis, C. "Nonextensive Statistics: Theoretical, Experimental and Computational Evidences and Connections." *Braz. J. Phys.* **29** (1999): 1–35.

[15] Tsallis, C. "Possible Generalization of Boltzmann-Gibbs Statistics." *J. Stat. Phys.* **52** (1988): 479–487.

[16] Winkler, R. G. "Extended-phase-space Isothermal Molecular-Dynamics—Canonical Harmonic-Oscillator." *Phys. Rev. A* **45** (1992): 2250–2255.

Nonequilibrium Systems

Robin Stinchcombe

Nonequilibrium system and behavior are briefly reviewed, with an emphasis on recent progress using minimal microscopic models and on implications for macroscopic descriptions.

1 INTRODUCTION

Our daily life continually confronts us with large systems whose internal processes or external influences are such that standard physical equilibrium descriptions of macroscopic behavior do not apply. Among complex examples are weather, crowds, traffic, financial markets, and so on, and at the other end of the spectrum are simple queuing, processing, and decision-making setups.

The most common, most interesting and most complex examples in nature are predominately collective stochastic systems, in which many constituents influence/interact with each other in some way, and the processes are probabilistic and dissipative. This is true of most examples given above, certainly of the first ones.

None of these achieves ordinary equilibrium states of the sort met in thermodynamics; such systems are generically called nonequilibrium systems (NES). In the case of weather, a reason for not going into standard equilibrium is the sun's continual heating of the earth's land surface, oceans, and atmosphere. A feeding mechanism like that also occurs in traffic systems through the entry and exit of vehicles. In addition, traffic transition rates are not set by thermodynamic balances. Traffic can achieve steady states of flow or jammed states. In common with most other nonequilibrium (NE) steady states, these are quite unlike the equilibrium states provided by the standard "general" macroscopic and microscopic descriptions of thermodynamics and statistical mechanics (see Boltzmann and Gibbs).

Nevertheless, NES show many similarities to collective equilibrium systems (ES), largely in behavior at a quantitative level. For example, NES and ES classes both include systems showing phase transitions (e.g., in the NE steady state, or in the thermal equilibrium state, respectively), whose phenomenology can typically be qualitatively interpreted in similar terms (using concepts of order parameter, scale invariance, and power laws, etc.).

Yet despite sharing many qualitative features, nonequilibrium and equilibrium phenomena are presently approached in fundamentally different manners, because no description of the generality and power of equilibrium thermodynamics and statistical mechanics presently exists for NES, nor is it known whether one is even, in principal, generally possible, though one hopes it might be.[1] This raises questions worthy of the widest possible dissemination and exploration.

Lacking the general Boltzmann-Gibbs approaches, the quantitative descriptions of the NE class are, at present, usually limited to treatments of specific microscopic models from a kinetic (Master Equation) point of view, which is the approach we have adopted [48].

Despite the *apparent* limitations of such an approach, great advances have been made, particularly in the last decade and particularly concerning mechanisms and characteristics. Very simple collective stochastic nonequilibrium (SNE) models [3, 17, 45, 48, 58] are now known to exhibit a wide range of the phenomena seen in real NES. Moreover, concepts of "universality" have been seen to apply, as in common types of collective equilibrium systems where, in "scalin up" from microscopic to macroscopic levels, all except a few "relevant" details scale away [59]. This means that the conclusions from the steady state of a simple model may be characteristic of the macroscopic behavior of a whole class, including real systems, giving great generality to the conclusions; the simplest model of a class is then sufficient.

Among the phenomena observed in NES and which can be captured by simple models are phase transitions, shocks, jamming, anomalous dynamics, coarsening, and sensitivity to boundary and initial conditions [3, 17, 39, 45, 48]. All

---

[1] "Optimism is an essential ingredient for innovation."—R. Noyce (from a plaque in the Santa Fe Institute).

these occur in weather, traffic, and many other NES, though the details will vary from class to class. The simplest system containing all these phenomena, and characteristic of one class, is a very idealized model [8, 17, 39, 45, 48] for traffic, compressible flow, and "Burgers" turbulence, which is discussed in section 3. Some of these phenomena are developed further in later sections (e.g., coarsening in section 5).

Other very interesting NE phenomena, about which a great deal is now understood through the study of simple models, include self-organization [2] and pattern formation [13, 24, 42], but we do not have space to discuss those further here. Our choice of models (see especially section 2) is decided by the aim to present in the simplest way, features (especially macroscopic ones) which emphasize the nonapplicability of standard thermodynamics/statistical mechanical approaches.

Sensitivity to boundary or to initial conditions appears in models discussed in sections 3 and 5, respectively. These features demonstrate the typically highly collective nature of NE states in space and time, which may also be reflected in anomalous size dependences and anomalous (nonexponential) probability distributions for extensive quantities (see sections 2 and 7).

Such anomalous features also occur in classical critical phenomena of equilibrium, and are there associated with the diverging controlling "correlation" length [59]. They also appear in ground-state behavior near quantum phase transitions [11]. A mapping of NE steady states to quantum ground states (section 4) for an important class of NE systems explains this particular correspondence; it also makes possible the application to NE systems of standard techniques for collective quantum systems [1, 5, 28, 29, 48]. Even without the mapping, many procedures like these for equilibrium systems can be developed and applied to NE systems. In particular, one can learn much, especially regarding anomalous features associated with long-range correlations, by developing and applying to NES techniques similar to those used for studying equilibrium critical phenomena [59] and quantum phase transitions [11], particularly scaling. Scaling approaches to NES are outlined in section 6, and their application to the investigation of anomalous distributions is included in section 7.

The study of states and correlation functions of simple NE models suggests that anomalous characteristics can arise from steady states having nonproduct measures, which is typically associated with their highly correlated nature. Correspondingly, the distributions of macroscopic variables differ from Boltzmann-Gibbs predictions [40] (see section 2).

The lack of a factorization of likelihood functions is one sense in which NES can be unlike standard equilibrium thermodynamic systems. Even more fundamental differences [16, 58] are reflected in the absence (except for a special case[2]) of any identified free energy function of macroscopic variables, or corresponding

---

[2]There has been progress in the recent provision of a free-energy functional (of microscopic variables) for the steady state of the traffic/flow model of section 3 [20]; see also section 7.

entropy which may be optimized, subject to constraints, to find the (steady) state.

The remaining sections attempt to present models, with particular emphasis on examples of special interest to practitioners in nonextensive thermodynamics [55, 57, 56]. It is hoped that the next sections will provide stimuli and challenges for those who see the possibility of providing an alternative description to the one currently used here, which relies on universality to extract generalities from detailed investigations of specific microscopic models.[3]

2   PRELIMINARIES AND MODELS

This section is largely concerned with the presentation of simple representative microscopic NE models, which are broadly applicable and "minimal" in the sense of being distilled as much as possible without losing their essence.

But we begin with some preliminary considerations which may help to distinguish the NE systems from simpler ones, and to address one feature that they share with systems already considered in the context of nonextensive thermodynamics.

One of the simplest systems is a dilute gas in thermal equilibrium. It differs fundamentally from NE systems such as the earth's atmosphere, and traffic, in its Maxwellian distribution of single particle energies, which results from the additivity of the energies and the factorization of probabilities into single-particle factors because the particles do not interact. That distribution should not be expected for single particle energies in any system with interactions, whether or not it is out of thermal equilibrium. But interacting equilibrium systems do usually have exponential distributions for the additive *macroscopic* variables such as the energy, because the factorization of probabilities applies to coarse-grained scales much larger than interaction-generated correlation lengths ξ, which are usually of microscopic size. An exception occurs in the critical region in the neighborhood of a phase transition where ξ diverges [59]. The nonexponential distribution there can be found by rescaling, using, for example, blocking of variables to coarser scales (see sections 6 and 7), and such procedures make it universal, applying for all the regimes or systems that flow into the particular fixed point. NE systems typically have (through interactions) states which are very strongly correlated, possibly across the whole system (as in familiar traffic situations such as jams, and as in the model discussed in section 3, which both have steady states without product measure [17, 45, 48]). Not surprisingly, they have nonexponential distributions of extensive variables. This is obviously similar to systems with long-range interactions (and, hence, long spatial correlations) or with long correlations in time through generation by a deterministic nonlinear map, to which nonextensive thermodynamics have been successfully applied [38,

---

[3]For a more extended discussion of the area see, for example, references [3, 17, 39, 48, 45].

43]. It may be of interest to note here that the long spatial correlations in the steady state of the model in section 3 are generated by a nonlinear map, in an approximation corresponding to the deterministic limit (see also section 8).

A continuously shaken system of granules, such as sand, might be expected to behave similarly to the dilute gas. In fact, the measured distribution of velocities v approximates to an exponential function of $v^{3/2}$. Though not understood in detail, this must arise at least partly from the agitation "feeding" mechanism. Such granular systems exhibit a remarkable range of NE phenomena [33], including self-organization and criticality, unusual static behavior including interesting force distributions, and glassy dynamics [37]. This last feature can be accounted for within a continuum version [49] of a simple class of minimal NE models to which we now turn.

These minimal models are very rich in their behavior, showing many important NE features, yet they are sometimes solvable [1, 17, 19, 26, 25, 27, 42]. They are, essentially, many-body systems which are collective and stochastic. And they are defined by their dynamic rules [45, 48]. Even in low dimensions, such as $1+1$ (one-space and one-time dimension), they can have highly nontrivial properties, including phase transitions [17, 39]. Any generality they might display comes at the macroscopic level because of universality.

Their dynamic processes are prescribed elementary moves, each involving only a few constituents ("particles") at a time. These occur stochastically, with transition probabilities per unit time (rates), which are not ratios of Gibbs probabilities. That choice, "detailed balance," would make the steady states into Gibbs states which is atypical in NES [58].

The other main ingredient is the interaction, the cause of collective properties. A very simple way of introducing interaction effects is to use a repulsive hard-core interaction, which keeps particles apart, and to limit particles to lattice sites separated by the hard-core repulsion diameter. The effect of the interaction is then to exclude particles from a site already occupied, in other words, to prevent double occupation of a site. In such lattice-based "exclusion" models the occupation n_l of site l is then 1 or 0 (one or no particles at site l). This corresponds to the well-known lattice gas, which is so important in the development of universal theories of equilibrium collective behavior, particularly phase transitions and critical phenomena [59]. The only difference is that now we have the added rule for the stochastic dynamic process (which does not satisfy detailed balance).

An example of such an exclusion model is the one-dimensional case (where l labels sites on a chain) having the following asymmetric hopping rule: a particle at site l hops with rate p to site $l+1$, provided it was empty, and with rate q to site $l-1$, provided this was empty. This model [8, 17, 20, 19, 28, 29, 45, 48], the one-dimensional asymmetrical exclusion process (ASEP), is the model for compressible flow or traffic, discussed in section 3. Other processes can involve "particle" annihilation or creation, singly or in pairs [5, 48] (see sections 4 and 5), in the bulk or at boundaries [17] (see also section 3). It should be emphasized that "particle" can have a variety of meanings. Creation or deposition of a "particle" at site l could correspond in a financial market model [12, 14] to placing an order at price l (see section 5), or in a chemical reaction to the appearance of a reaction product at position l, or in a growth model to incrementing the height difference at adjacent sites (see section 4), etc. The model just described has two states at each site (particle or vacancy; or, in other applications, A or B particles). Multistate/multispecies generalizations are straightforward but will not be discussed here.

The two-state case can be mapped to spin up or down, and that is (partly) how the equivalence to quantum spin models [1, 5, 28, 29] arises (see section 4).

Some emphasis has been placed on stochasticity, for example, on the importance of noise. This is known to have crucial effects in many contexts, such as in biological systems [7] or in any case where one has to allow for selective amplification of fluctuations in connection with instability or with the development of latent characteristics. Such amplifications obviously also require the collective/nonlinear behavior provided by interactions, as do any collective phenomena, such as phase transitions.

The discreteness of the site label can be removed by taking the continuum limit (section 3). If the noise is retained in this procedure, stochastic field equations are obtained, such as the noisy versions of the KPZ equation [35] and the Burgers equation [8] for scalar turbulence. These are ideal for investigation by field theoretic techniques [23]. The noiseless limits are relatively simple at the continuum level (section 3). The approach to the full stochastic discrete NE models is normally through the master equation [58], sometimes via the hierarchy of kinetic equations it gives rise to (section 3), or through reformulations involving operator algebras [19, 52] (section 3) or quantum Hamiltonians [1, 5, 28, 29] (section 4).

3   FLOW, TRANSITIONS, AND NON-GIBBSIAN STEADY STATES

This section illustrates many points which have already been referred to by considering the asymmetric exclusion process (ASEP) introduced in section 2. This is the simplest collective flow model, having "hard core" interaction effects mimicked by the exclusion of more than one particle from a site. That, combined with the bias in the hopping, causes strong nonlinearities in the model that are sufficient to give highly correlated non-Gibbsian steady states and transitions. The steady-state transitions are induced by forcing a current through the system, for instance, by boundary injection, and are analogous to jamming transitions in highway traffic flow (which allows intuitive interpretations).

The steady-state transitions are highly nontrivial but are (remarkably) solvable by techniques to be described later. But they are also amenable to a simple "mean field" treatment, which captures the essence of the problem. Further, the system is related by an exact mapping to an important growth model (see below and, especially, section 4).

We begin with a restatement of the model, followed by a mean-field treatment [18].

In its fully asymmetric one-dimensional version, particles can hop from site l of a chain to neighboring site $l + 1$, provided that it was unoccupied, with probability p (per time step). Denoting by $n_l = 1, 0$ the occupation number at site l, the average (over histories) of n_l, $\langle n_l \rangle \equiv \rho_l$, satisfies

$$\frac{\partial}{\partial t} \langle n_l \rangle = \langle J_{l-1,l} - J_{l,l+1} \rangle . \tag{1}$$

This is a continuity equation involving the current across a bond

$$J_{l,l+1} = p n_l (1 - n_l) . \tag{2}$$

Equation (1) is the first of a hierarchy of kinetic equations fully describing the model.

In the mean-field approximation, averages are factorized into products of single-site averages, so for example

$$\langle J_{l,l+1} \rangle \rightarrow p \rho_l (1 - \rho_{l+1}) . \tag{3}$$

This changes eq. (1) into an equation for the average site occupancies $\{\rho_l\}$. The mean-field neglect of correlations/fluctuations makes the equation noiseless (deterministic).

The steady-state version of this noiseless mean-field equation requires the average current across each bond to be the same and time independent. So the right-hand side of eq. (3) is constant, say, J. The ρ_l then satisfy the nonlinear map

$$\rho_{l+1} = 1 - \frac{J}{p \rho_l} . \tag{4}$$

The iteration of this map then determines the successive values of ρ_l, or, the density profile. For $J < p/4$ the map (4) has two fixed points, and any initial ρ between the two fixed-point values then iterates toward the larger (attractive) fixed point. In that way the profile can be seen to have a kink shape. For $J = p/4$ the two fixed points become coincident and for $J > p/4$ there are no fixed points. In this case the profile is monotonically decreasing. The equation $J_c \equiv p/4$ is the critical value of the current at which the steady-state transition occurs, dividing the phase with low current and kink profile from the high-current phase. The transition itself is marked by a diverging length (the kink width).

Instead of specifying the current, the state of the system can be specified by attempt rates α and β for particle entry, or exit at left and right boundaries of the system, respectively. This facilitates instructive interpretations from experience with highway traffic flow, and also emphasizes the fact that the internal state of the whole system is governed by boundary effects. This is an example of extreme correlations in SNE systems.

We next turn to the dynamics, still continuing with the mean-field approximation and adding to it a continuum limit in which derivatives with respect to x replace finite differences, in the fully time-dependent equation resulting from using eq. (3) in eq. (1). The result is the following noiseless version of the (nonlinear) Burgers equation [8]:

$$\frac{\partial \rho}{\partial t} = p \frac{\partial}{\partial x} \left(\frac{1}{2} \frac{\partial \rho}{\partial x} - \rho(1 - \rho) \right). \tag{5}$$

This equation is completely solvable either by linearization using the Cole-Hopf transformation [10, 31], or by using its Galilean invariance to deduce the full solution from the steady-state solution. The result is basically traveling profiles of the steady-state shapes, or decaying forms.

Using the transformation $\rho = \partial \ln h / \partial x$, the noiseless Burgers equation (5) can be seen to be equivalent to a noiseless KPZ equation [35] for a growing interface model with height $h(x)$. This mapping is further discussed in section 4, which includes an exact equivalence of the full ASEP to a discrete growth model.

The mean-field description can be compared to the exact steady-state solution, provided by an operator algebra approach [19], and to field theoretic [23] and Bethe ansatz [28, 29] solutions for a dynamic exponent $z = 3/2$. This exponent describes the broadening of kinks (and of the interface width in the related growth models), which is completely absent from the mean-field description. On the other hand, the mean-field description locates the steady-state transition exactly. However, it produces a factorizable (product measure) steady state, which is not the general result of the exact operator algebra solution, and it provides incorrect exponents because of neglect of fluctuations. The exact results imply nontrivial current and density distributions, which are further discussed in section 7.

For completeness and later reference we now give some details of the operator algebra approach [19, 52] referred to above. For the one-dimensional ASEP, it represents configurations along the chain as strings of letters D or E, representing particle or vacancy, respectively [19]. It can be shown that by taking the letters to be operators with an appropriate algebra, a matrix element, between special state vectors $\langle W|$ and $|V\rangle$, of the string of operators gives the probability of the configuration they represent, as it would be determined by the master equation.

For the fully dynamic problem the operator algebra can be shown to be [52]

$$\frac{d}{dt} D = [C^{-1}, \Lambda], \tag{6}$$

$$DC^{-1}\Lambda = \Lambda C^{-1} D, \tag{7}$$

where

$$C \equiv D + E, \quad \text{and} \quad \Lambda \equiv DE. \tag{8}$$

This includes, as a special case with no time dependence, the original steady-state algebra [19]

$$\Lambda \equiv C . \tag{9}$$

In this steady-state case, the boundary equations involving the special state vectors are

$$\langle W|E = x\langle W|, \quad D|V\rangle = y|V\rangle \tag{10}$$

where $x = p/\alpha$ and $y = p/\beta$. The steady-state transition was originally solved [19] by finding explicit matrix representations of the operators D and E.

Further advances made with the algebra involve scaling approaches (section 6) and the representation of the ASEP in terms of an effective free-energy function [20] (see also section 7). So far, this latter advance has not led to a thermodynamic approach to SNE systems.

4 MAPPINGS

Various mappings of SNE systems have resulted in new perspectives and much progress through the availability of techniques developed for other areas.

The most important ones, briefly outlined below, are mappings of (1) stochastic dynamics to quantum evolution [1], (2) particle exclusion models to quantum spin systems [1, 5, 28, 29], and (3) flow to growth.

The techniques used are, respectively,

1. writing the master equation for the rate of change of the configuration probability as a Heisenberg equation for a quantum state vector (or, equivalently, using a quantum evolution operator $\exp(-\mathcal{H}t)$);
2. associating a particle or vacancy at a site l with spin (σ_l^z) up or down there, and consequently associating removal or placement of a particle at site l with the action of a spin lowering or raising operator there $(\sigma_l^{\pm})$, so the Hamiltonian in the evolution operator becomes a quantum spin Hamiltonian; and
3. associating particle density with the gradient or finite difference of a height variable.

In (1), the Hamiltonian need not be Hermitian, but its eigenvalues will normally have non-negative real parts. Then the steady state will be the (zero energy) ground state of $\mathcal{H}$, since it does not evolve in time.

As an example of (2), a term $-p\sigma_l^- \sigma_{l+1}^+$ in $\mathcal{H}$ accomplishes the stochastic hopping, at rate p, of a particle from filled site l to empty site $l+1$, which is the evolution step of the ASEP. However, it is accompanied in $\mathcal{H}$ by a further term $pP_l^- P_l^+$, with $P_l^{\mp} = 1/2(1 \pm \sigma_l^z)$, to account for the other possible outcome of the stochastic process (which might have taken place but did not, with probability $1 - p$ per unit time). Summing over all l (i.e., over all bonds $l, l+1$) provides the quantum spin Hamiltonian for the ASEP [28, 29].

It can be rewritten in the form of a ferromagnetic Heisenberg pair interaction combined with a staggered Dzyaloshinsky-Moriya pair term and with single spin terms corresponding to complex bulk fields. In addition, the boundary injection process amounts to boundary field terms. So the steady-state transition discussed in section 3 corresponds to a "quantum transition" [11] related to the change in the symmetry of the ground state of the quantum spin system caused by changing the boundary fields. These can produce abrupt or gradual twists in the spin configuration. The highly correlated nature of such states, and the way the boundary terms can affect the whole of the interior state, are readily understood in this picture. This is even clearer in the classical spin limit, corresponding to mean-field theory: the noise effects correspond to quantum fluctuations. These can be thought of as giving interesting distributions to stochastic quantities, such as current and density (see especially section 7).

The mapping of exclusion models to quantum spin systems allows the use of standard techniques for the latter. Among these are:

(a) further mappings via the Jordan Wigner transformation [34] in one dimension to fermion models. In certain cases a pair interaction can become a biquadratic form in fermion operators, then the fermions are "free" (noninteracting) and exact solutions can be achieved [26, 25];

(b) Bethe ansatz; by such means the one-dimensional ASEP dynamic exponent $z = 3/2$ was obtained [21, 28, 29]; and

(c) bosonization (which has not yet been much explored).

The third mapping ((3)—flow to growth) is more restricted, but does relate two important processes. Consider (in the lowest dimensional version) an interface in two dimensions consisting of a chain of line segments, each sloping up or down. The center of each segment is projected onto sites of a horizontal chain, and a down-sloping (up-sloping) segment is associated with a particle (vacancy) at the projection site. Then the ASEP hop of a particle to an empty site to the right corresponds in the interface model to the inversion of a downward-pointing "vee" in the growth profile, which is a process for interface growth. So this growth process is equivalent to the ASEP. A slight variation of the description gives a growth model of Meakin et al. [28, 29]. And the continuum version of the mapping is the noisy generalization of the Burgers equation-to-KPZ mapping accomplished by $\rho = \partial h / \partial x$ in section 3. So interface width and kink width share, in these models, the same growth law $(t^{1/z})$ and dynamic exponent $z = 3/2$ in $d = 1$.

This is one example of growing scales. Other examples include the coarsening process discussed in the next section.

5   EVAPORATION AND DEPOSITION, DIVERSITY, SINGLE, AND MULTI-SCALE COARSENING

We begin by discussing a basic nonparticle-conserving process which, with seemingly innocuous modifications, progresses to a rich variety of stochastic behaviors, including strongly non-ergodic behavior and two-scale coarsening.

Evaporation and deposition processes were mentioned in section 2. These are important ingredients in models of chemical reactions and a wide variety of other systems, ranging to econophysics market models (e.g., for limit-order markets [12, 14]).

The simplest version is a system of particles, each of which can evaporate from filled sites and deposit on empty ones. Through the mapping introduced in the previous section, this is equivalent to a generalized noninteracting spin system, and is, therefore, trivially solvable and not cooperative [48]. It has simple distributions, equivalent to those of a paramagnet. So, for example, the total number $N\rho$ of particles on N sites of the system has a steady-state distribution related (through $N\rho = 1/2(M + N)$ and through correspondences between deposition and evaporation rates and probabilities p, q of spin up or down) to the distribution

$$P(M) = p^{\frac{1}{2}(M+N)} q^{\frac{1}{2}(N-M)} \, {}_N C_{\frac{1}{2}(N-M)} \tag{11}$$

for the magnetization $M = \sum\limits_{l=1}^{N} \sigma_l^z$. This distribution is asymmetric (for unequal rates) and Gaussian near its peak, crossing over through an exponential regime on each side of the peak to cut off at $M = \pm N$. In the large N limit, the relative fluctuations are of order $1/\sqrt{N}$, and the Gaussian is all that matters (as expected from central limit considerations, c.f. section 1).

The single-particle-evaporation-deposition system evolves "exponentially fast" toward the steady state. Starting from an initially empty system, this approach is seen as a growth of average size of domains of adjacent filled sites, as in coarsening, but rapidly saturating at the steady-state coverage.

Evaporation and deposition of dimers is a cooperative process [5]. For equal evaporation and deposition rates, a sublattice mapping establishes an equivalence to symmetric hard core diffusion. So the late-time evolution has long-time power-law tails. For the unequal rate case, on bipartite lattices, a Goldstone symmetry can be demonstrated and this again leads to power-law late-time dependences [54].

The Goldstone argument generalizes to the much more interesting case of evaporation and deposition of "k-mers" with $k \geq 3$ [28, 29, 54]. This process shares with diffusion of k-mers for $k \geq 2$ a set of properties sometimes called dynamic diversity [3, 4]. These arise from a highly non-ergodic character of these systems: their configuration space separates into sectors not connected by the dynamic evolution. The number of such sectors is exponential in system size. This fragmentation of configuration space is related to the existence of exponentially many extensive constants of motion and dynamic exponents. The asymptotic behavior is sector dependent and, consequently, dependent on the initial conditions.

We now develop ideas of coarsening and ordering kinetics further by combining the threads introduced above. First, consider coarsening in the Ising chain at very low temperatures. The final state reached will be the equilibrium state, in other words, the steady-state satisfying detailed balance at the temperature considered. In the zero field the Ising chain reduced Hamiltonian

$$-\beta\mathcal{H} = K \sum_l \sigma_l^z \sigma_{l+1}^z \tag{12}$$

maps to that for spins $\tau_l \equiv \sigma_l^z \sigma_{l+1}^z$ in reduced field K, in other words, to that for a paramagnet. Thus the energy has distribution of the form (11), and the pair correlation function is

$$\langle \sigma_l^z \sigma_{l+m}^z \rangle = \left\langle \prod_{s=1}^m \tau_{l+s-1} \right\rangle = (\tanh K)^m . \tag{13}$$

So the equilibrium correlation length is $\xi = [-\ln \tanh K]^{-1}$, which is very large at low temperatures (K large). At the temperature concerned, ξ is the average domain size to which the system "coarsens" from a more disordered state. This coarsening can be discussed using single-spin flip Glauber dynamics, satisfying detailed balance conditions. The processes involved are domain wall diffusion, and pair creation or annihilation of domain walls. Regarding the domain walls as particles, we then have a combination of single-particle diffusion and dimer deposition and evaporation processes, with rates which are functions of K set by the detailed balance condition. This system can be solved directly or by free-fermion techniques referred to in section 4 [22, 27]. In the infinite system limit at zero temperature, where $\xi \to \infty$, the coarsening continues forever to arbitrary long scales, which grow diffusively (like $t^{1/2}$). So there is an associated invariance under related inflation of time and length scales. This a generalization of the scale invariance seen at equilibrium phase transitions, and is amenable to scaling treatments.

The SNE models provide a variety of generalized coarsening behaviors. We briefly discuss one example which develops from processes discussed above. It is one of many examples posing challenges for description, especially concerning distribution functions.

The model [36] combines (in one particular form, and in $d = 1$) dimer diffusion [3] and dimer deposition [5] as follows. Using $(n_l, n_{l+1}, n_{l+2}, \ldots)$ to denote the particle ($n = 1$) or hole ($n = 0$) configuration of a group of adjacent sites, a dimer diffusion step to the right corresponds to $(110) \to (011)$; and dimer deposition is $(00) \to (11)$. Without this second process the model has the fragmented phase space and "dynamical diversity" [3] referred to above. With both processes, an initially empty system can be seen in simulations to coarsen, organizing itself so that the particle configuration exhibits two distinct sets of clusters (domains), "e" and "o," with characteristic lengths growing in time t like $L_e \propto t^{1/z_e}$ and $L_o \propto t^{1/z_o}$ respectively. Analytic considerations identify L_e and L_o with even and odd domains, as verified in the simulation. The associated distribution P of clusters of length L can be approximately fitted to "two-scale coarsening" forms

$$P_i(L) = t^{-1/z_i} f_i(Lt^{-1/z_i}), \quad i = e, o. \tag{14}$$

The functions f_i appear to have exponential tails at large arguments, which can be accounted for by an "independent interval approximation," involving factorization of probabilities between different clusters. At small arguments, f_e and f_o are quite different (power law) functions which is one of many aspects of this generalized coarsening process that is presently not well understood.

6 SCALING FOR STEADY-STATE TRANSITIONS, GROWTH, AND GLASSY DYNAMICS

SNE systems provide many examples of highly correlated states, of divergent and growing length and time scales, and of ubiquitous power-law dependencies at steady-state transitions and in coarsening. Diverging correlation lengths and scale invariance are encountered elsewhere, particularly in equilibrium critical phenomena. Scaling and renormalization procedures provided the breakthrough there, and it is not surprising that these approaches are very valuable, sometimes essential, in studies of SNE systems.

Here, as in studies of equilibrium phenomena [59], they provide many explanations, for example, of universality via basins of attraction of fixed points of scaling relations; they capture fluctuation phenomena, which can be extreme in SNE systems (as they often are in quantum spin models).

Field theoretical renormalization procedures are particularly powerful and have proved very effective for the SNE models described by noisy continuum field equations. An example is the KPZ system [35], where early field theoretic studies [23] gave the exact exponents in $d = 1$.

For insight into aspects which require a closer contact with intuition, less controlled but simpler scaling methods (such as real space renormalization or blocking [9]) can be very useful and informative [30, 50]. While quantitative results (e.g., for exponents) can be poor, they can provide valuable basic understanding and can approach features that are very difficult to study by more controlled methods, such as distributions (see section 7).

We give here two examples of the application of such simple scaling methods to SNE models. The first is a scaling, by blocking, of the ASEP flow model in $d = 1$, particularly for its steady-state transition; the other is a scaling for glassy dynamics in a primitive model [49]. Further applications are deferred to section 7.

The ASEP is perhaps the simplest SNE model exhibiting all of the features referred to in the first sentence of this section. A blocking scaling method for the SNE systems has recently been developed and exploited [50] which uses matching of fundamental quantities at two different scales. For the description of the steady state of the ASEP the simplest form is to match, by rescaling rate parameters, currents and densities for two primitive blocks differing in size by a scale factor b, using a "majority rule" (see below). The simplest parameters to rescale are the ratios $x = p/\alpha$ and $y = p/\beta$ of internal bulk rate to boundary injection rates (defined in section 3).

The densities ρ and currents J can be worked out for the blocks directly from the steady-state master equation or, more simply, using the corresponding steady-state algebra (section 3). For the smallest blocks, differing in size by the factor $b = 2$, the majority rule is (in the operator algebra formulation)

$$D' = DD + \frac{1}{2}(DE + ED), \quad E' = EE + \frac{1}{2}(DE + ED) \tag{15}$$

and the following matching result:

$$\rho'(x', y') = y'(x' + y')^{-1} = \rho(x, y) = \frac{g_2(x, y)}{h_2(x, y)}, \tag{16}$$

$$J'(x', y') = \frac{(x' + y')}{h_2(x', y')} = J(x, y) = \frac{g_3(x, y)}{h_4(x, y)}, \tag{17}$$

where g_n, h_n are nth-order polynomials (e.g., $g_2(x, y) \equiv y^2 + 1/2(x + y + xy)$, $h_2(x, y) \equiv y^2 + xy + x^2 + x + y$). From these equations, or from matchings using larger blocks and/or larger values of b, scaling equations of the following form result:

$$x' = X_b(x, y), \quad y' = Y_b(x, y). \tag{18}$$

The equations become more accurate for larger blocks and larger values of b.

The approach generalizes: (1) for other processes, such as dimer evaporation and deposition; (2) for higher dimensions; and (3) for disordered generalizations (see section 7). The dynamic generalization is obtained by using the full-time dependent master equation or, more easily in $d = 1$, using the dynamic operator algebra (section 3) [48]. This provides the dynamic scaling behavior of the ASEP (and its corresponding growth model) and can be applied to other nonequilibrium evolution problems.

The scaling equations are used in a standard way. For the steady-state ASEP, eq. (18) has (for any of the blockings) a fixed point at $x = y = 2$. Linearizing about the fixed point shows it to be unstable (for all blockings except the most primitive one, where eqs. (16) and (17) apply: that is only marginally unstable and we disregard it henceforth). The associated eigenvalues provide the length scaling exponent for the differences $\rho - \rho_c$, $J - J_c$ of density and current from their critical values $\rho_c = 1/2$, $J_c = 1/4$ (or, equivalently, correlation lengths as powers of $\rho - \rho_c$, $J - J_c$). All the blockings give the exact location for the transition, but larger blocks are required for acceptable exponents.

In practice, such scaling methods can be used to obtain steady-state distributions of fluctuating current and density. However, these distributions are actually determined directly by the operator algebra itself, and this will be discussed in the next section.

The dynamic generalization of the above scaling techniques also provides an understanding of glassy dynamics in simple constrained-dynamics models [32, 46].

A slightly different scaling approach has been applied to a model of compaction in granular systems, which exhibits glassy dynamics of the type mentioned in section 2. This model is a continuum deposition process where blocks of unit size park on a line if sufficient space exists at the attempted parking position, chosen at random anywhere along the line. The model has been solved exactly and by a (more generally applicable) time-scaling approach [49]. The scaling has a marginally unstable character (unit eigenvalue for the linearized scaling), which has a physical origin and is argued to be generic for glassy dynamics. It leads to universality classes characterized by Vogel-Fulcher, inverse-exponential-squared and other behaviors known in particular models.

7 DISTRIBUTIONS BY SCALING, ALGEBRAIC TECHNIQUES, AND ITERATIONS

In critical phenomena, distributions are given by scaling, which establishes the universality of their forms. These universal forms arise as the fixed-point functions satisfying the scaling equation for the distribution or, equivalently, are reached by iterating these equations many times.

The first examples of these were universal distributions arising for: (1) the conductance of disordered, particularly dilute networks [53]; and (2) blocked spins in the Ising model [6]. Since the ideas encountered in these examples apply also to SNE systems, we briefly present them now using the simplest illustration, namely example (1).

In such a case, one starts from a "composition rule" relating (here, through Kirchhoffs' laws) the (conductance) variable, σ, across a block to the individual variables $\{\sigma_l\}$ of bonds comprising the block. The resulting relation, say,

$$\sigma = f(\{\sigma_l\}),\tag{19}$$

would have

$$f(\{\sigma_l\}) = \frac{\sigma_1\sigma_2}{\sigma_1 + \sigma_2} + \frac{\sigma_3\sigma_4}{\sigma_3 + \sigma_4}\tag{20}$$

in the simple example of corner-to-corner conductance of a square block of four bonds, as encountered in the simplest two-dimensional "decimation" procedure [53]. The same equation occurs in the simplest Migdal-Kadanoff blocking. Suppose the values of σ_l are independently distributed, each with distribution

$P^{(0)}(\sigma_l)$. For example, $P^{(0)}(\sigma_l) = p\delta(\sigma_l - \sigma) + (1 - p)\delta(\sigma_l)$ applies for the case of dilute networks with probability p for each bond to be present. Then, under scaling, the distribution develops according to

$$P^{(n+1)}(\sigma) = \int \left(\prod_i P^{(n)}(\sigma_i) d\sigma_i \right) \delta(\sigma - f(\{\sigma_i\})). \tag{21}$$

In the case of the dilute network at the critical percolation concentration, $p = p_c$, many scalings take $P^{(n)}$ into a universal distribution, which determines the large-scale conductance properties of the network.

This type of procedure applies readily to the ASEP flow/traffic model in its generalization where the internal rates p_l are disordered. Then, for example, the approximate steady-state scaling equation (8) generalizes to scaling relations for $x_l \equiv p_l/\alpha$, $y_l \equiv x_l/\beta$ [50, 51]. Such analyses result in typically skewed universal distributions for rates, which are sufficient to provide approximate results for number (or density) and current distributions in the steady state. phenomena, distributions

For the pure steady-state ASEP, such distributions can be found exactly from the bulk operator algebra (9), combined with the open-boundary relations (10).

For example [15], the number distribution $P_L(N)$ (i.e., the probability of finding N particles in the ASEP, length L) can be obtained as the coefficient of λ^N in $\Gamma_{L,0}(\lambda)/\Gamma_{L,0}(1)$, where

$$\Gamma_{l,m}(\lambda) = \langle W|(E + \lambda D)^l E^m|V\rangle. \tag{22}$$

The steady-state algebra provides linear recurrence relations for $\Gamma_{l,m}(\lambda)$, which can be solved by means of a generating function. The resulting exact number distributions have asymmetric peaks (or a symmetric peak) in the low (or high) current phases respectively, and an intermediate critical form.

$P_L(N)$ can be written as $\exp[-LF(L, N, x, y)]$ where F is a "free energy" function whose minimization gives the most likely value of the macroscopic variable N. This provides a partial thermodynamic description for the special case of the steady-state ASEP. Derrida et al. [20] give a more complete free-energy description of this special case.

Such distributions include fluctuation effects that are not provided by mean-field theory, which corresponds to a deterministic description. But deterministic systems can have interesting distributions, such as the visitation probability under iterations of a nonlinear map. There are corresponding distributions in the mean-field ASEP: the visitation probability under iteration of the profile map (4) is the probability distribution $P(\rho)$ of the occupation values ρ in the steady state and this is actually the fluctuation-free approximation to $P_L(\rho L)$. This probability distribution can be determined analytically (because of the simple monotonic character of the map). For finite L (where only a finite number of iterations is involved) it takes the following (nonuniversal) forms (where a has the sign of $J - J_c$, so is positive, zero, or negative above, at, and below the transition, reflecting the different density profiles in these cases):

$$P(\rho) \propto \left| \left(\rho - \frac{1}{2} \right)^2 + a \right|^{-(\nu+1)/2\nu} \quad ; \tag{23}$$

ν is a mean-field exponent, $\nu = 1$.

In both the exact $P_L(N)$ and the mean-field distribution $P(\rho)$ the appearance of three forms is reminiscent of the blocked order parameter distribution of the Ising model in $d \geq 1$, where the three forms apply everywhere below, everywhere above, and at the transition. These are respectively binary (two delta functions), Gaussian, or a nontrivial critical distribution. The first two forms occur because of the attractive nature of the zero- and infinite-temperature fixed points, and all these forms are universal.

The mean-field result (23) can be extended to allow for fluctuations, since their effect on the steady-state density profiles is known from the exact operator algebra solution. The result is similar to eq. (23), but with the exact exponent $\nu = 1/2$ controlling the behavior in the wings of the distribution. phenomena, distributions

8 CONCLUSION

Stochastic nonequilibrium systems and phenomena have been briefly reviewed with an emphasis on strongly cooperative properties, especially complex nonlinear behavior in steady states and asymptotic (long-time) dynamics. The discussions and illustrations have been given in terms of minimal models of the stochastic lattice-based exclusion type, the study of which has, so far, been the source of most recent advances concerning nonequilibrium behavior.

Particular emphasis has been given to the challenges posed by situations in which highly correlated steady states occur because of bulk and/or boundary driving and through criticality, all of which are common in this area. In such situations, product measure does not typically apply and stationary distributions of key macroscopic variables are typically not exponential. Examples of such anomalous distributions have been given, some fully or qualitatively understood, and others where further study is needed. Anomalous distributions have also been seen to occur in evolving situations, such as, coarsening processes, and are also expected in processes (e.g., those with "dynamic diversity") where the exploration of state space is incomplete, hierarchical, or strongly nonuniform.

Scaling methods have been presented because they are particularly effective here. They deal with long-range correlations resulting from boundary and bulk driving, with steady state and dynamic behavior at nonequilibrium phase transitions, and with asymptotic dynamics. They show, through ideas of coarse-graining as RG blockings, that the distributions of macroscopic variables are universal (fixed point) distributions.

The models are presented in terms of their microscopic dynamic processes. In this sense they are like nonlinear deterministic dynamical systems, to which they reduce in their noiseless (mean field) limits. This point, and one in the previous paragraph, are relevant in connection with the following discussion, with which we conclude.

The absence of a general thermodynamic approach to such systems has been emphasized. The Tsallis formalism [55, 57, 56] is one candidate for consideration in this context. In that connection we make two particular suggestions whose relevance preliminary exploration has not excluded or established. The first relates to the work of Lyra et al. [38] and of Robledo [44], who by direct and scaling techniques have shown the applicability of the Tsallis formulation to two particular deterministic systems (the logistic and circle maps). Corresponding investigations of the noiseless limits of simple stochastic systems could at least test the applicability of the formulation there, and, if successful (a necessary condition), could lead into studies of noisy generalizations (the truly stochastic systems, or simply maps with noise). Secondly, because of general arguments that distributions of macroscopic variables should be universal, it would be of interest to find out, for instance, by scaling, whether the Tsallis distribution corresponds to a specially prevalent universality class (or to a crossover between behaviors characteristic of two prevalent classes).

It would certainly be important to know whether NES fit in any way into the framework of nonextensive thermodynamics. Whether or not they do, NES present many phenomena, many insights, and many challenges, and it is hoped that this introduction will have conveyed some of that.

REFERENCES

[1] Alcaraz, F. C., M. Droz, M. Henkel, and V. Rittenberg. *Ann. Phys. (N.Y.)* **230** (1994): 250.

[2] Bak, P. *How Nature Works: The Science of Self-Organized Criticality.* Oxford: Copernicus, 1997.

[3] Barma, M. *Non-equilibrium Statistical Mechanics in One-Dimension*, edited by V. Privman. Cambridge: Cambridge University Press, 1997.

[4] Barma, M., and D. Dhar. *Proceedings of the 19th IUPAP Conference on Statistical Physics*, edited by B.-L. Hao. Singapore: World Scientific, 1995.

[5] Barma, M., M. D. Grynberg, and R. B. Stinchcombe. *Phys. Rev. Lett.* **70** (1993): 1033.

[6] Bruce, A. D. *J. Phys. C* **14** (1981): 3667.

[7] Buiatti, M. "The Living State of Matter: Between Noise and Homeorrhetic Constraints." This volume.

[8] Burgers, J. M. *The Nonlinear Diffusion Equation.* Boston: Riedel, 1974.

[9] Burkhardt, T. W., and J. M. J. van Leeuwen, eds. *Real Space Renormalization*. Berlin: Springer-Verlag, 1982.

[10] Cole, J. D. *Quart: Appl. Math.* **9** (1951): 225.

[11] Chakrabarti, B. K., A. Dutta, and P. Sen. *Quantum Ising Phases and Transitions in Transverse Ising Models*. Berlin: Springer-Verlag, 1996.

[12] Challet, D., and R. B. Stinchcombe. "Analyzing ad Modeling $1 + 1d$ Markets." *Physica A* **300** (2001): 285.

[13] Chopard, B., and M. Droz. *Cellular Automata Modelling of Physical Systems*. Cambridge: Cambridge University Press, 1998.

[14] Daniels, M. G., J. D. Farmer, J. Iori, and E. Smith. "How Storing Supply and Demand Affects Price Diffusion." Dec. 2001. arXiv e-Print archive, Condensed Matter, Cornell University. December 2002. ⟨http://www.lanl.gov/abs/cond-mat/0112422⟩.

[15] Depken, M., and R. B. Stinchcombe. To be published.

[16] de Groot, S. R., and Mazur. *Non-equilibrium Thermodynamics*. North-Holland: Amsterdam, 1962

[17] Derrida, B., and M. R. Evans. *Non-equilibrium Statistical Mechanics in One-Dimension*, edited by V. Privman. Cambridge: Cambridge University Press, 1997.

[18] Derrida, B., E. Domany, and D. Mukhamel. *J. Stat. Phys.* **69** (1992): 667.

[19] Derrida, B., M. R. Evans, V. Hakim, and V. Pasquier. *J. Phys.* **A26** (1993): 1493.

[20] Derrida, B., J. Lebowitz, and A. Speer. *Phys. Rev. Lett.* **87(15)** (2001): 150601.

[21] Dhar, D. *Phase Transitions* **9** (1987): 51.

[22] Family, F., and J. G. Amar. *J. Stat. Phys.* **65** (1991): 1235.

[23] Forster, D., D. Nelson, and M. Stephen. *Phys. Rev.* **A16** (1977): 732.

[24] Grindrod, P. *The Theory and Application of Reaction-Diffusion Equations—Patterns and Waves*, 2d ed. Oxford University Press, 1996.

[25] Grynberg, M. D., and R. B. Stinchcombe. *Phys. Rev. Lett.* **74** (1995): 1242.

[26] Grynberg, M. D., and R. B. Stinchcombe. *Phys. Rev. Lett.* **76** (1996): 851.

[27] Grynberg, M. D., T. J. Newman, and R. B. Stinchcombe. *Phys. Rev.* **E50** (1994): 957.

[28] Gwa, L.-H., and H. Spohn. *Phys. Rev. Lett.* **68** (1992): 725.

[29] Gwa, L.-H., and H. Spohn. *Phys. Rev. A* **46** (1992): 844.

[30] Hooyberghs, J., and C. Vanderzande. *J. Phys. A* **33** (2000): 907.

[31] Hopf, E. *Commun. Pure Appl. Math.* **3** (1950): 201.

[32] Jäckle, J., and S. Eisinger. *Z. Phys. B* **84** (1991): 115.

[33] Jaeger, H. M., S. R. Nagel, and R. P. Behringer. *Phys. Today* **49(4)** (1996): 32.

[34] Jordan, P., and E. Wigner. *Z. Phys.* **47** (1982): 631.

[35] Kardar, M., G. Parisi, and Y.-C. Zhang. *Phys. Rev. Lett.* **56** (1986): 889.

[36] Krishnamurthy, S., and R. B. Stinchcombe. To be published.

[37] Liu, A. J., and S. Nagel, eds. *Jamming and Rheology*, London: Taylor & Francis, 2001.

[38] Lyra, M. L. "Nonextensive Entropies and Sensitivity to Initial Conditions of Complex Systems." This volume.

[39] Marro, J., and R. Dickman. *Non-equilibrium Phase Transitions in Lattice Models*. Cambridge: Cambridge University Press, 1999.

[40] McQuarrie, D. A. *Statistical Thermodynamics*. Mill Valley, CA: University Science Books, 1973.

[41] Meakin, P., P. Ramanlal, L. M. Sander, and R. C. Ball. *Phys. Rev. A* **34** (1986): 5091.

[42] Murray, J. D. *Mathematical Biology*. Berlin: Springer-Verlag, 1989.

[43] Rapisarda, A. "Nonextensive Effects in Hamiltonian Systems." This volume.

[44] Robledo, A. "Unifying Laws in Multidisciplinary Power-Law Phenomena: Fixed-Point Universality and Nonextensive Entropy." This volume.

[45] Schmittmann, B., and R. K. P. Zia. "Statistical Mechanics of Driven Diffusive Systems." In *Phase Transitions and Critical Phenomena*, edited by C. Domb and J. L. Lebowitz. New York: Academic, 1995.

[46] Sollich, P., and M. R. Evans. *Phys. Rev. Lett.* **83** (1999): 3238.

[47] Stinchcombe, R. B. "Dilute Magnetism." In *Phase Transitions and Critical Phenomena*, edited by C. Domb and J. Lebowitz, vol. 7. London: Academic Press, 1983.

[48] Stinchcombe, R. B. "Stochastic Non-equilibrium Systems." *Adv. Phys.* **50** (2001): 431.

[49] Stinchcombe, R., and M. Depken. "Marginal Scaling Scenario and Analytic Results for a Glassy Compaction Model." *Phys. Rev. Lett.* **88** (2002): 125701.

[50] Stinchcombe, R. B., and T. Hanney. "Exact and Scaling Approaches to Non-equilibrium Models." *Physica D* **168** (2002): 313.

[51] Stinchcombe, R. B., and R. J. Harris. To be published.

[52] Stinchcombe, R. B., and G. M. Schütz. *Europhys. Lett* **29** (1995): 663.

[53] Stinchcombe, R. B., and B. P. Watson. *J. Phys. C* **9** (1976): 3221.

[54] Stinchcombe, R. B., M. D. Grynberg, and M. Barma. *Phys. Rev.* **E47** (1993): 4018.

[55] Tsallis, C. *J. Stat. Phys.* **52(1-2)** (1998): 479–487.

[56] Tsallis, C. "need title of chapter." This volume.

[57] Tsallis, C., R. S. Mendes, and A. R. Plastino. *Physica A* **261(3-4)** (1998): 534–544.

[58] van Kampen, N. G. *Stochastic Processes in Physics and Chemistry*, 2d ed. Amsterdam: North-Holland, 1992.

[59] Wilson, K. G., and J. Kogut. *Phys. Rep.* **12C** (1974): 75.

Temperature Fluctuations and Mixtures of Equilibrium States in the Canonical Ensemble

Hugo Touchette

It has been suggested recently that "q-exponential" distributions, which form the basis of Tsallis' nonextensive thermostatistical formalism, may be viewed as mixtures of exponential (Gibbs) distributions characterized by a fluctuating inverse temperature. In this chapter, we revisit this idea in connection with a detailed microscopic calculation of the energy and temperature fluctuations present in a finite vessel of perfect gas thermally coupled to a heat bath. We find that the probability density related to the inverse temperature of the gas has a form similar to a χ^2 density, and that the "mixed" Gibbs distribution inferred from this density is non-Gibbsian. These findings are compared with those obtained by a number of researchers who worked on mixtures of Gibbsian distributions in the context of velocity difference measurements in turbulent fluids as well as secondary distributions in nuclear scattering experiments.

Nonextensive Entropy—Interdisciplinary Applications
edited by Murray Gell-Mann and Constantino Tsallis, Oxford University Press

1   INTRODUCTION

Most, if not all, textbooks on thermodynamics and statistical physics define temperature as being a quantity which, contrary to other thermodynamic observables like energy or pressure, does not admit fluctuations. Because of that, it is somewhat surprising to see papers with the expression "temperature fluctuations" in their titles appearing from time to time in serious scientific journals on subjects as various as particle physics and fluid dynamics (see, e.g., Ashkenazi and Steinberg [3], Ching [9], Chiu et al. [10], and Stodolsky [24]). Indeed, how can the temperature of a system, however small, fluctuate if one defines it "as equal to the temperature of a very large heat reservoir with which the system is in equilibrium and in thermal contact" [18]? Also, in the case of the reservoir, how can temperature be a fluctuating parameter if its definition requires one to assume the thermodynamic limit, in other words, to assume that the system acting as a reservoir is composed of an infinite number of particles or degrees of freedom? Presumably, the thermodynamic limit should rule out any fluctuations of thermodynamic quantities like the mean energy or the pressure, so that if temperature is related to these quantities, how can it fluctuate?

The solution to this conundrum is quite simple. First, the standard definition of temperature found in textbooks is too restrictive: there is not *one* but *many* definitions of temperature and of quantities analogous to temperature, as well as many physical (nonequilibrium) situations in the context of which these different definitions admit fluctuations [20]. Second, the standard definition of temperature involving the thermodynamic limit is only an idealization, a "purist" definition. Real physical bodies are always composed of a finite number of particles or degrees of freedom, which means that the concept of temperature must be applicable outside the idealized realm in which it is defined if experimentalists are indeed able to measure the temperature of real bodies in real laboratories. Physically, this means also that there must be a threshold number of particles or degrees of freedom (more or less precisely defined) above which a system can be measured or "felt" to have a temperature [15]. Well above this number, temperature is assured to be defined and likely to be constant, while close to this number it may be well defined, but may change in time or vary in space; that is, it may fluctuate!

All the studies concerned with temperature fluctuations exploit one of the above "indents" to the standard definition of temperature. That is, they either consider alternatives to the standard definition of temperature which do admit fluctuations or apply the "thermodynamic limit" definition of temperature in situations where the thermodynamic limit is assumed to be "effectively" reached without being reached "formally," so to speak. Our aim in this chapter is to review a number of the alternative definitions and situations, and to dispel, in doing so, some of the misunderstanding and misconceptions surrounding the notion of temperature fluctuations. We will be particularly interested in giving a detailed calculation of temperature fluctuations present in a system which is commonly thought to be at constant temperature, namely a system composed of a finite number of independent particles (basically a finite volume of perfect gas) thermally coupled to an infinite-size heat reservoir at constant temperature. In the following, we will show that if, instead of defining the temperature of the particle system simply as being equal to the temperature of the reservoir, we apply the statistical definition of temperature to the finite-sized system of particles (provided that the number of particles is sufficiently large), then we must come to the conclusion that the system's temperature is fluctuating, just as its internal energy is fluctuating because of the thermal coupling with the heat reservoir. The temperature of the particle system, in this case, is precisely related to its internal energy, and can be seen as a "microcanonical temperature" associated with the microscopic configurations of a system whose internal energy is held fixed during a period of time shorter than the energy fluctuations time scale.

Our motivation for studying the temperature fluctuations of a system of particles in a canonical ensemble setting, and moreover for presenting this study in a book about nonextensive statistical mechanics is threefold. First, a system composed of a bunch of independent particles coupled to a heat bath is one of the few thermodynamic systems for which the probability density describing the temperature fluctuations can be calculated directly using "first principle" or "microscopic" arguments. Second, for this specific system, the probability density of the temperature happens to be very similar to a class of χ^2 densities of temperature fluctuations recently introduced by Wilk and Włodarczyk [28, 29, 30], as well as by Beck [4, 5, 6], in the context of nonextensive statistical mechanics [1, 27]. Finally, the models of nonextensive behavior proposed by these authors are all based on the idea of "mixed equilibrium states," such as near-equilibrium states of systems characterized by fluctuating temperatures. In the context of the present study, this idea, as we will see, arises very naturally.

2   PHENOMENOLOGY OF TEMPERATURE FLUCTUATIONS

One can imagine many different systems exhibiting temperature fluctuations. The common characteristic of all of these systems is that they are nonequilibrium systems. Below, we list and briefly comment on four systems or, more precisely, four generic situations for which temperature can be defined and be thought to fluctuate. The list is far from being exhaustive: the first three situations are presented to give an idea of the physical phenomena involving temperature fluctuations, which have been discussed recently from the point of view of nonextensive statistical mechanics. The fourth and last case on the list, the particles and heat bath system, is the focus of this chapter (see section 3).

2.1   TEMPERATURE FLUCTUATIONS IN A GAS

A system with fluctuations of temperature "spread" over space can be constructed simply in the following way. Take a vessel of gas, and divide it in some number of compartments that are thermally insulated from one another. Bring the content of each compartment to a different temperature, and then remove the insulating partition. From the moment the partition is removed, a process of temperature relaxation will take place, whereby the particles forming the gas will collide and exchange energy until a state of uniform temperature is achieved.

The details of the relaxation process are quite complicated at the microscopic level, and depend on the nature and properties of the gas considered. But, at the macroscopic level, the net result of this experiment is simply described: between the time when the partition is removed and the time when the gas temperature is completely uniform, the temperature field of the gas will vary in space as well as in time. Thus, as a whole, the vessel of gas can be said to be in a state of fluctuating temperature. This is admittedly an expletive way to say that the temperature is not homogeneous in space, but the expression is nonetheless correct and widely used (e.g., when referring to the spatial temperature fluctuations of the cosmic background radiation).

Experimentally, there are various ways by which one can reconstruct the temperature field of the gas, apart from plunging a thermometer into it at different places. A simple method (conceptually, not experimentally) consists in measuring the momenta (along a fixed direction) of many particles of the gas at one point in space or, equivalently, sampling the momentum of a single particle over some period of time, and constructing from the measurements a histogram of the number of particles $L(x)$ having a momentum value between x and $x + \Delta x$. (Δx is the coarse-graining scale at which two particles are considered to have different momentum values.) If the sample of measurements is large enough, then it is expected that the form of $L(x)$ should be approximately Gaussian, as predicted by Maxwell and Boltzmann, with a variance proportional to the temperature T. Hence, fitting $L(x)$ with a Gibbs distribution proportional to $\exp(-\beta x^2)$ or calculating its variance or its half-width all constitute operational procedures for probing the temperature $T = (k_B \beta)^{-1}$ of the gas. It should be noted that the accuracy of any of these methods for obtaining β depends on (i) the number of measurements used to construct $L(x)$, which should be large, but not necessarily infinite! And (ii) the assumption that the gas is noninteracting (perfect) or weakly interacting. These two points are necessary to assume that the momenta of the particles are Gaussian distributed [23].

2.2   VELOCITY TEMPERATURE IN TURBULENT FLUIDS

It is common in turbulent flow experiments to define an analog of temperature by looking at the distribution $L(x)$ of particle velocity differences in a restricted region of a fluid using anemometry or interferometry equipment [17]. Just as in the case of the gas, temperature is defined for a fluid by fitting $L(x)$ with a Gibbs distribution of the form $e^{-\beta u(x)}$, where $u(x)$ is the one-particle energy function taken to be a quadratic or a nearly quadratic function of the velocity variable. This defines a local inverse temperature β which, it is important to note, does not represent the physical inverse temperature of the fluid. Rather, it is a correlate of the local rate of energy dissipation that takes place at the microscopic level over a time scale known as the Kolmogorov time [3, 4].

The Gibbsian character of $L(x)$ and the fluctuations of the velocity temperature in space, related to the spatial fluctuations of the local energy dissipation rate, have been observed in many experiments of weakly turbulent fluids (see Arimitsu and Arimitsu [2], Beck et al. [8], and references cited therein). However, for fluids at a high Reynolds number, such as highly turbulent fluids, a totally different behavior of $L(x)$ is observed. Indeed, recent experiments have demonstrated that $L(x)$ in strong turbulence regimes is not Gibbsian; instead, it takes the form of a power law which appears to be well fitted by a so-called q-exponential function

$$e_q^{-\beta_q u(x)} = [1 - (1-q)\beta_q u(x)]^{1/(1-q)}, \tag{1}$$

where β_q^{-1} is a fitting parameter analogous to temperature [2, 4, 8]. To account for this non-Gibbsian behavior, Beck has suggested interpreting q-exponential distributions as "mixed" distributions arising from an ensemble of exponential distributions $e^{-\beta u(x)}$ parametrized by a fluctuating inverse temperature β [4, 5, 6]. That is to say, if one assumes that what is probed in those experiments is not *one* velocity distribution $L(x)$ characterized by a fixed temperature, but a *continuum* of distributions $L(x)$ having different temperatures, then what should be observed physically is an *average* Gibbs distribution, the average being performed over the temperature fluctuations. In this context, the essential point made by Beck [4] is that, if the probability density $f(\beta)$ ruling the temperature fluctuations has the following form:

$$f(\beta) = \frac{1}{\Gamma\left(\frac{1}{q-1}\right)}\left[\frac{1}{(q-1)\beta_0}\right]^{\frac{1}{q-1}} \beta^{\frac{1}{q-1}-1} \exp\left[-\frac{\beta}{(q-1)\beta_0}\right], \tag{2}$$

where $\beta \geq 0$ and $q > 1$, then the mixed distribution obtained by averaging the Gibbs kernel $e^{-\beta u(x)}$ with $f(\beta)$ is q-exponential. Indeed, one can readily verify that

$$e_q^{-\beta_0 u(x)} = \int_0^\infty e^{-\beta u(x)} f(\beta) d\beta \tag{3}$$

using the above variant of the χ^2 or gamma density [16] for $f(\beta)$. This integral representation of the q-exponential function is sometimes referred to as Hilhorst's formula [22, 26].

2.3 NUCLEAR COLLISION TEMPERATURE

The basic idea involved in the definition of temperature in nuclear scattering experiments is to consider the set of particles produced during a collision (called the products) as forming a gas of particles which, at a first level of approximation, can be treated as being noninteracting (perfect gas approximation). From this point of view, a concept of "collision temperature" is defined essentially in the same way that temperature was defined for turbulent fluids, except that the precise physical property to look at in scattering experiments is not the shape of the momenta distribution itself, but the so-called exponential dependence of the distribution of secondaries with respect to transverse momentum [6, 24].

Since the number of particles probed during one scattering experiment is never very large ($\sim 10 - 1000$), one must sometimes collect the momenta of particles over many scattering experiments before the exponential shape of the secondaries distribution reveals itself. However, this is not always the case: in heavy-ion experiments at very high energy, for example, it is often observed that a single event, for example, only one scattering experiment, is sufficient for a thermostatistical analysis to be effective [24]. Also, what is often seen is that scattering events of the same nature, repeated over time, yield different collision temperatures, making it obvious that temperature is a fluctuating parameter.

Observations of "nonextensive" behavior in relation to this thermodynamic picture of scattering experiments have been reported so far on two different fronts. The first is related to the distribution of secondary, and, more precisely, to observed deviations of this distribution from its expected exponential form. Due to the limited space available here, we will not discuss this case, as it is quite involved. Let us only mention that Wilk and Włodarczyk [29] have advanced a "mixed exponential distribution" model of these deviations analogous to the one suggested by Beck.

The second case of "nonextensive" behavior concerns the absorption of cosmic ray particles in lead chambers [30, 29, 28]. This case was also studied by Wilk and Włodarczyk, who suggested for its explanation yet another variant of the χ^2 temperature fluctuations model (actually before Beck applied similar ideas to the study of turbulent fluids). The physics explained by their model is the following. The number N of hadronic particles absorbed in lead chambers is usually measured to be distributed as a function of the depth l according to

$$\frac{dN}{dl} \propto e^{-l/\lambda}, \tag{4}$$

where λ is the mean-free path parameter or mean penetration depth (an analog of temperature). This exponential distribution is, at least, what is observed at small penetration depths (~ 60 cm of lead); beyond that, what is observed is that dN/dl changes to a power law which can be fitted by a q-exponential with $q \simeq 1.3$. To account for this crossover, Wilk et al. simply conjectured that the λ parameter characterizing the long flying components (i.e., the deep penetration events) is subject to fluctuations, and, thus, that the q-exponential penetration profiles observed experimentally for these components are mixtures of exponential distributions. By assuming that the probability density of λ is a χ^2 density, they were effectively able to reproduce the nonexponential distributions measured in laboratories [19, 29, 30].

2.4   SYSTEM COUPLED TO A HEAT BATH

Our last example in the panorama of thermodynamic systems characterized by temperature fluctuations is the prototypical system defining the canonical ensemble: that is, a small system S in thermal contact with a larger system R acting as a heat reservoir. Following the standard textbook definition of the canonical ensemble, one should say that the temperature of system S at equilibrium is constant, and is equal to the temperature of system R; after all, this is how thermal equilibrium is defined. However, such a statement does not do justice to one important property of S, which is that the energy density of S fluctuates (because of its finiteness), while the energy density of R does not (by definition of a heat bath).

To make this statement more precise, suppose that S consists of n independent particles whose energy density or mean total energy is given by

$$U_n = \frac{1}{n} \sum_{i=1}^{n} u_i \,. \tag{5}$$

Since the particles are coupled to R, the values of u_i above are random variables, which means that U_n is also a random variable. Moreover, observe that U_n, for any finite n, has a non-negligible probability of assuming many different values because, in this case, the probability density $g_n(u)$ of U_n is not a Dirac-delta function. The Dirac density, formally, is only a limiting density which "attracts" g_n as $n \to \infty$. (This basically follows from the law of large numbers.) Thus, if we can associate an inverse temperature $\beta(u)$ to all energy states such that $U_n = u$, e.g., by applying the equipartition theorem or by fitting a distribution of energy levels with a Gibbs distribution as described earlier, then we must conclude that there are different values of β effectively realized "in" or "by" the particle system, so to speak. That is to say, the probability density $f_n(\beta)$ for β, obtained from $g_n(u)$ by a change of variables $u \to \beta(u)$, cannot be a Dirac-delta function if $g_n(u)$ is not itself a delta function. It is to be expected that $f_n(\beta) \to \delta(\beta - \beta_0)$, where β_0 is the inverse temperature of the heat bath, only in the thermodynamic limit where $n \to \infty$. These points are discussed in more mathematical detail in the next sections.

3   ENERGY AND TEMPERATURE FLUCTUATIONS IN THE CANONICAL ENSEMBLE

Our analysis of energy and temperature fluctuations of a system coupled to a heat bath will be presented in the context of the following model. Let a vessel of gas containing n independent (classical) particles be thermally coupled to a heat reservoir characterized by a fixed inverse temperature β_0. The state of each particle is represented by a random variable X_i, $i = 1, 2, \ldots, n$, to which is associated a (one-particle) energy $u(X_i)$. The set of outcomes of each of X_i (the one-particle state space) is denoted by $\mathcal{X}$. With these notations, the energy density or mean energy of the gas is written as

$$U_n(x^n) = \frac{1}{n} \sum_{i=1}^{n} u(x_i) = \sum_{x \in \mathcal{X}} L_n(x) u(x) \,, \tag{6}$$

where $x^n = x_1, x_2, \ldots, x_n$ is the joint state of the system or the state of the system as a whole. Note that in the above expression we have defined $L_n(x)$ as the relative number of particles which are in state x as

$$L_n(x) = \frac{\#(\text{particles} : X_i = x)}{n} \,. \tag{7}$$

It should be noted that the vector L_n is nothing but the histogram of one-particle states referred to previously when we discussed temperature fluctuations. Indeed, in the case where x represents a momentum variable, the quantity $nL(x)$ precisely counts the number of particles having a momentum value equal to x. (We assume throughout that the values of X_i are discrete random variables; the continuous case can be treated with minor modifications.)

Now, owing to the fact that the gas is treated in the canonical ensemble, in the sense that it is coupled to a heat bath, we have

$$P_n(x^n) = P_n(X_1 = x_1, X_2 = x_2, \ldots, X_n = x_n) = \frac{e^{-\beta_0 n U_n(x^n)}}{Z_n(\beta_0)} \tag{8}$$

as the joint probability distribution over the states x^n, where

$$Z_n(\beta_0) = \sum_{x^n \in \mathcal{X}^n} e^{-\beta_0 n U_n(x^n)} \tag{9}$$

is the n-particle partition function. Of course, since all the particles are assumed to be independent (perfect gas assumption), as well as individually coupled to the same heat bath, we can also write

$$P_n(x^n) = p(x_1)p(x_2) \cdots p(x_n) = \frac{e^{-\beta_0 u(x_1)}}{Z(\beta_0)} \frac{e^{-\beta_0 u(x_2)}}{Z(\beta_0)} \cdots \frac{e^{-\beta_0 u(x_n)}}{Z(\beta_0)} \tag{10}$$

with $Z(\beta_0) = Z_1(\beta_0)$ (one-particle partition function). These equations make obvious the fact that what we are dealing with is a system of independent and identically distributed (IID) random variables.

The first quantity that we are interested in calculating at this point is the probability distribution or probability density $g_n(u)$ associated with the outcomes $U_n = u$. A priori, finding an exact expression for $g_n(u)$ is not an easy task, even though U_n is the simplest sum of random variables that one can imagine, that is, one involving IID random variables. Fortunately, there exists a general method by which one can obtain a very accurate approximation of $g_n(u)$ for $n \gg 1$ without too much effort. This method is based on the theory of large deviations [12], and proceeds by observing that probability densities of normalized sums of IID random variables, such as the one defining U_n, satisfy two basic properties: (i) they decay exponentially with the number n of random variables involved; and (ii) the rate of decay is a function of the value (outcome) of the sum alone. In the present context, this means specifically that $g_n(u)$ has the form

$$g_n(u) \asymp e^{-nD(u)}. \tag{11}$$

The sign "$\asymp$" above is there to emphasize that the large deviation approximation of the density $g_n(u)$ is "exponentially tight" with n, in other words, that it is exact up to $O(n^{-1}\ln n)$ marginal corrections to the rate of decay $D(u)$. This rate of decay or *rate function* is itself calculated as the Legendre transform of the quantity

$$\lambda(k) = \ln E[e^{ku(X)}] = \ln \sum_{x \in \mathcal{X}} p(x)e^{ku(x)} \tag{12}$$

which is the *cumulant generating function* of the probability distribution $p(x)$ associated with the IID random variables. The result of this transform is

$$D(u) = uk(u) - \lambda(k(u)), \tag{13}$$

$k(u)$ being the solution of

$$\left.\frac{d\lambda(k)}{dk}\right|_{k(u)} = u. \tag{14}$$

A proof of this result can be found in Dembo and Zeitouni [12], Ellis [13, 14], and Oono [21] (see also the notes contained in Dembo and Zeitouni [12] for a historical account of the development of the theory of large deviations together with a list of the founding papers of this theory).

Physicists who are not familiar with the formalism of large deviations will probably look at the above formula for calculating $g_n(u)$ as being quite formal if not fancy (in a pejorative way). For them, we offer the following alternative derivation of $g_n(u)$. Consider the density $\Omega_n(u)$ of states x^n having the same energy $U_n(x^n) = u$. Following the thermostatistics of Gibbs and Boltzmann, this density of states must be an exponential function of n taking the form

$$\Omega_n(u) \asymp e^{nH(u)}, \tag{15}$$

where $H(u)$ is the entropy of the system at energy density u. As is well known, the function $H(u)$ is also obtained by a Legendre transform, this time involving the logarithm of the one-particle partition function or free energy. Now, using the above approximation for $\Omega_n(u)$, and the fact that all states x^n such that $U_n(x^n) = u$ have the same probability

$$P_n(x^n : U_n(x^n) = u) = \frac{e^{-\beta_0 n u}}{Z(\beta_0)}, \tag{16}$$

we can write

$$g_n(u) = \Omega_n(u) P_n(x^n : U_n(x^n) = u) \asymp e^{-n[u\beta_0 + \ln Z(\beta_0) - H(u)]}. \tag{17}$$

Thus, we arrive at

$$D(u) = u\beta_0 + \ln Z(\beta_0) - H(u). \tag{18}$$

One can verify that the above expression for the rate function is totally equivalent to the one found in the context of large deviation theory. Both expressions are, in fact, related by the transformation $\beta(u) = \beta_0 - k(u)$, where the equation

$$\beta(u) = \frac{dH(u)}{du} \tag{19}$$

is the usual thermostatistical definition of the inverse temperature. The proof of this equivalence result follows, essentially, by noting that

$$\lambda(k) = \ln \sum_{x \in \mathcal{X}} \frac{e^{-\beta_0 u(x)}}{Z(\beta_0)} e^{ku(x)} = \ln Z(\beta_0 - k) - \ln Z(\beta_0), \tag{20}$$

and by using the familiar expression $H(u) = u\beta(u) + \ln Z(\beta(u))$ for the entropy. (The complete verification of the result is left as an exercise to the reader.)

Let us now turn to the matter of defining an inverse temperature β for our system of IID particles, and to the complement matter of inferring the probability density $f_n(\beta)$. Following our discussion of temperature fluctuations, it should be expected that there are many ways in which one can assign a temperature to the microcanonical set of states defined by

$$M_n(u) = \{x^n : U_n(x^n) = u\}. \tag{21}$$

Also, it is to be expected that one definition of temperature may not necessarily coincide with another in the case of finite-size ($n < \infty$) systems. We illustrate this possibility by comparing below three different definitions or "flavors" of temperature.

- **Derivative of entropy or free energy.** An obvious way to associate a temperature to the states in $M(u)$ is to take the energy derivative of the microcanonical entropy $H(u)$ as in eq. (19). Equivalently, one can solve the equation

$$-\frac{d\ln Z(\beta)}{d\beta} = u \tag{22}$$

for β, or compute the function $k(u)$ from eq. (14) and use the relation $\beta(u) = \beta_0 - k(u)$. The inverse temperature obtained by any of these methods will be denoted by $\beta_{th}(u)$ to emphasize that it is based on *intensive* thermodynamic potentials which do not depend on n.

- **Derivative of the density of state.** A slightly different definition of inverse temperature is obtained by taking the "logarithmic derivative" of $\Omega_n(u)$ with respect to the total energy nu

$$\beta_\Omega(u) = \frac{1}{\Omega_n(u)}\frac{d\Omega_n(u)}{d(nu)} = \frac{d\ln\Omega_n(u)}{d(nu)} \tag{23}$$

in lieu of the derivative of the entropy exponent as in eq. (19). This defines another inverse temperature $\beta_\Omega(u)$ which differs from $\beta_{th}(u)$ by a term of order $O(n^{-1}\ln n)$ which vanishes as $n \to \infty$.

- **Gibbsian distribution of states.** An inverse temperature $\beta_L(u)$ can be defined from a phenomenological point of view by fitting a given distribution of states $L_n(x)$ of mean energy $U_n = u$ with a Gibbs distribution of the form

$$L^u(x) = \frac{e^{-\beta_L(u)u(x)}}{Z(\beta_L(u))} . \tag{24}$$

We have described this definition of temperature earlier (see section 2), and have noted that it is accurate when n is large. To be more precise, it is accurate in a probabilistic sense because, in theory, there is always a possibility that non-Gibbsian distributions $L_n(x)$ of mean energy $U_n = u$ can be observed. However, the probability associated with such a possibility is very small and vanishes rapidly as $n \to \infty$. To see why, let us consider all the states x^n and their corresponding distributions L_n present in the energy "box" $M(u)$. What we want to show is that the probability $P_n(L^u)$ that L^u is observed in $M(u)$ is overwhelmingly large compared to the probability $P_n(L)$ of observing any other distribution $L \neq L^u$. To show this, we use another result of the theory of large deviations [13, 14, 21] which states that

$$\frac{P_n(L)}{P_n(L^u)} \asymp e^{-n[H(L^u)-H(L)]} = e^{-n\Delta H} , \tag{25}$$

where

$$H(L) = -\sum_{x \in \mathcal{X}} L(x)\ln L(x) \tag{26}$$

is the Boltzmann-Gibbs-Shannon entropy, and $\Delta H = H(L^u) - H(L)$. Using the fact that L^u is a maximum entropy distribution under the constraint $U_n = u$, it is easy to see that $\Delta H \geq 0$ with equality if and only if $L = L^u$, so that $P_n(L)/P_n(L^u) \to 0$ as $n \to 0$. Moreover, the discrepancy between the two probabilities is exponentially large in n. Thus, for n large it can be said that any distribution L_n picked at random in $M(u)$ will be such that $L_n \simeq L^u$. As this holds for any $M(u)$, this implies that any measured distribution related to some randomly chosen state x^n with $n \gg 1$ ought to be a Gibbs distribution or be very close to a Gibbs distribution with a probability nearly equal to 1.

The preceding paragraphs show that there is some arbitrariness in defining the concept of temperature for systems composed of a finite number of particles or degrees of freedom. In theory, there is some indeed; however, if n is large, then defining the temperature in any of the ways described above should have little effect on the actual value of the temperature inferred. Thus, for all practical purposes, we can assume that $\beta_{th}(u) \simeq \beta_\Omega(u) \simeq \beta_L(u)$ for $n \gg 1$. In view of what was said in section 2, it should be noted that the particular approximation $\beta_{th}(u) \simeq \beta_L(u)$ is of deep consequence: if we look at the distributions $L_n(x)$ associated with the states $x^n \in \mathcal{X}^n$, then we are likely to realize that the majority of these distributions, those which have the most probability of being observed, form a set of Gibbs distributions L^u parametrized by a fluctuating inverse temperature $\beta(u)$ (from now on we do not distinguish between the different flavors of inverse temperature). This means that for $1 \ll n < \infty$, all the statistical and thermodynamic properties of our system can be described in an effective manner, using an ensemble of Gibbs distributions with a fluctuating temperature. The probability density $f_n(\beta)$ ruling the inverse temperature fluctuations must, in this case, be given by

$$f_n(\beta) = g_n(u(\beta)) \left| \frac{du(\beta)}{d\beta} \right| , \tag{27}$$

where $u(\beta)$ is the inverse function of $\beta(u)$. Using this density, one then defines a "mixed" or "average" Gibbs distribution of one-particle states as follows:

$$\tilde{L}(x) = \int L^{u(\beta)} f_n(\beta) d\beta = \int \frac{e^{-\beta u(x)}}{Z(\beta)} f_n(\beta) d\beta . \tag{28}$$

This integral is a definite integral which must be evaluated over the range of definition of β. Equivalently, the average can be taken over the energy coordinate:

$$\tilde{L}(x) = \int L^u g_n(u) du = \int \frac{e^{-\beta(u)u(x)}}{Z(\beta(u))} g_n(u) du . \tag{29}$$

In the above equation, be sure to distinguish the energy function $u(x)$ from the value u of the mean energy U_n. Also note the slight difference between these mixed distributions and those proposed by Wilk et al. and Beck: in our version of mixed distributions, we take the average over the Gibbs factor $e^{-\beta u(x)}$ *normalized* by the partition function, which is itself a function of β (compare eqs.(3) and (28)).

4 THE CASE OF THE PERFECT GAS

As an application of the large deviation formalism, we carry out in this section the complete calculation of $g_n(u)$ and $f_n(\beta)$ for $u(x) = x^2/2$. By using this form of energy, we assume that the particles composing the gas have a unit mass, and that their momentum x_i, $i = 1, 2, \ldots, n$, is confined to one dimension ($\mathcal{X}$ is the real line extending from $-\infty$ to $+\infty$). We also abstract out the position's of the particles from the analysis, since the mean energy U_n of the n particles does not depend on the position's degree of freedom.

To find $g_n(u)$, we first calculate the rate function $D(u)$ using the Legendre transform method. The cumulant generating function associated with the quadratic energy function is calculated to be

$$\lambda(k) = \ln \int_{-\infty}^{\infty} \frac{e^{-\beta_0 x^2/2}}{Z(\beta_0)} e^{kx^2/2} dx = \frac{1}{2} \ln \frac{\beta_0}{\beta_0 - k}. \tag{30}$$

From this equation, we find the "translated" inverse temperature $k(u)$ by solving

$$\frac{d\lambda(k)}{dk}\bigg|_{k(u)} = \frac{1}{2} \frac{1}{\beta_0 - k(u)} = u. \tag{31}$$

The solution is $k(u) = \beta_0 - (2u)^{-1}$, so that

$$D(u) = uk(u) - \lambda(k(u)) = u\beta_0 - \frac{1}{2} - \frac{1}{2} \ln 2\beta_0 u. \tag{32}$$

Thus,

$$g_n(u) \asymp u^{n/2} e^{-nu\beta_0}. \tag{33}$$

This form of density is a variant of the χ^2 or gamma density mentioned previously with n as the number of degrees of freedom [16]. Note that this density for the mean energy can be derived directly by noting that U_n, for $u(x) = x^2/2$, is a normalized sum of squares of n Gaussian random variables. In statistics, this is usually how the χ^2 density is introduced [16].

At this point, the density $f_n(\beta)$ describing the fluctuations of β is readily deduced from the expression of $g_n(u)$ found above by calculating the physical inverse temperature $\beta(u)$. To this end, we can use the fact that $\beta(u) = \beta_0 - k(u)$ or use the equipartition theorem to find in both cases that $\beta(u) = (2u)^{-1}$. Hence, following eq. (27), $f_n(\beta)$ must have the form

$$f_n(\beta) \asymp \frac{1}{\beta^{n/2}} e^{-\frac{n\beta_0}{2\beta}} \frac{1}{\beta^2}. \tag{34}$$

Normalizing this expression for $\beta \in [0, \infty)$ yields

$$f_n^{\mathrm{ld}}(\beta) = \frac{\beta_0}{\Gamma(\frac{n}{2})} \left(\frac{n\beta_0}{2}\right)^{n/2} \beta^{-n/2-2} e^{-\frac{n\beta_0}{2\beta}} \tag{35}$$

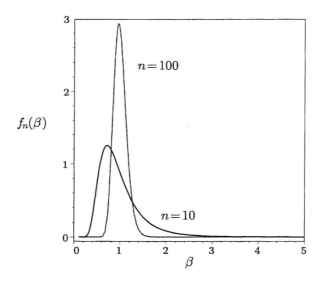

FIGURE 1 Probability densities $f_n^{\mathrm{ld}}(\beta)$ characterizing the β fluctuations of $n = 10$ and $n = 100$ free particles thermally coupled to a heat bath with $\beta_0 = 1$. Each density is defined for $\beta > 0$, and shows a maximum at $\beta_0 n/(4+n)$.

as the large deviation (ld) approximation of $f_n(\beta)$. A plot of this density for two values of n (10 and 100) is shown in figure 1 with $\beta_0 = 1$. The plot corresponding to $n = 10$ has no real physical significance, since the large deviation approximation is not expected to be effective in this case. However, it is presented to illustrate the skewness (to the right) of $f_n(\beta)$, which disappears as $n \to \infty$. The maximum value of $f_n(\beta)$ is given by $\beta_{\max} = \beta_0 n/(4+n)$. As expected, $f_n(\beta)$ converges (in a uniform sense) to the thermodynamic-limit density $f_\infty(\beta) = \delta(\beta - \beta_0)$ when $n \to \infty$; this is partially seen by looking at figure 1. By virtue of the law of large numbers, $g_n(u)$ must also converge in the same limit to a δ density, taking this time the form $g_\infty(u) = \delta(u - u(\beta_0))$ where

$$u(\beta_0) = E[u(X)] = \int_{-\infty}^{\infty} \frac{e^{-\beta_0 x^2/2}}{Z(\beta_0)} \frac{x^2}{2} dx = \frac{1}{2\beta_0} . \tag{36}$$

We now come to the main point of our study, which is to compare the $f_n(\beta)$ density obtained here and the χ^2 probability density of eq. (2), which was "postulated" by Wilk et al. and Beck in their studies of mixed distributions (see section 2). To establish this comparison, we present in figure 2 two plots of $f_n^{\mathrm{ld}}(\beta)$ for two different values of n and a variant of the χ^2 β-density proposed by Beck

$$f_n^B(\beta) = \frac{1}{\Gamma(\frac{n}{2})} \left(\frac{n}{2\beta_0} \right)^{n/2} \beta^{n/2-1} e^{-\frac{n\beta}{2\beta_0}} , \tag{37}$$

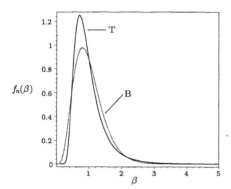 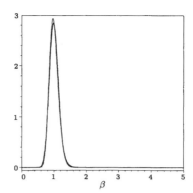

FIGURE 2   Comparison of the χ^2 β-density proposed by Beck (B) [4] and the one proposed in this work (T) for $n = 10$ (left) and $n = 100$ (right) particles (see text). For $n = 100$, the two densities are quasi-indistinguishable. The maximum value of the β-density, in the case of Beck, is located at $\beta_0(n-2)/n$.

which results from identifying $1/(q-1)$ in eq. (2) with $n/2$ [4]. The plots are presented again for $n = 10$ and $n = 100$. A rapid inspection of the expressions of f_n^{ld} and f_n^B reveals that these densities are not at all the same. The density f_n^B can, in fact, be viewed as emerging from the sum of the squares of n Gaussian random variables, whereas, in our case, the sum of squared Gaussian random variables arises as the mean energy, and so as β^{-1} modulo some constant. This explains why performing the change of variables $\beta \to \beta^{-1}$ in f_n^{ld} yields f_n^B modulo some constant and a Jacobian term arising from the change of variables. In spite of this important difference, the second plot of figure 2 shows that both densities are remarkably similar as n gets large. This, at first, does not seem surprising as both densities converge to the delta density $f_\infty(\beta)$ in the thermodynamic limit $n \to \infty$. However, it is to be noted that each of them gives totally different mixed distributions when they are used in eq. (28). Indeed, it can be shown [25] that the mixed distribution associated with f_n^{ld} has the asymptotic form

$$\tilde{L}_n^{\mathrm{ld}}(x) \sim e^{-|x|} \tag{38}$$

for $|x| \gg 1$, instead of

$$\tilde{L}^B(x) \propto e_q^{-x^2}, \tag{39}$$

where $q = 1 - 2/n$. Both of these results should be compared with the (pure) Gibbsian distribution $L^G(x) \propto e^{-x^2}$ which is the limiting distribution of $\tilde{L}_n^{\mathrm{ld}}$ and $\tilde{L}_n^B$ in the thermodynamic limit ($n \to \infty$).

The above scaling relationships clearly indicate that choosing between $f_n^{\mathrm{ld}}(\beta)$ or $f_n^B(\beta)$ has a dramatic consequence on the functional form of the mixed distribution calculated even for $n \gg 1$. Does that imply that our model of temperature fluctuations cannot serve as a model of the non-Gibbsian distributions which have been observed in turbulent fluid experiments as well as in nuclear scattering experiments? The answer is not as straightforward as one would think. First, it is not at all clear that turbulent fluids can actually be treated in the canonical ensemble and/or that the perfect gas assumption is a valid approximation in this case. These points call for further justifications. Second, the extreme events $|x| \gg 1$ needed to validate either one of two temperature densities compared in this work are often very difficult to detect experimentally in a reliable way. Surely, additional calculations and experimental data would be welcome in order to test the validity of our approach, and to confront it with that of the authors mentioned in the present study. This seems to be especially true for nuclear scattering experiments, which are usually thought to fit perfectly well into the canonical ensemble picture [11, 24].

5 CONCLUDING REMARKS

Our treatment of the energy and temperature fluctuations of a system coupled to a heat bath has focused mainly on the perfect gas. However, it is worth noting that the large deviation approach presented in this chapter for calculating the energy and temperature probability densities in the canonical ensemble is very general. It can be applied independently of the form of the energy function $u(x)$, which defines the mean energy U_n, and can also be generalized without too much difficulty to cases involving other forms of probability distribution for $P_n(x^n)$ (e.g., q-exponential distributions). In this context, an obvious extension of our work could be to consider different forms for $u(x)$, and to look at the mixed distributions which result from the corresponding temperature fluctuations. This line of thought has been followed recently by Beck and Cohen [7], who derived a number of "superstatistical" mixed distributions (sometimes unphysical ones) by assuming different forms of temperature fluctuations. Another problem could be to solve the following "inverse problem": for which $u(x)$ is $f_n(\beta)$ the same χ^2 density as the one suggested by Beck? Finally, note that a large deviation calculation of $g_n(u)$ and $f_n(\beta)$ can also be carried out for systems involving dependent random variables. Unfortunately, the calculations leading to the specific forms of $g_n(u)$ and $f_n(\beta)$ in this case are likely to be tedious. Also, the concept of mixed distribution does not generalize easily to the case of interacting particles because the eqs.(25) and (26), which were used to prove that Gibbs distributions are the only distributions likely to be observed in large systems, are valid for sequences of IID random variables only. It is, in fact, a long-standing open problem of large deviation theory to generalize these equations to sequences of dependent random variables. Solving this problem would have direct consequences in statistical physics, for it implies *ipso facto* a generalization of the maximum entropy principle to systems of interacting particles.

ACKNOWLEDGMENTS

It is a pleasure for me to thank M. Gell-Mann and C. Tsallis for the kind invitation to participate to the Workshop on Interdisciplinary Applications of Nonextensive Statistical Mechanics. I also want to thank S. Lloyd, who shares a great part of the responsibility for my visit to the Santa Fe Institute. This work has been supported in part by the Natural Sciences and Engineering Research Council of Canada, and the Fonds québécois de la recherche sur la nature et les technologies.

REFERENCES

[1] Abe, S., and Y. Okamoto, eds. *Nonextensive Statistical Mechanics and Its Applications*. New York: Springer-Verlag, 2001.

[2] Arimitsu, T., and N. Arimitsu. "PDF of Velocity Fluctuation in Turbulence by a Statistics based on Generalized Entropy." *Physica A* **305** (2002): 218.

[3] Ashkenazi, S., and V. Steinberg. "Spectra and Statistics of Velocity and Temperature Fluctuations in Turbulent Convection." *Phys. Rev. Lett.* **83** (1999): 4760.

[4] Beck, C. "Dynamical Foundations of Nonextensive Statistical Mechanics." *Phys. Rev. Lett.* **87** (2001): 180601.

[5] Beck, C. "Non-additivity of Tsallis Entropies and Fluctuations of Temperature." *Europhys. Lett.* **57** (2002): 329.

[6] Beck, C. "Nonextensive Methods in Turbulence and Particle Physics." *Physica A* **305** (2002): 209.

[7] Beck, C., and E. G. D. Cohen. "Superstatistics." May 2002. arXiv.org e-Print Archive, Condensed Matter, Cornell University. December 2002. ⟨http://xxx.lanl.gov/abs/cond-mat/0205097⟩.

[8] Beck, C., G. S. Lewis, and H. L. Swinney. "Measuring Nonextensitivity Parameters in a Turbulent Couette-Taylor Flow." *Phys. Rev. E* **63** (2001): 035303.

[9] Ching, E. S. C. "Probability Densities of Turbulent Temperature Fluctuations." *Phys. Rev. Lett.* **70** (1993): 283.

[10] Chui, T. C. P., D. R. Swanson, M. J. Adriaans, J. A. Nissen, and J. A. Lipa. "Temperature Fluctuations in the Canonical Ensemble." *Phys. Rev. Lett.* **69** (1992): 3005.

[11] Cottinghman, W. W., and D. A. Greenwood. *An Introduction to Nuclear Physics*, 2d ed. Cambridge: Cambridge University Press, 2001.

[12] Dembo, A., and O. Zeitouni. *Large Deviations Techniques and Applications*, 2d ed. New York: Springer-Verlag, 1998.

[13] Ellis, R. "An Overview of the Theory of Large Deviations and Applications to Statistical Mechanics." *Scand. Actuar. J.* **1** (1995): 97.

[14] Ellis, R. "The Theory of Large Deviations: From Boltzmann's 1877 Calculation to Equilibrium Macrostates in 2D Turbulence." *Physica D* **133** (1999): 106.

[15] Feshbach, H. "Small Systems: When Does Thermodynamics Apply?" *Phys. Today* **40** (1987): 9.

[16] Freund, J. E. *Mathematical Statistics*, 5th ed. New Jersey: Prentice Hall, 1992.

[17] Johnson, R. W., ed. *The Handbook of Fluid Dynamics*. New York: CRC Press, 1998.

[18] Kittel, C. "Temperature Fluctuation: An Oxymoron." *Phys. Today* **41** (1988): 93.

[19] Korus, R., St. Mrówczyński, M. Rybczyński, and Z. Włodarczyk. "Transverse Momentum Fluctuations due to Temperature Variation in High-Energy Nuclear Collisions." *Phys. Rev. C* **64** (2001): 054908.

[20] Mandelbrot, B. B. "Temperature Fluctuation: A Well-Defined and Unavoidable Notion." *Phys. Today* **42** (1989): 71.

[21] Oono, Y. "Large Deviation and Statistical Physics." *Prog. Theor. Phys. Suppl.* **99** (1989): 165.

[22] Prato, D. "Generalized Statistical Mechanics: Extension of the Hilhorst Formula and Application to the Classical Ideal Gas." *Phys. Lett. A* **203** (1995): 165.

[23] Reif, R. *Fundamentals of Statistical and Thermal Physics*. New York: McGraw-Hill, 1965.

[24] Stodolsky, L. "Temperature Fluctuations in Multiparticle Production." *Phys. Rev. Lett.* **75** (1995): 1044.

[25] Touchette, H. Unpublished manuscript, 2002 (in preparation).

[26] Tsallis, C. "Extensive Versus Nonextensive Physics." In *New Trends in Magnetism, Magnetic Materials, and Their Applications*, edited by J. L. Morán-López and J. M. Sanchez. New York: Plenum Press, 1994.

[27] Tsallis, C. "Possible Generalization of Boltzmann-Gibbs Statistics." *J. Stat. Phys.* **52** (1988): 479.

[28] Wilk, G., and Z. Włodarczyk. "Application of Nonextensive Statistics to Particle and Nuclear Physics." *Physica A* **305** (2002): 227.

[29] Wilk, G., and Z. Włodarczyk. "The Imprints of Nonextensive Statistical Mechanics in High-Energy Collisions." *Chaos, Solitons & Fractals* **13** (2002): 581.

[30] Wilk, G., and Z. Włodarczyk. "Interpretation of the Nonextensivity Parameter q in Some Applications of Tsallis Statistics and Lévy Distributions." *Phys. Rev. Lett.* **84** (2000): 2770.

On the Role of Non-Gaussian Noises on Noise-Induced Phenomena

Horacio Wio

Most of the studies of noise-induced phenomena assume that the noise source is Gaussian (either white or colored). Here we present recent results of some of those noise-induced phenomena when driven by a noise source taken as colored and non-Gaussian, generated by a nonextensive q-distribution. In all the cases analyzed we have found that the response of the system is strongly affected by a departure of the noise source from the Gaussian behavior, showing an enhancement and/or a marked broadening of the corresponding system's response. The general result is that the value of the parameter q, optimizing the system's response, results in $q \neq 1$ (where $q = 1$ corresponds to a Gaussian distribution). These results are of great relevance for many technological applications as well as for some situations of medical interest, like the noisy control of Wenckebach rhythms.

1 INTRODUCTION

Fluctuations (or noise) have had a changing role in the history of science. We can identify three different stages. During the a first one, which lasted until the end of nineteenth century, noise was considered a nuisance to be avoided or eliminated. In the second stage, which started at the beginning of the twentieth century, it was possible to extract more information from a physical system through the study of fluctuations via Onsager, fluctuation-dissipation, and other related relations. The third stage corresponds to the last few decades of the twentieth century, with the recognition that in many situations noise can actually play a driving role that induces new phenomena. Some examples are noise-induced phase transitions [18, 27, 28], noise-induced transport [2, 29, 34, 35], stochastic resonance [17], and noise-sustained patterns [18].

Most of the studies on the noise-induced phenomena indicated above assume that the noise source is Gaussian (either white or colored). In addition to the intrinsic interest in the study of non-Gaussian noises, there is some experimental evidence, particularly in sensory and biological systems [3, 16, 20, 30, 32, 39], indicating that in at least some of these phenomena the noise sources could be non-Gaussian. The use of non-Gaussian noises in such studies is rare, mainly due to the difficulties of handling them and to the possibility of obtaining some analytical results when working with Gaussian (particularly white) noises.

Here we present recent results on some of those noise-induced phenomena when driven by a noise source taken as colored and non-Gaussian, generated by a Tsallis nonextensive q-distribution [10]. The phenomena we discuss here correspond to: stochastic resonance, gated trapping processes, Brownian motors, and, finally, the possibility of noisy control and elimination of Wenckebach rhythms [13].

In all the cases that were analyzed we have found that the response of the system is strongly affected by a departure of the noise source from the Gaussian behavior. For instance, in stochastic resonance we found an enhancement of the response (that is a larger maximum in the curve for the signal-to-noise ratio [SNR]) and a marked broadening of the SNR curve indicating a larger degree of independence on the precise value of the noise intensity needed to tune the external signal. The general result is that the value of the parameter q optimizing the system's response is $q \neq 1$ ($q = 1$ corresponding to a Gaussian distribution). These results have strong implications for technological and medical (cardiology) applications as well as for explaining some experimental and theoretical analyses on sensory systems.

In the next section we briefly discuss the form and properties of the non-Gaussian noise source. After that we present the results for the different noise-induced phenomena we have analyzed. In the last section we draw some conclusions.

2   NON-GAUSSIAN NOISE AND ITS PROPERTIES

We start considering the following general form of a Langevin equation

$$\dot{x} = -\frac{\partial U}{\partial x} + \eta(t).$$ (1)

However, at variance with other studies, we assume that the noise term $\eta(t)$ has a non-Gaussian distribution. Although we believe that our results are quite general, for concreteness and motivated by the work in Borland [4] based on a nonextensive thermostatistics distribution [10], we consider that the noise term is a Markovian process generated as the solution of the following Langevin equation

$$\dot{\eta} = -\frac{1}{\tau}\frac{d}{d\eta}V_q(\eta) + \frac{1}{\tau}\xi(t),$$ (2)

where $\xi(t)$ is a standard Gaussian white noise of zero mean and correlation $\langle\xi(t)\xi(t')\rangle = D\delta(t - t')$, and

$$V_q(\eta) = \frac{D}{\tau(q-1)}\ln\left[1 + \frac{\tau}{D}(q-1)\frac{\eta^2}{2}\right].$$ (3)

The stationary properties of the noise η, including the time correlation function, have been studied in Fuentes et al. [15]. The stationary probability distribution (pdf) is given by

$$P_q^{st}(\eta) = \frac{1}{Z_q}\left[1 + \frac{\tau}{D}(q-1)\eta^2\right]^{\frac{-1}{q-1}},$$ (4)

where Z_q is the normalization factor. This distribution can be normalized only for $q < 3$. The first moment, $\langle\eta\rangle = 0$, is always equal to zero, and the second moment, $\langle\eta^2\rangle = 2D/\tau(5 - 3q) \equiv D_q$, is finite only for $q < 5/3$. Clearly, when $q \to 1$ we recover the limit of η being a Gaussian colored noise (Ornstein-Uhlenbeck or OU process). Furthermore, for $q < 1$, the distribution has a cut off and it is only defined for $|\eta| < \sqrt{2D/\tau(1 - q)}$. Finally, the correlation time τ_q of the process η diverges near $q = 5/3$ and it can be approximated over the whole range of values of q as $\tau_q \approx 2\tau/(5 - 3q)$.

In Fuentes et al. [15], the above-indicated stochastic processes were analyzed in detail. Also, an effective Markovian approximation via a path integral procedure was obtained, making it possible to get quasi-analytical results for the mean-first-passage time or transition rate. Such results and their dependence on the different parameters in the case of a double-well potential were compared with extensive numerical simulations with excellent agreement.

In order to have an idea of the form of the stationary pdf for η, in figure 1 we show it for different values of q, with $\beta = \tau/D$.

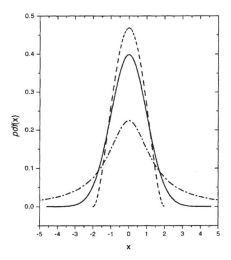

FIGURE 1 The stationary pdf given by eq. (4) with $\beta = 1$. Solid line: Gaussian case ($q = 1$); dashed line: bounded distribution ($q = 0.5$); dashed-dotted line, wide distribution ($q = 2$).

3 SOME APPLICATIONS

3.1 STOCHASTIC RESONANCE

Stochastic resonance (SR) has attracted considerable interest due, among other aspects, to its potential technological applications for optimizing the output signal-to-noise ratio (SNR) in nonlinear dynamical systems, as well as to its connection with some biological mechanisms. The phenomenon shows the counterintuitive role played by noise in nonlinear systems as it enhances the response of a system subject to a weak external signal [17]. A tendency shown in recent papers, and determined by the possible technological applications, points toward achieving an enhancement of the system response (that is, obtaining a larger output SNR) by means of the coupling of several SR units in what forms an *extended medium* [5, 7, 8, 22, 23, 41].

A majority of studies on SR have been made analyzing a paradigmatic system: a bistable, one-dimensional, double-well potential. In almost all descriptions, and particularly within the *two-state model* (TST) [31], the transition rates between the two wells are estimated as the inverse of the *mean first-passage time* [31], which is evaluated using standard techniques, and most specifically through the Kramers approximation [14, 15]. In all cases the noises are assumed to be Gaussian.

Here we present the results of analyzing SR when the noise source is non-Gaussian. We consider the following problem

$$\dot{x} = f(x,t) + \eta(t)\,, \tag{5}$$

$$\dot{\eta} = -\frac{1}{\tau}\frac{d}{d\eta}V_q(\eta) + \frac{1}{\tau}\xi(t)\,, \tag{6}$$

where $\xi(t)$ is a Gaussian white noise as in eq. (2), and $V_q(\eta)$ is given as in eq. (3). The function $f(x,t)$ is derived from a potential $U(x,t)$, consisting of a double well potential and a linear term modulated by $S(t) \sim F\cos(\omega t)$ ($f(x,t) = -dU/dx = -U_0' + S(t)$). This problem corresponds (for $\omega = 0$) to the case of diffusion in a potential $U_0(x)$, induced by η, a colored non-Gaussian noise.

The details about the form of the effective Markovian Fokker-Planck equation can be found in Fuentes et al. [14, 15]. There we calculated the stationary probability density, and derived the expression for the first-passage time. We also exploited the TST approach [31] in order to obtain the power spectral density (psd) and the SNR.

In figure 2 we depict R vs. D, for a fixed value of the time correlation τ ($\tau = 0.1$) and various q. The general trend is that the maximum of the SNR curve increases when $q < 1$, this is when the system departs from the Gaussian behavior. Figure 3 again shows R vs. D, but for a fixed value of q ($q = 0.75$) and several values of τ. The general trend agrees with the results for colored Gaussian noises [17], where it was shown that the increase of the correlation time induces a decrease of the maximum of SNR as well as its shift toward larger values of the noise intensity. The latter fact is a consequence of the suppression of the switching rate with increasing τ. Both qualitative trends are confirmed by Monte Carlo simulations of eq. (5). We have integrated eqs. (5) and (6) numerically using the Heun method. In all cases the results were obtained averaging over 2000 trajectories (5000 trajectories for $\tau = 0$).

Figure 4 shows the simulation results for the same situation and parameters indicated in figure 2. Here, in addition to the increase of the maximum of the SNR curve for values of $q < 1$, we see also an aspect that is not well reproduced or predicted by the effective Markovian approximation. It is the fact that the maximum of the SNR curve flattens for lower values of q, indicating that the system, when departing from Gaussian behavior, does not require a fine tuning of the noise intensity in order to maximize its response to a weak external signal. Figure 4 shows the simulation results for the same situation and parameters indicated in figure 2. Again, we found an agreement with the behavior found for colored Gaussian noises [17].

Our numerical and theoretical results indicate that:

1. for a fixed value of τ, the maximum value of the SNR increases with decreasing q;
2. for a given value of q, the optimal noise intensity (the one that maximizes SNR) decreases with q and its value is approximately independent of τ; and

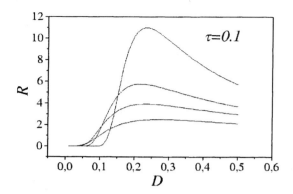

FIGURE 2   Theoretical value of SNR vs. D, for $\tau = 0.1$ and the following values of $q = 0.25, 0.75, 1.0, 1.25$ (from top to bottom).

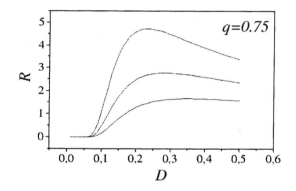

FIGURE 3   Theoretical value of SNR vs. D, for $q = 0.75$ and the following values of $\tau = 0.25, 0.75, 1.5$ (from top to bottom).

3. for a fixed value of the noise intensity, the optimal value of q is independent of τ and, in general, it turns out that $q_{op} \neq 1$.

In Castro et al. [9] we analyzed the case of SR when the noise source is non-Gaussian, but from an experimental point of view. We studied an experimental setup similar to the one used in Fauve and Heslot [12], but we used a non-Gaussian noise source that was built to exploit the form of noise introduced above, particularly white noise. Those results confirmed the predictions indicated above.

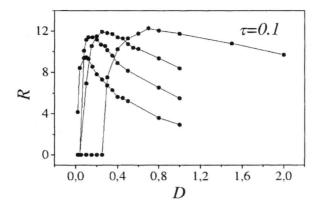

FIGURE 4 Simulation results of SNR vs. D, for $\tau = 0.1$ and the following values of $q = 0.25, 0.75, 1.0, 1.25$ (from top to bottom).

3.2 GATED TRAPS

As indicated above, stochastic resonance has been found to play a relevant role in several problems in biology. In particular, there is an experiment related to the measurement of the current through voltage-sensitive ion channels in a cell membrane [3]. These channels switch (randomly) between open and closed states, thus controlling the ion current. This and other related phenomena have stimulated several theoretical studies of the problem of ionic transport through biomembranes, using different approaches, as well as different ways of characterizing stochastic resonance in such systems [16, 20, 32].

In Sánchez et al. [37] we have studied a toy model, prompted by the work in Bezrukov and Vodyanoy [3], sketching the behavior of an ion channel. Among other factors, the ion transport depends on the membrane electric potential (which plays the role of the barrier height) and can be stimulated by both *dc* and *ac* external fields. This included the simultaneous action of a deterministic and a stochastic external field on the trapping rate of a gated imperfect trap. Rather than attempting a precise modeling of the behavior of an ionic channel, we proposed a simple model of dynamical trap behavior. Our main result was that even such a simple model of a gated trapping process shows SR-like behavior. In that initial study we assumed that the stochastic external field was a Gaussian white noise. Here, we sketch the main results, obtained analyzing the same model, but using a correlated non-Gaussian noise source, as detailed in Wio et al. [40].

In Sánchez et al. [37], the study was based on the so-called *stochastic model* for reactions [1, 36], generalized in order to include the internal dynamics of traps. The dynamical process consists of the opening or closing of the traps

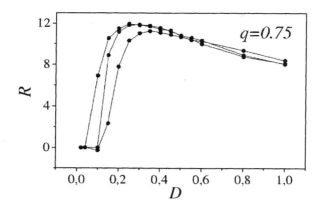

FIGURE 5 Simulation results of SNR vs. D, for $q = 0.75$ and the following values of $\tau = 0.25, 0.75, 1.5$ (from top to bottom).

according to an external field. Such a field has two contributions, one that is periodic with a small amplitude, and the other that is stochastic, the intensity of which will be (as usual) the tuning parameter. Here, to be consistent with the work in Sánchez et al. [37, 40], we change notation and use the same one as in those works. The starting model equation was

$$\partial_t \rho(x,t) = D \partial_x^2 \rho(x,t) - \gamma(t)\delta(x)\rho(x,t) + n_u\,, \tag{7}$$

where γ is a stochastic process that represents the absorption probability of the trap, and ρ is the particle density (particles that have not yet been trapped). For a given realization of γ, here D is the diffusion coefficient, x is the coordinate over the one-dimensional system, and n_u is a source term that represents a constant flux of ions. The injection of ions can be at a trap position or at any other position. In this last case the ion can diffuse to the trap position. This diffusion coefficient represents an effective diffusion through the volume rather than a diffusion over the membrane surface.

The absorption is modeled as $\gamma(t) = \gamma^*\theta[B\sin(\omega t) + \xi - \xi_c]$, where $\theta(x)$, the Heaviside function, determines when the trap is open or closed. The trap works as follows: if the signal, composed of the harmonic part plus ξ (the noise contribution), reaches a threshold ξ_c, then the trap opens; otherwise, it is closed. We are interested in the case where $\xi_c > B$; that is, without noise the trap is always closed. When the trap is open the particles are trapped with a given frequency (probability per unit time) γ^*. In other words the open trap is represented by an "imperfect trap." Finally, in order to complete the model, we must give the statistical properties of the noise ξ. In Sánchez et al. [37] we assumed that ξ is an uncorrelated Gaussian noise of intensity ξ_0. In [40] we used a "colored" non-Gaussian noise given by the same eq. (6), where $\xi(t)$ is the non-Gaussian noise, ξ_w is the white noise intensity and V_q is given by $V_q(\xi) = 1/\beta(q-1)\ln[1 + \beta(q-1)\xi^2/2]$, where $\beta = \tau/\xi_w$.

We defined the current through the trap as $J(t) = \langle \gamma_j(t)\rho(jl,t)\rangle$. The brackets mean averages over all realizations of the noise. In Sánchez et al. [37], that is, in the case of $\xi(t)$ being a Gaussian white noise, we have obtained some analytical results and solved the equation numerically. However, for the non-Gaussian case [40] we should resort only to Monte Carlo simulations.

As in Sánchez et al. [37], we choose to quantify the SR-like phenomenon by computing the amplitude of the oscillating part of the absorption current given by $\Delta J = J|_{\sin(\omega t)=1} - J|_{\sin(\omega t)=-1}$. The qualitative behavior of the system can be explained as follows. For small noise intensities the current is low (remember that $\xi_c > B$); hence, ΔJ is small, too. For a large noise intensity, the deterministic (harmonic) part of the signal becomes irrelevant and the ΔJ is also small. Therefore, there must be a maximum at some intermediate value of the noise.

The details of the way the simulations were done can be found in Sánchez et al. [37] and Wio et al. [40]. All simulations shown in the figures correspond to averages over 1000 realizations. We have plotted all results as functions of the non-Gaussian noise intensity ξ_0. It is related to ξ_w by $\xi_0 = 2\xi_w/(5-3q)$ [36].

In figure 6 we show the amplitude of the absorption current $\Delta J(t)$ as a function of the noise intensity ξ_0 for: (a) different values of q and fixed τ and observational time (t), and (b) for three different τ and fixed values of q and t. The results are in agreement with those found in the case of Gaussian white noise. In the first case we see that the system response increases when $q < 1$, and there is a shift of the maximum of $\Delta J(t)$ to larger noise values for increasing q. In the second case, the curves also show a shift of the maximum to larger noise intensities as τ increases. The shift of the $\Delta J(t)$ maximum with τ to larger values of ξ_0 is in agreement with a similar effect in "usual" SR [14, 15]. In that case, it was associated with the suppression of the switching rate with increasing τ. In figure 7 we depict the maximum of $\Delta J(t)$ (which are averaged over the τ values for each q) for two different observational times as a function of the parameter q. For each observational time, the values are scaled with the corresponding maximum. We observe the existence of a new resonant-like maximum as a function of the parameter q. This implies that we can find a region of q where the maximum ΔJ reach optimal values (*corresponding to a bounded and non-Gaussian pdf*), yielding the largest system response.

The present results show that the use of non-Gaussian noises in the simple trapping process defined by eq. (7) produces significant changes in the system response when compared with the Gaussian case. In particular, we want to emphasize that we have found a double resonance-like phenomenon indicating that, in addition to an optimal noise intensity, there is an optimal q value which yields the larger enhancement of the system response. The remarkable fact is that it corresponds to $q < 1$ indicating that this enhancement occurs for a non-Gaussian

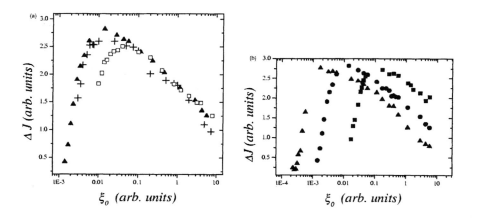

FIGURE 6 Value of ΔJ (amplitude of the oscillating part of the absorption current) as a function of ξ_0 for a given observational time ($t = 1140$). (a) different values of q (triangles $q = 0.5$, crosses $q = 1.0$, squares $q = 1.5$) and a fixed value of τ ($\tau = 0.1$). (b) different values of τ (triangles $\tau = 0.01$, circles $\tau = 0.1$, squares $\tau = 1.0$) and a fixed value of q ($q = 0.5$).

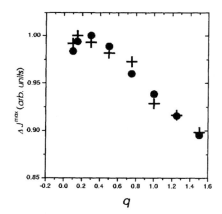

FIGURE 7 Dependence of $\Delta J_{\max}$, the value of ΔJ at the maximum, as a function of q and different observational times: circles $t = 633$, crosses $t = 1140$.

and *bounded* distribution. The increase in the system response (here of $\Delta J(t)$) and the reduction in the need to tune the noise of ξ_0, can be understood as similar to the case of the usual SR [14, 15]. The bounded character of the pdf for $q < 1$ contributes positively to the rate of overcoming the threshold ξ_c, and such a rate remains of the same order for a larger range of values of ξ_0.

3.3 BROWNIAN MOTORS

The study of noise-induced transport by "ratchets" has attracted the attention of an increasing number of researchers due to its biological interest and also to its potential technological applications [2, 34]. A recent new aspect has been to relax the requirement of a built-in bias: a system of periodically coupled nonlinear phase oscillators in a symmetric "pulsating" environment has been shown to undergo a noise-induced nonequilibrium phase transition, wherein the spontaneous symmetry breakdown of the stationary probability distribution function gives rise to an *effective* ratchetlike potential. Some of the striking consequences of this fact are the appearance of *negative* (absolute) zero-bias conductance in the disordered phase, but near the phase-transition line (for small values of the bias force F, the particle current $\langle \dot{X} \rangle$ opposes F), and *anomalous hysteresis* in the strong-coupling region of the ordered phase (the $\langle \dot{X} \rangle$ vs. F cycle runs *clockwise*, as opposed, for instance, to the B vs. H cycle of a ferromagnet) [29, 35].

Here, in line with the work in Bouzat and Wio [6], we analyze the effect of the class of colored non-Gaussian noise introduced before on the transport properties of Brownian motors. We start considering the general system

$$m\frac{d^2x}{dt^2} = -\gamma\frac{dx}{dt} - V'(x) - F + \xi(t) + \eta(t)\,, \tag{8}$$

where m is the mass of the particle, γ the friction constant, $V(x)$ the ratchet potential, F is a constant "load" force, and $\xi(t)$ the thermal noise satisfying $\langle \xi(t)\xi(t') \rangle = 2\gamma T\delta(t - t')$. Finally, $\eta(t)$ is the time-correlated forcing (with zero mean) that keeps the system out of thermal equilibrium by allowing rectification of the motion. For this type of ratchet model several different kinds of time-correlated forcing have been considered in the literature [2, 34]. The main characteristic introduced by the non-Gaussian form of the forcing we consider here is the appearance of arbitrary strong "kicks" with relatively high probability when compared, for example, with the Gaussian OU process. As we shall see, in a general situation (without fine tuning of the parameters), this leads to the above-indicated enhancement effects.

We will consider the dynamics of $\eta(t)$ as described by the Langevin equation (2). As discussed before, for $1 < q < 3$, the probability distribution decays more slowly than a Gaussian, as a power law. Hence, keeping D constant, the width or dispersion of the distribution increases with q, meaning that, the higher the q, the stronger the "kicks" that the particle will receive.

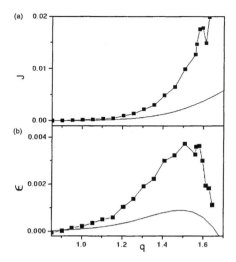

FIGURE 8   Current (a) and efficiency (b) as functions of q. The solid line corresponds to an adiabatic approximation, the line with squares shows results from simulations. All calculations are for $m = 0, \gamma = 1, T = 0.5, F = 0.1, D = 1$, and $\tau = 100/(2\pi)$.

For the ratchet potential we will first consider the same form as in Magnasco [26] (with period 2π) $V(x) = V_1(x) = -\int dx(\exp[\alpha \cos(x)]/J_0(i\alpha) - 1)$, with $\alpha = 16$. The integrand in $V(x)$ is the ratchet force $(-V'(x))$ appearing in eq. (8).

First, we have analyzed the overdamped regime setting $m = 0$ and $\gamma = 1$. Our intention was to analyze the dependence of the mean current $J = \langle dx/dt \rangle$ and the efficiency ε on the different parameters, in particular, their dependence on the parameter q. As defined, ε is the ratio of the work (per unit time) done by the particle "against" the load force F $(1/T_f \int_{x=x(0)}^{x=x(T_f)} F dx(t))$, to the mean power injected into the system through the external forcing η $(1/T_f \int_{x=x(0)}^{x=x(T_f)} \eta(t) dx(t))$. For the numerator we got $F\langle dx/dt \rangle = FJ$, while, for the denominator, we obtained $1/T_f \int_0^{T_f} \eta(t) dx/dt dt = 1/\gamma T_f \int_0^{T_f} \eta(t)^2 dt = 2D[\gamma T_f \tau(5-3q)]^{-1}$. Interesting and complete discussions on the thermodynamics and energetics of ratchet systems can be found in Parrondo et al. [33] and Sekimoto [38].

In figure 8, we show typical analytical results for the current and the efficiency as functions of q, together with results coming from numerical simulations. Calculations have been done in a region of parameters similar to the one studied in Magnasco [26], but they consider (apart from the difference provided by the non-Gaussian noise) a non-zero load force that leads to a nonvanishing efficiency. As can be seen, although there is not a quantitative agreement between theory and simulations, the adiabatic approximation predicts qualitatively very well the behavior of J (and ε) as q is varied. As shown in the figure, the current grows monotonously with q (at least for $q < 5/3$), while there is an optimal value of q (> 1) which gives the maximum efficiency. This fact is interpreted as follows: when q is increased, the width of the $P_q(\eta)$ distribution grows and high values of the non-Gaussian noise become more frequent, which leads to an improvement of the current. Although the mean value of J increases monotonously with q, the grow of the width of $P_q(\eta)$ leads to an enhancement of the fluctuations around this mean value. This is the origin of the efficiency's decay that occurs for high values of q: in this region, in spite of having a large (positive) mean value of the current, for a given realization of the process, the transport of the particle toward the desired direction is far from being assured. Hence, our results show that the transport mechanism becomes more efficient when the stochastic forcing has a non-Gaussian distribution with $q > 1$.

Now we turn to the $m \neq 0$ case, that is, the situations in which the inertia effects are relevant. The results that we found [6] imply that separation of masses (particles with different masses moving in opposite directions) occurs, and that this happens in the absence of load force. In view of the results discussed above, it is reasonable to expect that non-Gaussian noises may improve the capability of mass separation in ratchets. Lindner et al. [24] was one of the primary works discussing mass separation by ratchets. There, the authors analyzed a ratchet system like the one described by eq. (8), considering OU noise as external forcing (in our case it corresponds to $q = 1$). They studied (both numerically and analytically) the dynamics for different values of the correlation time of the forcing τ, finding that there is a region of parameters where mass separation occurs. This means that the direction of the current is found to be mass dependent: the "heavy" species moves in the negative sense while the "light" one does so in the positive sense.

In order to compare results, we analyzed the same system studied in Lindner et al. [24, 25], but consider the non-Gaussian forcing. Hence, we studied eq. (8) with $V(x) = V_2(x) = -[\sin(2\pi x) + 0.25\sin(4\pi x)]/(2\pi)$ as the ratchet potential. We focused on the region of parameters where, in Lindner et al. [24] (for $q = 1$), separation of masses was found. We fixed $\gamma = 2, T = 0.1, \tau = 0.75$, and $D = 0.1875$ and considered the values of the masses $m = m_1 = 0.5$ and $m = m_2 = 1.5$ as in Lindner et al. [24]. Our main result was that the separation of masses is enhanced when a non-Gaussian noise with $q > 1$ is considered. In figure 9(a) we show J as function of q for $m_1 = 0.5$ and $m_2 = 1.5$. It can be seen that there is an optimum value of q that maximizes the difference of currents. This value, which is close to $q = 1.25$, is indicated with a vertical double arrow. Another double arrow indicates the separation of masses occurring for $q = 1$ (Gaussian OU forcing). We have observed that, when the value of the load force is varied, the difference between the curves remains approximately constant, but both are shifted together to positive or negative values (depending on the sign of the variation of the loading). By controlling this parameter it is possible to

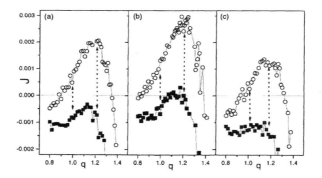

FIGURE 9 Separation of masses: results from simulations for the current as a function of q for particles of masses $m = 0.5$ (hollow circles) and $m = 1.5$ (solid squares). Calculations for three different values of the load force: (a) $F = 0.025$, (b) $F = 0.02$, and (c) $F = 0.03$.

achieve, for example, the situation shown in figure 9(b), where, for the value of q at which the difference of currents is maximal, the heavy "species" remains static on average (has $J = 0$), while the light one has $J > 0$. Also, the situation shown in figure 9(c), occurs when, for the optimal q, the two species move in the opposite direction with equal absolute velocity.

4 FINAL REMARKS

We have presented the results of a series of studies on noise-induced phenomena in which the noise source was non-Gaussian. In all cases, we have found that the system's response leads to enhancement of values of the parameter q, which indicates a departure from Gaussian behavior; that is, the optimum response happens for $q \neq 1$.

For the case of SR in a bistable system, we observed that the SNR, as we depart from Gaussian behavior (with $q < 1$), shows two main aspects: first, its maximum as a function of the noise intensity increases; and second, it becomes less dependent on the precise value of the noise intensity. Both aspects are of great relevance for technological applications [17]. In addition to the increase in the response (SNR), the reduction in the need of *tuning* a precise value of the noise intensity is of particular relevance in order to understand how a biological system can exploit this phenomenon. The present results indicate that the noise model used here offers an adequate framework for analyzing such a problem. An analogous result occurs for the case of SR in the trap model. Finally, for the Brownian motor, what we have found is that a departure from Gaussian behavior, given now by a value of $q > 1$, induces a remarkable increase in the current together with an enhancement of the motor efficiency. The latter shows, in addition, an optimum value for a given degree of non-Gaussianity. When inertia is taken into account, we also find a considerable increment in the mass separation capability.

Let us conclude by commenting on some related, although still preliminary, results. The problem that we want to address is associated with the so-called *Wenckebach rhythms* [19, 21]. In its classical manifestation, the Wenckebach phenomenon is characterized by a succession of electrocardiographic complexes in which atrioventricular (AV) conduction time (or P-R interval) increases progressively in decreasing increments until transmission failure occurs [19, 21]. In a recent contribution we analyzed the way Wenckebach rhythms arise in an excitable system due to the presence of defects (necrotic tissue) [13]. Afterward, we studied the possibility of eliminating or controlling the appearance of such rhythms through the injection of noise after or at the defect. Our intention was to determine the dependence of such rhythms on q, the parameter that controls the non-Gaussianity, and to analyze the optimum form of the noise to reach the control indicated above.

We used the simple FitzHugh model which involves only the membrane potential u and a recovery variable v. The variable v lumps together the time-dependent activation of the potassium current with the time-dependent inactivation of the sodium current, in other words, the two slow variables of the Hodgkin-Huxley equation. In order to simulate the propagation of a wave train through the medium, induced by a periodic forcing, we have imposed a time-dependent boundary condition for the variable u at $x = 0$, and have used initial conditions as discussed in Fuentes and Wio [13].

Even though our results are, so far, only preliminary and further studies are required, it is worth remarking here on what we have found: for a fixed (and low) noise intensity, the optimal form of noise for controlling Wenckebach rhythm production appears to be results to be non-Gaussian (i.e., $q \neq 1$).

ACKNOWLEDGMENTS

I wish to thank S. Bouzat, F. Castro, M. Fuentes, M. Kuperman, J. Revelli, A. Sánchez, R. Toral, and C. Tsallis for their collaboration and/or useful discussions. Partial support from CONICET and ANPCyT, both Argentine agencies, is also acknowledged. I also want to thank Iberdrola S. A., Spain, for an award within the *Iberdrola Visiting Professor Program in Science and Technology*, and the IMEDEA and Universitat de les Illes Balears, Palma de Mallorca, Spain, for the kind hospitality extended to me.

REFERENCES

[1] Abramson, G., A. Bru Espino, M. A. Rodriguez, and H. S. Wio. *Phys. Rev. E.* **50** (1994): 4319.

[2] Astumian, R. D. *Science* **276** (1997): 917.

[3] Bezrukov, S. M., and I. Vodyanoy. *Nature* **378** (1995): 362.

[4] Borland, L. *Phys. Lett. A* **245** (1998): 67.

[5] Bouzat, S., and H. S. Wio. *Phys. Rev. E* **59** (1999): 5142.

[6] Bouzat, S., and H. S. Wio. "Strong Enhancement of Current, Efficiency and Mass Separation in Brownian Motors Driven by Non-Gaussian Noises." *Phys. Rev. Lett.* submitted.

[7] Castelpoggi, F., and H. S. Wio. *Europhys. Lett.* **38** (1997): 91.

[8] Castelpoggi, F., and H. S. Wio. *Phys. Rev. E* **57** (1998): 5112.

[9] Castro, F. J., M. N. Kuperman, M. A. Fuentes, and H. S. Wio. *Phys. Rev. E* **64** (2001): 051105.

[10] Curado, E. M. F., and C. Tsallis. *J. Phys. A* **25** (1992): 1019.

[11] Dialynas, T. E., K. Lindenberg, and G. P. Tsironis. *Phys. Rev. E* **56** (1997): 3976.

[12] Fauve, S., and F. Heslot. *Phys. Lett. A* **97** (1983): 5.

[13] Fuentes, M. A., and H. S. Wio. *Physica A* **286** (2000): 345.

[14] Fuentes, M. A., R. Toral, and H.S. Wio. *Physica A* **295** (2001): 114–122.

[15] Fuentes, M. A., Horacio S. Wio, and Raúl Toral. *Physica A* **303** (2002): 91–104.

[16] Fuliński, A. *Phys. Rev. Lett.* **79** (1997): 4926.

[17] Gammaitoni, L., P. Hänggi, P. Jung, and F. Marchesoni. *Rev. Mod. Phys.* **70** (1998): 223.

[18] García-Ojalvo, J., and J. M. Sancho. *Noise in Spatially Extended Systems.* New York: Springer-Verlag, 1999.

[19] Glass, L., P. Hunter, and A. McCulloch, eds. *Theory of the Heart.* New York: Springer-Verlag, 1991.

[20] Goychuk, I., and P. Hänggi. *Phys. Rev. E* **61** (2000): 4272.

[21] Keener, J., and J. Sneyd. *Mathematical Physiology.* New York: Springer-Verlag, 1998.

[22] Lindner, J. F., B. K. Meadows, W. L. Ditto, M. E. Inchiosa, and A. Bulsara. *Phys. Rev. Lett.* **75** (1995): 3.

[23] Lindner, J. F., B. K. Meadows, W. L. Ditto, M. E. Inchiosa, and A. Bulsara. *Phys. Rev. E* **53** (1996): 2081.

[24] Lindner, R., L. Schimansky-Geier, P. Reimann, and P. Hänggi. In *Applied Linear Dynamics and Stochastic Systems Near the Millenium*, edited by J. B. Kadtke and A. Bulsara, 309. AIP, 1997.

[25] Lindner, R., L. Schimansky-Geier, P. Reimann, P. Hänggi, and M. Nagaoka. *Phys. Rev. E* **59** (1999): 1417.

[26] Magnasco, M. *Phys. Rev. Lett.* **71** (1993): 1477.

[27] Mangioni, S., R. Deza, H. S. Wio, and R. Toral. *Phys. Rev. Lett.* **79** (1997): 2389.

[28] Mangioni, S. R. Deza, R. Toral, and H. S. Wio. *Phys. Rev. E* **61** (2000): 223.

[29] Mangioni, S., R. Deza, and H. S. Wio. *Phys. Rev. E* **63** (2001): 041115.

[30] Manwani, A. Ph.D. thesis, California Institute of Technology, Pasadena, CA, 2000.

[31] McNamara, B., and K. Wiesenfeld. *Phys. Rev. A* **39** (1989): 4854.

[32] Nozaki, D., D. J. Mar, P. Griegg, and J. D. Collins. *Phys. Rev. Lett.* **72** (1999): 2125.

[33] Parrondo, J. M. R., J. M. Blanco, F. J. Cao, and R. Brito. *Europhys. Lett.* **43** (1998): 248.

[34] Reimann, P. *Phys. Rep.* **361** (2002): 57.

[35] Reimann, P., R. Kawai, C. Van den Broeck, and P. Hänggi. *Europhys. Lett.* **45** (1999): 545.

[36] Sánchez, A. D., E. M. Nicola, and H. S. Wio. *Phys. Rev. Lett.* **78** (1997): 2244.

[37] Sánchez, A. D., J. A. Revelli, and H. S. Wio. *Phys. Lett. A* **277** (2000): 304.

[38] Sekimoto, K. *Prog. Theor. Phys. Supl.* **130** (1998): 17.

[39] Wiesenfeld, K., D. Pierson, E. Pantazelou, Ch. Dames, and F. Moss. *Phys. Rev. Lett.* **52** (1994): 2125.

[40] Wio, H. S., J. A. Revelli, and A.D. Sánchez. *Physica D* (2002): in press. [cond-mat/0109454].

[41] Wio, Horacio S., B. Von Haeften, and S. Bouzat. Special Issue: Proceedings of the Twenty-First IUPAP International Conference on Statistical Physics, STATPHYS21. *Physica A* **306C** (2002): 140–156.

A Dripping Faucet as a Nonextensive System

T. J. P. Penna
J. C. Sartorelli
R. D. Pinto
W. M. Gonçalves

Here we present our attempt to characterize a time series of drop-to-drop intervals from a dripping faucet as a nonextensive system. We found a long-range anticorrelated behavior as evidence of memory in the dynamics of our system. The hypothesis of faucets dripping at the edge of chaos is reinforced by results of the linear rate of the increase of the nonextensive Tsallis statistics. We also present some similarities between dripping faucets and healthy hearts.

1   INTRODUCTION

Many systems in Nature exhibit complex or chaotic behaviors. Chaotic behavior is characterized by short-range correlations and strong sensitivity to small changes of the initial conditions. Complex behavior is characterized by the presence of long-range power-law correlations in its dynamics. In the latter, the sensitivity to a perturbation of the initial condition is weaker than in the former.

Nonextensive Entropy—Interdisciplinary Applications
edited by Murray Gell-Mann and Constantino Tsallis, Oxford University Press   195

Because the probability densities are frequently described as inverse power laws, the variance and the mean often diverge.

Although it is hard to predict the long-term behavior of such systems, it is still possible to get some information from them and even to find similarities between two apparently very distinct systems. Tools from statistical physics are frequently used because the main task here is to deal with diverse macroscopic phenomena and to try to explain them, starting with the microscopic interactions among many individual components. The microscopic interactions are not necessarily complicated, but the collective behavior can determine a rather intricate macroscopic description.

Nonextensive statistical mechanics, since its proposal in 1988 [27], has been applied to an impressive collection of systems in which spatial or temporal long-range correlations appear. Hence, it can also become a useful tool to characterize such systems. Here, we present an attempt of using such formalism to try to understand the intriguing behavior of an apparently simple system: a dripping faucet.

2   EXPERIMENTAL TIME SERIES

Robert Shaw [24], in 1984, was the first to propose that the dripping faucet could be a prototype of a chaotic system. Using real taps and microphones, he measured the intervals between successive drops for many flow rates. Hence, he was able to show the great variety of attractors corresponding to different dynamical behaviors for various flow rates. Further, his group also proposed a theoretical mass-spring model that reproduces several of the return maps that they have found in the experiments [13]. They also performed a mutual information investigation in order to confirm the chaotic behavior of the dripping faucets. The mass-spring model was somewhat improved by Fuchikami, Kyono, and Ishioka [3, 6].

Other groups soon became interested in the dripping faucet problem and new experiments were performed that confirm the curious behavior of leaky faucets [29, 31]. Dreyer and Hickey [1] measured the time series on tiny faucets ($\approx 3\,\mathrm{mm}$). They found that return maps are different than those from the real-sized faucets. Therefore, the size and shape of the faucets are also important to the proper characterization of a leaky faucet time series. It is also worthwhile to cite some experiments with a leaky faucet that are not related to the dynamical behavior of its time series, but to its shape [25, 14].

From a previous lattice drop model by Manna, Herrmann, and Landau [12], for the shape of a water drop pending from a vertical wall, Oliveira and Penna [15, 16] were also able to reproduce many of the return maps of real faucets, in particular the tiny ones of Dreyer and Hickey. Larger simulations and comparisons with other dynamical aspects of a leaky faucet were done [9], which are in agreement with the experimental data. A recent review on lattice models for dripping

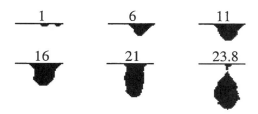

FIGURE 1   Time evolution of a two-dimensional drop following the lattice model (courtesy of A. R. Lima). The last snapshot was taken right after the disconnection from the faucet. No sort of average has been done to smooth the boundary of the drop since we are not interested in the shape, but in the drop-to-drop interval. The non-integer part of the time of the last snapshot is the fraction of the number of relaxations performed before the drop is disconnected.

faucets is available [17]. The lattice model was successfully applied to other physical phenomena such as nuclear fragmentation, mercury drops fragmentation, and magnetic hysteresis. Although we are not going into the details of this model, it has some remarkable features. It is a two-dimensional Ising model with next- and nearest-neighbors following mass-conserving Kawasaki dynamics (the extension to three dimensions is straightforward, but CPU time is excessive). It is easily implementable in a computer. Its dynamics are based on an energy balance between the attraction of the fluid molecules and the gravitational field. In figure 1, we show a sequence of snapshots of a simulation on a small lattice.

The quality of data from leaky faucet experiments has been considerably improved in recent years by using lasers, new sensors, high-speed and high-resolution cameras, fast electronics, and computers for data aquisition. Longer and more stable time series are relatively easy to obtain. The ability to measure higher flow rates makes the results from longer time series more reliable and less subject to fluctuations. New experiments were performed and new results were found on this system: crisis and intermittence [20, 23], Hopf bifurcation [4], homoclinic tangencies [22, 21], and others. These new phenomena seem to corroborate Shaw's original hypothesis of chaos in the dripping faucet experiment. It will be ultimately confirmed by a precise determination of the Lyapunov exponent λ, which is still missing. Currently available data give $\lambda \approx 0$. Traditional approaches for determination of the highest Lyapunov exponent fail (or do not give an accurate result [30]) at low positive values of λ. This is the main motivation of this work.

At this point, it is worth noticing that time series of beat-to-beat intervals for human hearts are similar [30]. However, long and precise time series for hearts are much harder to obtain than those for leaky faucets. That is why some authors suggest that a human heart behaves as a highly nonstationary and out-

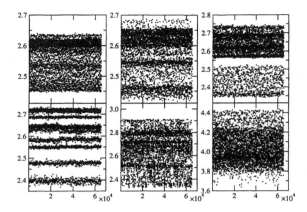

FIGURE 2   Series of drop-to-drop time intervals. The abscissa is the rank of the drop on time. Interval time in an arbitrary scale is the ordinate. Each series contains 65,536 (2^{16}) intervals. Although they are quite different, all series are stationary ones.

of-equilibrium system [5]. In figure 2, we show some time series for the dripping faucet experiment, with six different flow rates. The series are clearly stationary. Let us stress that this time series can be obtained on minutes, which are roughly the same length as a time series obtained from a subject trying to be at rest for almost 24 hours!

An easy and more comprehensive way to visualize the differences between chaotic or complex time series is to reconstruct the attractor of the system using return maps. The first return map is obtained by plotting $B(n + 1)$ vs. $B(n)$, where $B(n)$ is the nth interval of the time series. Different attractors are obtained, varying the flow rates. Even for small changes in the flow rate, it is possible to have remarkable differences in the attractors. Figure 3 is an impressive example of this property. Time is an arbitrary scale—it depends only on the electronic device used during the data acquisition: a very simple calibration can be performed if we need time given in milliseconds, for example.

3   TIME SERIES ANALYSIS

We analyze more than twenty different time series. Each set has at least 8,192 intervals, with the longest ones having up to 65,536 intervals. A set at the highest flow rates took only a few minutes to obtain, which minimizes the problems

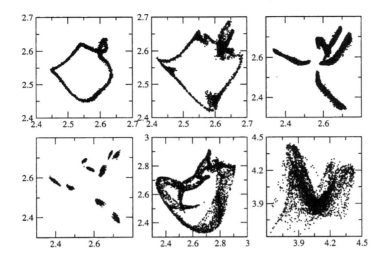

FIGURE 3   First return map for the time series shown in figure 2. For the sake of clarity, points are plotted at each ten drops. The attractors are much more dense than the ones shown above.

with the long-term stability of the series. We refer the reader interested in the experimental setup to previous work. [21, 23]. We also analyze the time series from computer simulations of the lattice model. We have obtained up to one million interdrop intervals.

We follow the same approach as Peng et al. [18] for heartbeat series. We also tried the *detrended fluctuation analysis* (DFA) [19] on both interdrop intervals and interdrop interval increments with no difference in the results. For the heartbeat problem, the interbeat interval increments series $I(n)$ is important because it exhibits power-law anticorrelations [18]. This series $I(n)$ is stationary, allowing the use of standard spectral analysis techniques [5]. On the other hand, the interbeat interval $B(n)$, is nonstationary and fluctuates in a healthy subject, even at rest.

The presence of long-range anticorrelations on the interbeat interval increments can also be tested by the power spectrum. Peng et al. [18], and further confirmed by Ivanov et al. [5], found that for waking and sleeping dynamics the log-log plot of the power spectrum $S(f)$ vs. f is linear, implying

$$S(f) \sim f^{-\beta}. \tag{1}$$

The exponent β is related to the mean fluctuation exponent α by

$$\beta = 2\alpha - 1. \tag{2}$$

The exponent α is obtained from the fluctuation function

$$F(n) = \overline{|B(n' + n) - B(n')|} \tag{3}$$

where the horizontal bar denotes an average over all values of n' (for a peda-gogical description of how to characterize sequences as time series and random walks, see Stanley et al. [26]). If the sequence is a true random walk or the correlations are local (short-ranged), then $F(n) \sim n^{1/2}$ and if there is no char-acteristic length, $F(n) \sim n^{\alpha}$, with $\alpha \neq 1/2$. Moreover, if $\alpha > 1/2$ the dynamics is called "persistent," i.e., the system tends to stay at the same mean behavior. If $\alpha < 1/2$ it is said to be "antipersistent," which means an attempt "to escape" the mean behavior. Antipersistent dynamics provide more variability in the time series and, therefore, it can appear on evolutive processes. Another important conclusion that we can obtain from this analysis is, if the mean fluctuation de-cays as a power law, then the correlation function is also described by a power law $(C(n) \sim (1/n)^{\gamma})$ [2, 26].

The log-log plots of $F(n)$ vs. n for two in the series in figure 3 are presented in figure 4. For all sequences that we have studied we found α ranging between -0.08 and 0.09. One impressive finding is the value of α and, consequently, β and γ are found to be *faucet-width- and flow-rate-independent* even though the attractors are quite different. Even more surprising is the fact that such different systems such as dripping faucets and hearts display the same behavior. By the way (!), Peng et al. [18] have found that only interbeat interval increments series from healthy hearts display long-range anticorrelations. Our very simplified model of a two-dimensional faucet in a lattice also displays the same exponents! Hence, we could say that not only two, but three different dynamical systems are characterized by the same set of exponents. Our lattice model does not take into account the vibrations on the drop surface, a crucial element on the mass-spring model.

We also have the power spectrum of dripping faucets. It is presented in figure 5, although the direct measure of α is much more precise. However, the relation $\beta = 2\alpha - 1$ is fulfilled.

The results we present here suggest that dripping faucets behave more like a complex system at the edge of chaos than a chaotic one, as previously supposed. Again, because the traditional methods of nonlinear analysis fail at low values of the Lyapunov exponents, we are going to use the newly developed tools of nonextensive physics in order to get some insight into the dynamics of such an

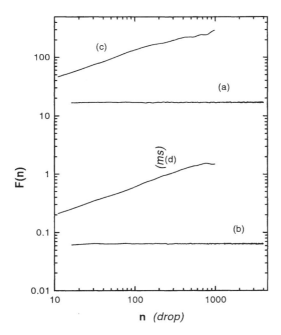

FIGURE 4   Log-log plot of $F(n)$ vs. n for two of the time series from dripping faucets. From regression we have found respectively $\alpha = 0.001$ (a) and -0.001 (b). Curves (c) and (d) correspond to a shuffle in the ordering of the events of original time series (a) and (b), respectively. For these new shuffled series we have $\alpha \approx 0.5$ and therefore they are uncorrelated, as expected.

amazing system. Nonextensive analysis seems to be suitable to address our main question: is a dripping faucet a truly chaotic system or is it working at the edge of chaos?

4   NONEXTENSIVE ANALYSIS

The nonextensive analysis we are going to present in this section is based on the Tsallis q-entropy [27]

$$S_q = \frac{1 - \sum_{i=1}^{W} p_i^q}{q - 1} \quad (q \in \mathbb{R}).$$ (4)

For $q = 1$, we have $S_1 = -\sum_{i=1}^{W} p_i \ln p_i$, the Boltzmann-Gibbs entropy.

Here we are interested in the sensitivity to the initial conditions [28]. The separation between nearby trajectories $\xi(t)$ diverges in time as an exponential

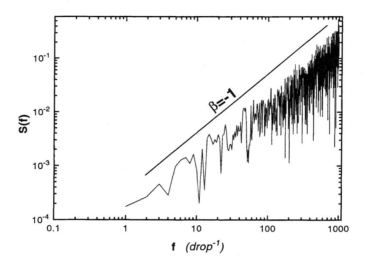

FIGURE 5   Power spectra $S(f)$ for the interval increments for the time series presented in figure 1. A straight line corresponding to $\beta = -1$ curve is presented for comparison.

$\exp(\lambda_1 t)$ with $(\lambda_1 > 0)$, if the system is chaotic. At the edge of chaos, the separation goes as a power law

$$\xi(t) \propto [1 + (1-q)\lambda_q t]^{1/(1-q)} . \tag{5}$$

The edge of chaos for the logistic map has $\lambda_1 = 0$ and the separation is well described by $q = 0.2445$, for example. This behavior is called *weak sensitivity* as opposed to the *strong sensitivity* (exponential) of the chaotic systems. An alternative way to determine q for systems that present multifractality is described in Lyra and Tsallis [11]. A more detailed description of the latter approach is the work by Lyra [10].

We decided to apply the method recently proposed by Latora et al. [8]. In a previous work, Latora and Baranger found the connection between the Kolmogorov-Sinai (KS) entropy rate and the Boltzmann-Gibbs-Shannon (BGS) entropy [7]. To summarize their results: BGS entropy grows at a linear rate in the chaotic regime; at the edge of chaos, there is a particular value of $q \neq 1$ for which the Tsallis entropy grows linearly with time.

The numerical procedure for obtaining q from a time series is easily implemented for maps such as the logistic map, although it demands a good statistic and, consequently, a reasonable computational effort. The phase space is divided into W cells of equal measure. One of these W is chosen and N initial conditions *inside* the cell are considered. As t evolves, the N points spread out the phase space by occupying different cells. The probability, at time t of the ith cell to be occupied is given by

$$p_i \equiv \frac{N_i(t)}{N} \tag{6}$$

where $N_i(t)$ gives the number of points inside the cell i. At $t = 0$, all points are in the same cell, $p_i = 0$ but one that gives $S_q(0) = 0$. As t evolves, $S_q(t)$ tends to increase, and we need to find which value of $q = q^\star$ gives the constant rate of increase:

$$\kappa_q \equiv \lim_{t\to\infty} \lim_{W\to\infty} \lim_{N\to\infty} \frac{S_q}{t}. \tag{7}$$

The procedure briefly described above has been used to find $q^\star$ for the logistic map, using $N = 10^6$ initial conditions and $W = 10^5$. This statistic is impossible to obtain from experimental data (sets of 10^{11} points!). Therefore, we know that we will not be able to obtain the same precision when measuring q for dripping faucets. However, as we stressed before, we are interested in the signal of the Lyapunov exponent: If it is positive, the BGS grows linearly with time and the system is really chaotic. Otherwise, the system could be poised on the edge of chaos.

We cannot generate drop-to-drop intervals inside a cell at will. To circumvent this limitation we adopt the following procedure:

- Take the whole time series (as long a series as possible). Divide the phase space into W cells. W must be smaller than the number of intervals, of course.
- Choose one of the cells (preferably the one which is the most visited during the whole experiment).
- Every time an interval is found to fall inside the chosen cell, consider it as a new initial condition.
- After these steps, we have N initial conditions taken from only one cell.

As pointed out in the original work [8], the entropy fluctuates too much at the edge of chaos. To minimize the effect of the fluctuations, we chose to study the accumulated entropy, which should grow as t^2.

Our result is presented in figure 6. We show the following values of $q = 0.5, 0.7$, and 1. Because no linear rate can be found at the very first steps for $q = 1$, this value should be discarded. The other limit $q = 0.5$ is to be discarded as well, because it always seems to grow faster than it does as a linear rate. The best value we found is $q \approx 0.7$, although it is far from precise. The important point here is to discard the $q = 1$.

5   CONCLUSIONS

We have used the rate of Tsallis entropy increase to show that dripping faucets should not be considered as chaotic systems. Their intricate dynamics seem to be better explained in terms of long-range power-law correlations. This work can

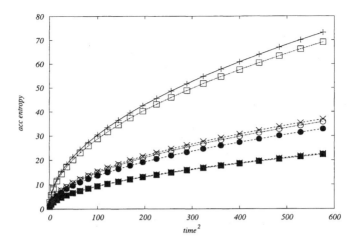

FIGURE 6   Accumulated entropy as a function of t^2. The lowest curve (full box) corresponds to $q = 1$. Full circles and $\times$ are $q = 0.7$ for two experimental series and one simulation. Notice the agreement suggesting that they have the same value of $q^\star$. The remaining curves ($\square$ and $+$) are for $q = 0.5$ as a lower bound for $q^\star$.

be extended by studying the possible multifractality, if any, of drop-to-drop time series. Because our results are not very precise and also because this approach demands a huge amount of data, it cannot presently be applied to hearts in order to test the similarities between the dynamics of both systems. We are waiting with great enthusiasm for a more suitable tool to deal with reasonably short time series and nonextensive systems.

ACKNOWLEDGMENTS

TJPP is grateful to M. Lyra and R. Stinchcombe for encouragement and discussions during the meeting. He would like to thank to C. Tsallis and M. Gell-Mann for their kind invitation and the opportunity to be at SFI during such a profitable meeting.

REFERENCES

[1] Dreyer, K., and F. R. Hickey. "The Route to Chaos in a Dripping Water Faucet." *Am. J. Phys.* **59** (1991): 619–627.
[2] Feder, J., ed. *Fractals*. New York: Plenum Press, 1988.

[3] Fuchikami, N., S. Ishioka, and K. Kyiono. "Simulation of a Dripping Faucet." *J. Phys. Soc. Jpn.* **68** (1999): 1185–1196.

[4] Gonçalves, W. M., R. D. Pinto, J. C. Sartorelli, and M. J. de Oliveira. *Physica* **A257** (1998): 385–389.

[5] Ivanov, P. Ch., L. A. Nunes Amaral, A. Goldberger, S. Havlin, M. G. Rosenblum, H. E. Stanley, and Z. R. Struzik. *Chaos* **11** (2001): 641.

[6] Kyiono, K., and N. Fuchikami. *J. Phys. Soc. Jpn.* **68** (1999): 3259.

[7] Latora, V., and M. Baranger. *Phys. Rev. Lett.* **82** (1999): 520.

[8] Latora, V., M. Baranger, A. Rapisarda, and C. Tsallis. *Phys. Lett.* **A273** (2000): 97.

[9] Lima, A. R., T. J. P. Penna, and P. M. C. de Oliveira. *Intl. J. Mod. Phys.* **C8** (1997): 1073.

[10] Lyra, M. L. "Nonextensive Entropies and Sensitivity to Initial Conditions of Complex Systems." This volume.

[11] Lyra, M. L., and C. Tsallis. *Phys. Rev. Lett.* **80** (1998): 53.

[12] Manna, S. S., H. J. Herrmann, and D. P. Landau. *J. Stat. Phys.* **66** (1992): 1155.

[13] Martien, P., S. C. Pope, P. L. Scott, and R. Shaw. *Phys. Lett.* **A110** (1985): 339.

[14] Nagel, S. R. *Am. J. Phys.* **67** (1999): 17.

[15] de Oliveira, P. M. C., and J. T. P. Penna. *J. Stat. Phys* **73** (1993): 789.

[16] de Oliveira, P. M. C., and T. J. P. Penna. *Intl. J. Mod. Phys.* **C5** (1994): 997.

[17] de Oliveira, P. M. C., T. J. P. Penna,A. R. Lima, J. S. Sá Martins, C. Moukarzel, and C. A. F. Leite. *Trends Stat. Phys.* **3** (2000): 137.

[18] Peng, C.-K., J. Mietus, J. M. Hausdorff, S. Havlin, H. E. Stanley, and A. L. Goldberger. *Phys. Rev. Lett.* **70** (1993): 1343.

[19] Peng, C.-K., S. V. Buldyrev, S. Havlin, M. Simons, H. E. Stanley, and A. L. Goldberger. *Phys. Rev.* **E49** (1994): 1685.

[20] Pinto, R. D., W. M. Gonçalves, J. C. Sartorelli, I. L. Caldas, and M. S. Batista. *Phys. Rev.* **E58** (1998): 4009.

[21] Pinto, R. D., and J. C. Sartorelli. *Phys. Rev.* **E61** (2000): 342.

[22] Pinto, R. D., J. C. Sartorelli, and W. M. Gonçalves. *Physica* **A291** (2001): 244.

[23] Sartorelli, J. C., W. M. Gonçalves, and R. D. Pinto. *Phys. Rev.* **E49** (1994): 3963.

[24] Shaw, R. *Dripping Faucet as a Model Chaotic System.* Santa Cruz, CA: Aerial Press, 1984.

[25] Shi, X. D., M. P. Brenner, and S. R. Nagel. *Science* **265** (1994): 219.

[26] Stanley, H. E., S. V. Buldyrev, A. L. Goldberger, Z. D. Goldberger, S. Havlin, R. N. Mantegna, S. M. Ossadnik, C.-K. Peng, and M. Simons. *Physica* **A205** (1994): 214.

[27] Tsallis, C. *J. Stat. Phys.* **52** (1988): 479.

[28] Tsallis, C., A. R. Plastino, and W.-M. Zheng. *Chaos, Solitons and Fractals* **8** (1997): 885.

[29] Yépes, H. N. N., A. L. S. Brito, C. A. Vargas, and L. A. Vicente. *Eur. J. Phys.* **10** (1989): 99.

[30] West, B. J., R. Zhang, R., A. W. Sanders, S. Miniyar, J. H. Zucherman, and B. D. Levine. *Physica* **270** (1999): 552.

[31] Wu, X., and A. Schelly. *Physica* **D40** (1989): 433.

Power-Law Persistence in the Atmosphere: Analysis and Applications

Armin Bunde
Jan Eichner
Rathinaswamy Govindan
Shlomo Havlin
Eva Koscielny-Bunde
Diego Rybski
Dmitry Vjushin

We review recent results on the appearance of long-term persistence in climatic records and their relevance for the evaluation of global climate models and rare events. The persistence can be characterized, for example, by the correlation $C(s)$ of temperature variations separated by s days. We show that, contrary to previous expectations, $C(s)$ decays for large s as a power law, $C(s) \sim s^{-\gamma}$. For continental stations, the exponent γ is always close to 0.7, while for stations on islands $\gamma \cong 0.4$. In contrast to the temperature fluctuations, the fluctuations of the rainfall usually cannot be characterized by long-term power-law correlations but rather by pronounced short-term correlations. The universal persistence law for the temperature fluctuations on continental stations represents an ideal (and uncomfortable) test-bed for the state-of-the-art global climate models and allows us to evaluate their performance. In addition, the presence of long-term correlations leads to a noval approach for evaluating the statistics of rare events.

Nonextensive Entropy—Interdisciplinary Applications
edited by Murray Gell-Mann and Constantino Tsallis, Oxford University Press   207

1   INTRODUCTION

The persistence of weather states on short terms is a well-known phenomenon: a warm day is more likely to be followed by a warm day than by a cold day and vice versa. The trivial forecast that the weather of tomorrow is the same as the weather of today was, in previous times, often used as a "minimum skill" forecast for assessing the usefulness of short-term weather forecasts. The typical time scale for weather changes is about one week, a time period which corresponds to the average duration of so-called "general weather regimes" or "Grosswetterlagen," so this type of short-term persistence usually stops after about one week. On larger scales, other types of persistence occur, one of them is related to circulation patterns associated with blocking [5]. A blocking situation occurs when a very stable high-pressure system is established over a particular region and remains in place for several weeks. As a result, the weather in the region of the high remains fairly persistent throughout this period. Furthermore, transient low-pressure systems are deflected around the blocking high so that the region downstream of the high experiences a larger than usual number of storms. On even longer terms, a source for weather persistence might be a slowly varying external (boundary), forcing sea surface temperatures and anomaly patterns, for example. On the scale of months to seasons, one of the most pronounced phenomena is the El Niño Southern Oscillation (ENSO) event, which occurs every three–five years and which strongly affects the weather over the tropical Pacific as well as over North America [21].

The question is, *how* the persistence, that might be generated by very different mechanisms on different time scales, decays with time s. The answer to this question is not simple. Correlations, and in particular long-term correlations, can be masked by trends that are generated, for instance, by the well-known urban warming phenomenon. Even uncorrelated data in the presence of long-term trends may look like correlated data, and, on the other hand, long-term correlated data may look like uncorrelated data influenced by a trend.

Therefore, in order to distinguish between trends and correlations, one needs methods that can systematically eliminate trends. Those methods are available now: both wavelet techniques (WT) [1] and detrended fluctuation analysis (DFA) [2] can systematically eliminate trends in the data and thus reveal intrinsic dynamical properties such as distributions, scaling, and long-range correlations that are often masked by nonstationarities.

In recent studies [6, 15, 16, 22] we have used DFA and WT to study temperature and precipitation correlations in different climatic zones on the globe. The results indicate that the temperature variations are long-range power-law correlated above some crossover time that is on the order of 10 days. Above $10\,\mathrm{d}$, the persistence, characterized by the autocorrelation $C(s)$ of temperature

---

[1]See e.g., Arneodo et al. [1].
[2]See, e.g., Peng et al. [20] and Bunde et al. [3]

variations separated by s days, decays as

$$C(s) \sim s^{-\gamma}, \tag{1}$$

where, most interestingly, the exponent γ has roughly the same value $\gamma \cong 0.7$ for all continental records. For small islands the correlations are more pronounced, with γ around 0.4. This value is close to the value obtained recently for correlations of sea-surface temperatures [17]. In marked contrast, for most stations the precipitation records do not show indications of long-range temporal correlations on scales above 6 months. Our results are supported by independent analysis by several groups [18, 19, 23].

The fact that the correlation exponent varies only very little for the continental atmospheric temperatures, presents an ideal test-bed for the performance of the global climate models, as we will show below. We present an analysis of the two standard scenarios (greenhouse gas forcing only and greenhouse gas plus aerosols forcing) together with the analysis of a control run. Our analysis points to the clear deficiencies of the models. For further discussions we refer to Govindan et al. [8]. Finally, we review a recent approach to determine the statistics of rare events in the presence of long-term correlations.

The chapter is organized in four sections. In Section 2, we describe one of the detrending analysis methods, the detrended fluctuation analysis (DFA). In Section 3, we review the application of this method to both atmospheric temperature and precipitation records. In Section 4, finally, we describe how the "universal" persistence law for the atmospheric temperature fluctuations on continental stations can be used to test the three scenarios of the state-of-the-art climate models. In Section 5, finally, we describe how the common extreme value statistics is modified in the presence of long-term correlations.

2   THE METHODS OF ANALYSIS

Consider, for example, a record T_i, where the index i counts the days in the record, $i = 1, 2, \ldots, N$. This record T_i may represent the maximum daily temperature or the daily amount of precipitation, measured at a certain meteorological station. For eliminating the periodic seasonal trends, we concentrate on the departures of the T_i, $\Delta T_i = T_i - \overline{T}_i$, from their mean daily value $\overline{T}_i$ for each calendar date i, say the 1st of April, which has been obtained by averaging over all years in the record.

Quantitatively, correlations between two ΔT_i values separated by n days are defined by the (auto)correlation function

$$C(n) \equiv \langle \Delta T_i \Delta T_{i+n} \rangle = \frac{1}{N-n} \sum_{i=1}^{N-n} \Delta T_i \Delta T_{i+n}. \tag{2}$$

If the ΔT_i are uncorrelated, $C(n)$ is zero for n positive. If correlations exist up to a certain number of days $n_\times$, the correlation function will be positive up to

$n_\times$ and vanish above $n_\times$. A direct calculation of $C(n)$ is hindered by the level of noise present in the finite records, and by possible nonstationarities in the data.

To reduce the noise we do not calculate $C(n)$ directly, but instead study the "profile"

$$Y_m = \sum_{i=1}^{m} \Delta T_i \,. \tag{3}$$

We can consider the profile Y_m as the position of a random walker on a linear chain after m steps. The random walker starts at the origin and performs, in the ith step, a jump of length ΔT_i to the right, if ΔT_i is positive, and to the left, if ΔT_i is negative. The fluctuations $F^2(s)$ of the profile, in a given time window of size s, are related to the correlation function $C(s)$. For the relevant case (1) of long-term power-law correlations, $C(s) \sim s^{-\gamma}$, when $0 < \gamma < 1$, the mean-square fluctuations $\overline{F^2(s)}$, obtained by averaging over many time windows of size s (see below), asymptotically increase by a power law [2],

$$\overline{F^2(s)} \sim s^{2\alpha}, \quad \alpha = 1 - \gamma/2 \,. \tag{3}$$

For uncorrelated data (as well as for correlations decaying faster than $1/s$), we have $\alpha = 1/2$.

For the analysis of the fluctuations, we employ a hierarchy of methods that differ in the way the fluctuations are measured and possible trends are eliminated (for a detailed description of the methods we refer to Kantelhardt et al. [14]).

1. In the simplest type of fluctuation analysis (FA) (where trends are not going to be eliminated), we determine the difference of the profile at both ends of each window. The square of this difference represents the square of the fluctuations in each window.
2. In the *first-order* detrended fluctuation analysis (DFA1), we determine in each window the best linear fit of the profile. The variance of the profile from this straight line represents the square of the fluctuations in each window.
3. In general, in the nth-order DFA (DFAn) we determine in each window the best nth-order polynomial fit of the profile. The variance of the profile from these best nth-order polynomials represents the square of the fluctuations in each window.

By definition, FA does not eliminate trends similar to the Hurst method and the conventional power spectral methods [7]. In contrast, DFAn eliminates trends of order n in the profile and $n - 1$ in the original time series. Thus, from the comparison of fluctuation functions $F(s)$ obtained from different methods, one can learn about long-term correlations and types of trends, which cannot be achieved by the conventional techniques.

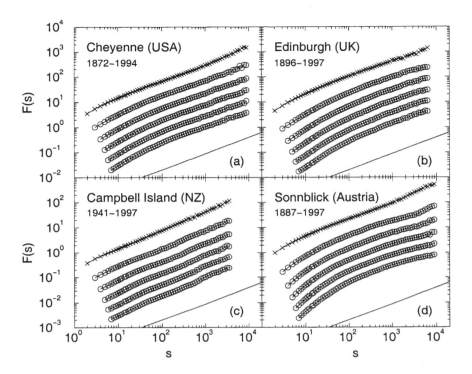

FIGURE 1 Analysis of daily temperature records of four representative weather stations. The four figures show the fluctuation functions obtained by FA, DFA1, DFA2, DFA3, DFA4, and DFA5 (from top to bottom) for the four sets of data. The scale of the fluctuation functions is arbitrary. In each panel, a line with slope 0.65 is shown as a guide to the eye.

3   ANALYSIS OF TEMPERATURE AND PRECIPITATION RECORDS

Figure 1 shows the results of the FA and DFA analysis of the maximum daily temperatures T_i of the following weather stations (the length of the records is written within the parentheses): (a) Cheyenne (USA, 123 y), (b) Edinburgh (UK, 102 y), (c) Campbell Island (New Zealand, 57 y), and (d) Sonnblick (Austria, 108 y). The results are typical for a large number of records that we have analyzed so far (see Eichner et al. [6] and Koscielny-Bunde et al. [15, 16]). Cheyenne has a continental climate, Edinburgh is on a coastline, Campbell Island is a small island in the Pacific Ocean, and the weather station of Sonnblick is on top of a mountain.

In the log-log plots, all curves are (except at small s values) approximately straight lines. For both the stations inside the continents and along coastlines, the slope is $\alpha \cong 0.65$. There exists a natural crossover (above the DFA crossover) that can be best estimated from FA and DFA1. As can be verified easily, the crossover occurs roughly at $t_c = 10d$, which is the order of magnitude for a typical Grosswetterlage. Above t_c, there exists long-range persistence expressed by the power-law decay of the correlation function with an exponent $\gamma = 2 - 2\alpha \cong 0.7$. These results are representative for the large number of records that we have analyzed. They indicate that the exponent is "universal," i.e., does not depend on the location and the climatic zone of the weather station. Below t_c, the fluctuation functions do not show universal behavior and reflect the different climatic zones.

However, there are exceptions from the universal behavior, and these occur for locations on small islands and on top of large mountains. In the first case, the exponent can be considerably larger, $\alpha \cong 0.8$, corresponding to $\gamma \cong 0.4$. In the second case, on top of a mountain, the exponent can be smaller, $\alpha \cong 0.58$, corresponding to $\gamma \cong 0.84$.

Next, we consider precipitation records. Figure 2 shows the results of the FA and DFA analysis of the daily precipitation P_i of the following weather stations (the length of the records is written within the parentheses): Cheyenne (USA, 117 y, fig. 2(a)), Edinburgh (UK, 102 y, fig. 2(b)), Campbell Island (New Zealand, 57 y, fig. 2(c)), and Sonnblick (Austria, 108 y, fig. 2(d)). The results are typical and represent a large number of records that we have analyzed so far [22].

In the log-log plots, all curves are (except at small s values) approximately straight lines at large times, with a slope close to 0.5. If there exist long-term correlations, then they are very small. Some exceptions are again stations on top of a mountain, where the exponent might be around 0.6, but this happens only very rarely. In most cases, the exponent is between 0.5 and 0.55, pointing to uncorrelated or weakly correlated behavior at large time spans. Unlike the temperature records, the exponents actually do not depend on specific climatic or geographic conditions.

Figure 3 summarizes the results for exponents α for (a) temperature records and (b) precipitation records. Different climatological conditions are marked in the histograms. First we concentrate on the temperature records (fig. 3(a)). One can see clearly that for stations that are neither on islands nor on summits, the average exponent is close to 0.65, with a variance of 0.03. For the islands (where only few records are available), the average value of α is 0.78, with quite a large variance of 0.08. The variance is large, since stations on larger islands, like Wrangelija, behave more like continental stations, with an exponent close to 0.65. For the precipitation records (fig. 3(b)), the average exponent α is close to 0.54, with a variance close to 0.05, and does not depend significantly on the climatic conditions around a weather station.

Since for the temperature records the exponent for continental and coastline stations does not depend on the location of the meteorological station and its

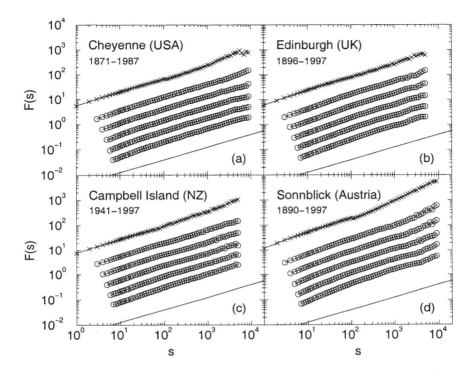

FIGURE 2 Analysis of daily precipitation records of four representative weather stations. The four figures show the fluctuation functions obtained by FA, DFA1, DFA2, DFA3, DFA4, and DFA5 (from top to bottom) for the four sets of data. The scale of the fluctuation functions is arbitrary. In each panel, a line with slope 0.5 is shown as a guide to the eye.

local environment, the power-law behavior can serve as an ideal test for climate models where regional details cannot be incorporated and, therefore, regional phenomena like urban warming cannot be accounted for. The power-law behavior seems to be a global phenomenon and, therefore, should also show up in the simulated data of the global climate models (GCM).

4 TEST OF GLOBAL CLIMATE MODELS

The state-of-the-art climate models that are used to estimate future climate are coupled atmosphere-ocean general circulation models (AOGCMs) [10, 12]. The models provide numerical solutions of the Navier Stokes equations devised for simulating meso-scale to large-scale atmospheric and oceanic dynamics. In

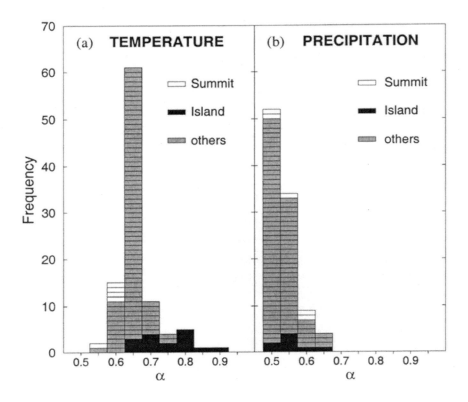

FIGURE 3   Histogram of the values of the fluctuation exponents α (a) for daily temperature records and (b) for daily precipitation records.

addition to the explicitly resolved scales of motions, the models also contain parametrization schemes representing the so-called subgrid-scale processes, such as radiative transfer, turbulent mixing, boundary layer processes, cumulus convection, precipitation, and gravity wave drag. A radiative transfer scheme, for example, is necessary for simulating the role of various greenhouse gases such as CO_2 and the effect of aerosol particles. The differences among the models usually lie in the selection of the numerical methods employed, the choice of the spatial resolution,[3] and the subgrid-scale parameters.

Three scenarios have been studied by the models, and the results are available, for four models, from the IPCC Data Distribution Center [13]. The first scenario represents a control run where the CO_2 content is kept fixed. In the

---

[3] A typical climate module within the overall AOGCM machinery will have a grid spacing of 300–500 km and 10–20 vertical layers as compared to a weather forecasting model with a grid spacing of 100 km or less and 30–40 layers.

second scenario, one considers only the effect of greenhouse gas forcing (GHG). The amounts of greenhouse gases are taken from the observations until 1990 and then increased at a rate of 1% per year. In the third scenario, the effect of aerosols (mainly sulfates) in the atmosphere is also taken into account. Only direct sulfate forcing is considered; until 1990, the sulfate concentrations are taken from historical measurements, and are increased linearly afterward. The effect of sulfates is to mitigate and partially offset the greenhouse gas warming. Although this scenario represents an important step toward comprehensive climate simulation, it introduces new uncertainties—regarding the distributions of natural and anthropogenic aerosols and, in particular, regarding indirect effects on the radiation balance through cloud-cover modification, etc. [11].

For the test, we consider the monthly temperature records from those four AOGCMs. Data for these three scenarios are available from the Internet: CSIRO-Mk2 (Melbourne), CCSR/NIES (Tokyo), ECHAM4/OPYC3 (Hamburg), and CGCM1 (Victoria, Canada). We extracted the data for six representative sites around the globe (Prague, Kasan, Seoul, Luling [Texas], Vancouver, and Melbourne). For each model and each of the three scenarios, we selected the temperature records of the four grid points closest to each site, and bilinearly interpolated the data to the location of the site.

Figure 4 shows representative results of the fluctuation functions, calculated using DFA3, for two sites (Kasan [Russia] and Luling [Texas]) for the four models and the three scenarios. As seen in figure 4 most of the DFA curves approach the slope of 0.5. However, the control runs seem to show a somewhat better performance, i.e., many of them have a slope close to 0.65 (e.g., Luling (CSIRO-Mk2)), and the "greenhouse gas forcing only" scenario shows the worst performance. The actual long-term exponents α for the three scenarios of the four models for the six cities are summarized in figure 5(a)–(c). Each histogram consists of 24 blocks and every block is specified by the model and the city.

For the control run (fig. 5(a)) there is a peak at $\alpha \cong 0.65$, but more than half of the exponents are below $\alpha \cong 0.62$. For the greenhouse gas only scenario (fig. 5(b)), the histogram shows a pronounced maximum at $\alpha = 0.5$. For best performance, all models should have exponents α close to 0.65, corresponding to a peak of height 24 in the window between 0.62 and 0.68. Actually, more than half of the exponents are close to 0.5, while only 3 exponents are in the proper window between 0.62 and 0.68. Figure 5(c) shows the histogram for the greenhouse gas plus aerosol scenario, where, in addition to the greenhouse gas forcing, the effects of aerosols are also taken into account. For this case, there is a pronounced maximum in the α window between 0.56 and 0.62 (more than half of the exponents are in this window), while again only three exponents are in the proper range between 0.62 and 0.68. This shows that although the greenhouse gas plus aerosol scenario is also far from reproducing the scaling behavior of the real data, its overall performance is better than the performance of the greenhouse gas scenario. The best performance is observed for the control run, which points to remarkable deficiencies in the way the forcings are introduced into the models.

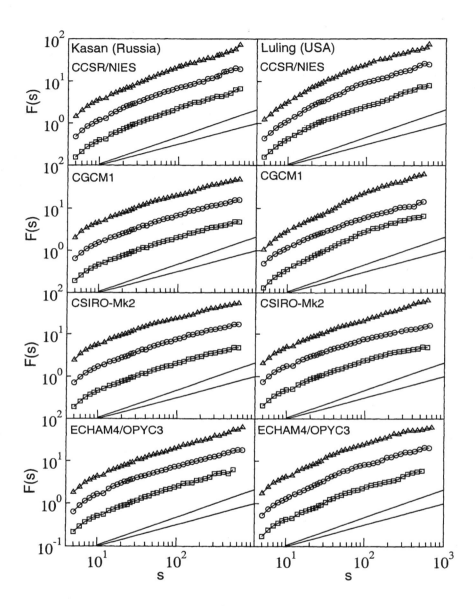

FIGURE 4 Comparison of the scaling performance of the three scenarios: control run (△), greenhouse gas forcing only (○), and greenhouse gas plus aerosols (□). All curves are obtained by applying DFA3 to the monthly mean of the daily maximum temperatures generated by the four AOGCMs. The lines with slopes 0.65 and 0.5 are shown as guides to the eye. For details of the records, we refer to IPCC [13].

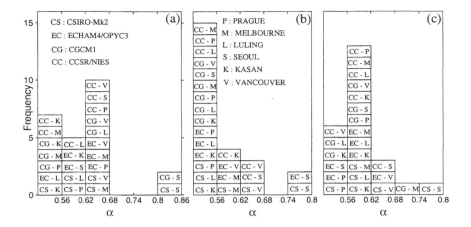

FIGURE 5   Histogram of the values of the fluctuation exponent (α) obtained from the simulations of the four AOGCMs (listed in (a)), for six sites (listed in (b)). The three panels are for the three scenarios: (a) control run, (b) greenhouse gas forcing only, and (c) greenhouse gas plus aerosol forcing. The entries in each box represent "Model–Site."

5   EXTREME VALUE STATISTICS

The long-term correlations in a record $\{T_i\}$ effect strongly the statistics of the extreme values in the record, as has been shown recently in Bunde [4]. The central quantity extreme value statistics (EVS) is the return time r_q between two events of size greater or equal to a certain threshold q. The basic assumption in conventional EVS is, that the events are uncorrelated(at least when the time lag between them is sufficiently large). In this case, on ecan obtain the mean return time R_q simply from the probability W_q that an event greater or equal to Q occurs, $R_q = 1/W_q$. Since the events are uncorrelated, also the return intervals are uncorrelated and follow the Poisson-statistics; i.e., their distribution function $P_q(r)$ is a simple exponential, $P_q(r) \sim (-r/R_q)$.

For long-term correlations, it has been shown in Bunde [4] that the distribution function $P_q(r)$ changes into a *stretched* exponential function,

$$P_q(r) \sim \exp[-\text{const}(\frac{r}{R_q})^\gamma)]\,, \tag{1}$$

for γ between zero and one. For γ above one, in the case of short-term correlations, P_q reduces to the Poisson distribution.

In addition, the return intervals become long-term correlated, with an exponent that is approximately identical to γ [4] This is seen in figure 6, where the fluctuation functions $F(s)$ of the return intervals (obtained by DFA2) are shown for two artificial long-term correlated records with $\gamma = 0.4$ and 07. Two values

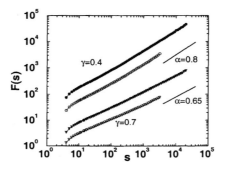

FIGURE 6 fluctution functions $F(s)$ for the record of return intervals obtained from two artificial long-term correlated records $\{T_i\}$ with $\gamma = 0.4$ (upper curves) and $\gamma = 0.7$ (lower curves). The distribution of the T_i values has been chosen as Gaussian, with zero mean and variance one. For the return intervals, the thresholds $q = 1.5$ and 2.5 have been considered. The straight lines in the figure have slopes $\alpha = 1 - \gamma/2$, suggesting that the return intervals are long-term correlated in the same way as the T_i.

of thresholds $q = 1.5$ and 2.5 have been considered for each value of γ. The distribution of the T_i values has been chosen as Gaussian, with zero mean and variance one. In the double logarithmic plot, all the curves are straight lines with slopes $\alpha = 1 - \gamma/2$, suggesting that the return intervals are long-term correlated in the same way as the T_i.

As a consequence, small return intervals are more likely to be followed by small intervals and large intervals are more likely to be followed by large intervals. Accordingly, for long-term correlated records it is more likely than for uncorrelated records that a sequence of large return times is followed by a sequence of short-return times.

This fact may be relevant for the occurrence of floods. It is well known that river flows are long-term correlated with exponents γ between 0l3 and 0.9, in most cases close to 0.4. In the last decades, the frequency of large floods in Europe has increased. It is possible, that this increase is due to global warming, but it is also possible that it has been triggered by the long-term correlations.

ACKNOWLEDGMENTS

We are grateful to Prof. H.-J. Schellnhuber, Dr. J. Kantelhardt, and Prof. S. Brenner for very useful discussions. We wish to acknowledge financial support by the Deutsche Forschungsgemeinschaft and the Israel Science Foundation.

REFERENCES

[1] Arneodo, A., Y. d'Aubenton-Carafa, E. Bacry, P. V. Graves, J. F. Muzy, and C. Thermes. "Wavelet-Based Fractal Analysis of DNA Sequences." *Physica D* **96** (1996): 291–320.

[2] Bunde, A., and S. Havlin, eds. *Fractals in Science.* New York: Springer-Verlag, 1995.

[3] Bunde, A., S. Havlin, J. W. Kantelhardt, T. Penzel, J. H. Peter, and K. Voigt. "Correlated and Uncorrelated Regions in Heart-Rate Fluctuations during Sleep." *Phys. Rev. Lett.* **85** (2000): 3736–3739.

[4] Bunde, A., J. Eichner, S. Havlin, and J. W. Kantelhardt. "Extreme Value Statistics in the Presence of Long-Term Correlations." *Physica A* (2003): in press.

[5] Charney, J. G., and J. G. Devore. "Multiple Flow Equilibrium in the Atmosphere and Blocking." *J. Atmos. Sci.* **36** (1979): 1205–1216.

[6] Eichner, J., A. Bunde, S. Havlin, E. Koscielny-Bunde, and H.-J. Schellnhuber. "Power-Law Persistence and Trends in the Atmosphere: A Detailed Study of long Temperature Records." *Phys. Rev. E* (2003): submitted.

[7] Feder, J. *Fractals.* New York: Plenum, 1989.

[8] Govindan, R. B., D. Vjushin, S. Brenner, A. Bunde, S. Havlin, and H.-J. Schellnhuber. "Global Climate Models Violate Scaling of the Observed Atmospheric Variability." *Phys. Rev. Lett.* **89** (2002): 028501.

[9] Govindan, R. B., D. Vjushin, D. Brenner, A. Bunde, S. Havlin, and H.-J. Schellnhuber. "Long-Range Correlations and Trends in Global Climate Models: Comparison with Real Data." *Physica A* **294** (2001): 239.

[10] Hasselmann, K. "Multi-pattern Fingerprint Method for Detection and Attribution of Climate Change, Multi-fingerprint Detection and Attribution Analysis of Greenhouse Gas, Greenhouse Gas-Plus-Aerosol and Solar Forced Climate Change." *Climate Dynamics* **13** (1997): 601–634.

[11] Houghton, J. T., ed. *Climate Change 2001: The Scientific Basis, Contribution of Working Group I to the Third Assessment Report of the Intergovernmental Panel on Climate Change (IPCC).* Cambridge: Cambridge University Press, 2001.

[12] Intergovernmental Panel on Climate Change. *The Regional Impacts of Climate Change. An Assessment of Vulnerability,* edited by R. T. Watson, M. C. Zinyowera, and R. H. Moss. Cambridge: Cambridge University Press, 1998.

[13] IPCC Data Distribution Center. ⟨http://ipcc-ddc.cru.uea.ac.uk/dkrz/dkrz_index.html⟩.

[14] Kantelhardt, J. W., E. Koscielny-Bunde, H. A. Rego, S. Havlin, and A. Bunde. "Detecting Long-Range Correlations with Detrended Fluctuation Analysis." *Physica A* **295** (2001): 441–454.

[15] Koscielny-Bunde, E., A. Bunde, S. Havlin, and Y. Goldreich. "Analysis of Daily Temperature Fluctuations." *Physica A* **231** (1996): 393–396.

[16] Koscielny-Bunde, E., A. Bunde, S. Havlin, H. E. Roman, Y. Goldreich, and H.-J. Schellnhuber. "Indication of a Universal Persistence Law Governing Atmospheric Variability." *Phys. Rev. Lett.* **81** (1998): 729–732

[17] Monetti, R. A., S. Havlin, and A. Bunde. "Long-Term Persistence in the Sea Surface Temperature Fluctuations." *Physica A* **320** (2003): 581–588.

[18] Pelletier, J. D. "Analysis and Modeling of the Natural Variability of Climate." *J. Climate* **10** (1997): 1331–1342.

[19] Pelletier, J. D., and D. L. Turcotte. "Long-Range Persistence in Climatological and Hydrological Time Series: Analysis, Modeling and Application to Drought Hazard Assessment." *J. Hydrol.* **203** (1997): 198–208.

[20] Peng, C.-K., S. V. Buldyrev, S. Havlin, M. Simons, H. E. Stanley, and A. L. Goldberger. "Mosaic Organization of DNA Nucleotides." *Phys. Rev. E* **49** (1994): 1685–1689.

[21] Philander, S. G. "El Niño, La Niña and the Southern Oscillation." *Int'l. Geophys.* **46** (1990).

[22] Rybski, D., A. Bunde, S. Havlin, and H. J. Schellnhuber. "Detrended Fluctuation Analysis of Precipitation Records: Scaling and Multiscaling." Preprint (2002).

[23] Talkner, P., and R. O. Weber. "Power Spectrum and Detrended Fluctuation Analysis: Application to Daily Temperatures." *Phys. Rev. E* **62** (2000): 150–160.

[24] Vjushin, D., R. B. Govindan, S. Brenner, A. Bunde, S. Havlin, and H.-J. Schellnhuber. "Lack of Scaling in Global Climate Models." *J. Phys. C* **14** (2002): 2275.

The Living State of Matter: Between Noise and Homeorrhetic Constraints

Marcello Buiatti
Patrizia Bogani
Claudia Acquisti
Giuseppe Mersi
Leone Fronzoni

The living state of matter, while obviously obeying physical and chemical laws, does exhibit some peculiar features which, taken altogether, distinguish it from the rest of the natural world. As discussed at length elsewhere [16, 18], these include, among others, hierarchical organization, individuality, invention, and homeostasis. One key category which was introduced by Waddington [94] is homeorrhesis. Homeorrhesis is the dynamical version of homeostasis and it may be defined as the capacity of living systems to maintain their harmonic dynamical networks through continuous change in response to external and internal noise.

In the present chapter, we will discuss, in the light of recent interdisciplinary data, the relationship between noise and self-organization. We propose, as a new feature specific to living systems, the capacity to integrate into flexible and versatile networks a portion of noise of internal and external origin (correlated information), and to utilize it to counteract the disrupting effect of the remaining noise. The analysis will be carried out at the levels of DNA and of organized networks, and will also be tackling the problem from the point of view of the relationship between increases in organismal complexity and entropy levels at the evolutionary scale.

1   THE DNA LEVEL

1.1   GENOME SIZE VARIATION

Present-day genomes are the results of evolution, and DNA may be considered as an archive of past events. We use as a working hypothesis the almost unanimously accepted concept of interplay between selection as it fixes "useful" sequences, and mutation-genetic drift forces as they generate noise and continuously shuffling of genomes.

Studies on structure and function of genomes have been increasing at a very fast rate following the extraordinary progress in sequencing techniques and, consequently, in GenBank whole genome sequences are available of a number of bacteria, Archaea, viruses, and eukaryotes. DNA content per cell has been increasing during evolution, more recent species having, in general, larger genomes than ancestral ones. This, however, is only a rough approximation of reality. A better analysis of the data tells us that while prokaryotes (bacteria and Archaea) have small genomes ranging from 500,000 base pairs (bp) to 5 million bp, an abrupt increase has occurred in eukaryotes. In this large group, a unicellular organism like *Saccharomyces cerevisiae* has a 12-million-bp genome, while multicellular organisms range from around 100 million to several billion bp.

Within eukaryotes, the minimum genome size *per phylum* seems to have been increasing almost exponentially throughout evolution, but, in many cases, genome size distributions of different *phyla* are very wide and show overlapping values. The increase in genome size observed in eukaryotes is mainly due to noncoding sequences, whereas the number of genes increases at a much lower rate. To give some examples, gene density is around 2,000/Mbp[1] in *Escherichia coli*, but drops to 500/Mbp in *Saccharomyces cerevisiae*, and to 12/Mbp in man. We, the humans, have around 35,000–40,000 genes, while the bacterium *Escherichia coli* has 2,000 genes. Therefore, the increase in number of genes has been only 20-fold from *E. coli* to humans, with a 1000-fold difference in genome size. So, coding sequences cover about 90% of prokaryote genomes, but only 2% of human DNA.

The structure of genomes is also dramatically different in prokaryotes and eukaryotes. Prokaryote genes are uninterrupted and separated by relatively short tracts of intergenic, noncoding DNA. In eukaryotes, coding sequences (exons) alternate with noncoding ones (introns). Consequently, prokaryote genes are relatively short, while some eukaryote ones may span over 100,000 bp. In these cases, short exons are literally "embedded" into the noncoding DNA. A num-

---

[1]Mbp stands for millions of base pairs.

ber of spontaneous mechanisms contributed to the increase in size of eukaryote genomes and are still operating. Plants are the eukaryotes with higher genome size variation, which is probably also due to the widespread occurrence of polyploidy (multiplication of whole genomes) and aneuploidy (gain or loss of single chromosomes). In plants, but not in animals, new species often derive from interspecific crosses and subsequent polyploidization. Other important mechanisms of increase in DNA amounts per genome are gene duplication, mobile elements replication and distribution throughout the genomes, specific sequence amplification and tandem integration, and low-complexity sequence lengthening. Most processes do not occur in prokaryotes (with the exceptions of transposon replication and of gene duplication).

Although genome sizes are not correlated with organismal complexity, their distributions are by no means random, and seem in many cases subject to selection pressure due to the effects of the increase of DNA contents on key phenotypic characters. For instance, genome size is inversely correlated with metabolic rates [45]. Consequently, as birds and bats, due to their flying habits, need high metabolic rates, they tend to have smaller genomes in comparison with earthbound organisms [46]. Larger genomes, on the other hand, are found in plants living at low temperatures [9]. In plants, the existence of "optimal" genome sizes for different organisms has been experimentally proven by a number of authors [23], the evidence being derived from the common experience that artificial induction of genome multiplication through polyploidy leads to increases in cell and plant size but to a decrease in cell solute concentrations. The ratio between cell size and gene expression levels becomes consequently limiting when it exceeds organism-specific levels over which further increases in genome size become disadvantageous. However, single processes, like genome number and size increase, can have contrasting and balancing effects. For instance, as shown by Lewis [55], genome multiplication leads to increased redundancy and allows a higher level of heterozygosity, which increases homeostatic capacity and, hence, better adaptation to unstable environments. That selection is continuously operating on genome sizes has been recently shown by Petrov et al. [72], who found that the prevalence of deletions over insertion is 40 times lower in the 11 times larger Laupala cricket genome than in *Drosophila*.

Genome size distributions throughout evolution seem, therefore, to have been determined by the balance between size-increase generating processes and the internal (physiological) and external (environmental) contexts, giving rise to directional selection pressure. Finally, it should be mentioned that genome size distributions could be both continuous and discontinuous. Discontinuous genome size distributions are frequent in plants probably for the impact in that kingdom of polyploidy and aneuploidy as shown by the "doubling law" proposed by Sparrow and Naumann [83]. In the case of animals, discontinuous variation seems to prevail in invertebrates, while continuous variation prevails in birds, mammals, and teleost fish [38]. A 1/4 power law correlates genome with intron size [92], and cell doubling time with genome size [80] thus increasing the vast number of cases in which this exponent or a multiple of it has been described.

1.2   CONSTRAINTS IN NUCLEOTIDE DISTRIBUTIONS

Since the discovery in the early 1970s of the widespread presence of repeated sequences in eukaryotic DNA, debate has been opened on their putative functional role. Noncoding DNA, in which most of the repeats are localized, has been defined as "junk DNA" [66] and "selfish DNA" [28, 67], although at least one multiplicative "nonselfish" process (DNA amplification) has been discovered in animals [74] and in plants [68]. Evolutionary theories have been built on concepts such as the "selfish gene" theory by Dawkins [24] or the "concerted evolution" by Dover [29]. In the second half of the 1980s, however, the exponential increase of available GenBank sequences and the discovery that noncoding sequences cover a large part of eukaryotic genomes prompted a vast series of statistical studies on nucleotide distributions in DNA, along with experimental work on the possible structural and functional meaning of "junk DNA." The main approaches used in this area have been essentially the search for repeated, short "words," the analysis of periodicities (the "hidden codes" proposed by Trifonov [90]), and the study of long-range correlations and, in general, of the global statistical structure of genomes.

Periodicities have been assessed with positional autocorrelation functions and mutual information analyses (see for instance Liò et al. [59] and Grosse et al. [39]), while different methods of linguistic studies have been utilized in the search for frequent and putatively relevant "words" in the genomes. Different methods have been used to analyze global statistical properties like the variance of "DNA walks" [84], the detrended fluctuation analysis (DFA) [70], wavelet transforms [6], average mutual information [39], conditional entropy and order q entropy [44], the diffusion entropy method (DEM) [78], etc. DNA walks are the walks generated by translating a DNA sequence into a binary one, where, for instance, a purine corresponds to 1 and a pyrimidine to $x_i - 1$ (see fig. 1).

Parallel to these mathematical and computational studies, a very large number of experiments has been carried out with the aim of identifying specific, short, DNA sequences, generally located in noncoding DNA, exerting critical roles in the regulation of the gene functions. These studies generally utilize some sort of "knock out" method based on the observation of the effects of deletions or changes in nucleotide sequences located in upstream or intervening regions. Through this work a series of "motifs" has been identified and used by bioinformatics students for "annotation" purposes. By annotation we mean a series of bioinformatics practices which allows the distinction between different functional DNA regions in anonymous isolated sequences.

Without entering into the details of the data obtained in these fields, we will shortly report on a series of converging results obtained. Periodic behaviors have been found particularly in eukaryotic genomes [91]. A 3-base periodicity has been observed in coding sequences. The general codon rule RNY (R, purine, N, any base, and Y, pyrimidine) had been proposed on theoretical grounds by Eigen and Schuster [33]. Other periodic behaviors found were the frequent occurrence of G in the first positions of triplets and its avoidance in the third. A different rule (G in the third position) was found in our laboratory to occur in bacteria living in extreme conditions [59] thus showing the presence of environment-led selection. Physical and functional causes of periodicities have been suggested a number of times. A 10.5-base periodicity of AA.CT dinucleotides has been related to the B-DNA helical repeat per turn and to their high bendability needed for the superhelical DNA winding in the nucleosomes. Indeed, chromatin structure and its constraints seem to be one of the main reasons for higher-order periodicities like the periodic behavior at periods 200 and 400 shown by many authors and recently emphasized by the Arneodo groups extensive work (see for instance Audit et al. [7]). These authors found that the scale of 100–200 bp separates two different power-law correlations in eukaryotes and Archaea but not in bacteria. Although these are fairly general rules, at least in eukaryotes selection operates in terms of "inner" functional constraints even if the environment may also have at least a local influence as in the quoted case of G in the third triplet position in extreme environments contradicting the general RNY principle.

Although, as discussed before, many different methods have been used for obtaining global estimates of deviations from randomness, the results obtained do allow some general conclusions to be drawn. Firstly, noncoding strings show very different statistical features from coding ones, both at short and long range. At short range, an example of the results obtained is the work by Grosse et al. [39]. In this paper the authors studied the decay of a positional mutual information function with the distance between nucleotide positions. The function used, as the authors remind us, measures the difference between the sum of the entropies of two subsystems (two positions on the string) and that of the compound system. This function showed a fast decay to zero when the distance between two nucleotides was higher than ten in noncoding sequences and a periodic behavior (period $= 3$) in coding ones. As far as long-range correlations are concerned, a large number of investigations have been focused on the detection of scaling. With all methods (see table 1), noncoding sequence distributions were found to follow power laws, while a discussion remains open on coding ones, although several authors [6, 98] confirmed the presence of weak correlations also in this case. Particularly, it has been shown that coding DNA statistics is characterized by two different scaling behaviors, being almost random at short range and of the Lévy kind at long range [3, 4, 78].

Similar results have been obtained with a linguistic complexity measure method by Gabrielian and Bolshoy [36] who found that least complex fragments were preferentially located in noncoding sequences.

The more constrained nature of noncoding regions and the presence of a higher percentage of these sequences in eukaryotes result in the logical consequence of a less-than-linear increase in sequence entropy with increasing genome

TABLE 1   Some examples of correlations in DNA.

Methods	Results
Variance of DNA walk	Long-range correlations higher in noncoding DNA [69].
Purine/pyrimidine cluster distribution	Power-law decay in noncoding DNA [5].
Detrended fluctuation analysis on DNA walk	Long-range correlations in noncoding DNA [70].
Modified standard deviation of block length	Higher deviation from randomness in noncoding sequences [5].
Linguistic sequence complexity	Low-complexity sequences higher in noncoding and in eukaryotes [36].
Two- and tri-dimensional DNA walk	Long-range correlations in coding and noncoding sequences [57].
Mapping a random Cantor set	Noncoding/coding fractal partition only in eukaryotes [73].
Second-moment, kurtosis, and diffusion entropy of DNA walk	Lévy statistics in prokaryotes [3, 4, 78].

length. However, in most cases the methods so far utilized only allow the differentiation between bacteria, Archaea and eukaryotes, but no parameter has been found yet which may reveal any evolutionary trend within these large groups.

A more experimental approach to the problem of evolutionary dynamics of complexity levels is, however, the analysis of density and distributions of low-complexity DNA tracts. An approach of this kind was attempted by Provata and Almirantis [73], who analyzed the size distributions of purine or pyrimidine clusters and found an exponential decay in coding ones and a power-law decay in noncoding ones. A strong suggestion that low-complexity tracts are a significant component of the observed deviations from randomness of nucleotide distributions, came from the observation [58] that the removal of long purine and pyrimidine stretches from human sequences strikingly reduced the level of long-range correlations, estimated through the distribution's second moment, approaching the quasi-randomness of bacterial genomes.

These observations prompted us to carry out an analysis of the impact of low-complexity tracts throughout evolution with the aim of gaining information on a possible evolutionary trend. An algorithm was then used to evaluate the density of low-complexity tracts (homopurine, homopyrimidine; homo-AT; homo-GC simple sequence repeats) in a large number of genomes from virus, bacteria, Archaea, and eukaryotes [19]. A clear difference was observed in the distributions of an impact coefficient of low-complexity sequences between coding and noncoding regions and between prokaryotes and eukaryotes, noncoding and eukaryote sequences showing large high-value tails. The analysis was then directed toward tract-length distributions of single classes of homogeneous sequences: purines, pyrimidines, A, T, G, C; "strong" (GC); "weak" (AT); (AC); and (GT). In this work more than 30 complete genomes were screened and the following results have been obtained. Prokaryote distributions generally showed exponential decay with the exception of the Archeoglobus genome where a small power-law component was observed. This component became more evident in Archaea and very clear in eukaryotes. In eukaryotes moreover, a slower type of decay was observed for longer tracts and, in general, maximum values of lengths were attained. The slower component was generally not present in Saccharomyces with the exception of AC rich tracts and was not present in Arabidopsis in the case of AT, AC, GT, GC, and strings. Longer lengths were always preferentially localized in noncoding sequences and interesting differences were observed between the homogeneous tracts of different composition. Power-law components were constantly larger and their coefficient was found to be lower for AT, GT, purines, pyrimidines, than for GC and AC and, in general, in noncoding than in coding regions, with stretches being in these last cases much shorter and having a higher power coefficient, thus suggesting the presence of a possible selective pressure against these last sequences. This hypothesis was also supported by the finding that intron sequences were less constrained than intergenic ones [1]. The general conclusion of the data reported in this paragraph may be that constraints to randomness are clearly higher in noncoding and in general in eukaryote sequences and these constraints suggest a possible functional role of low-complexity tracts.

1.3   THE POSSIBLE FUNCTIONAL ROLES OF CONSTRAINED SEQUENCES

A functional role of noncoding sequences is clearly contradictory to the concept of "junk DNA," prevalent until only a few years ago. It should be mentioned here that a tentative nonselective explanation of correlations was offered by Dokholyan et al. [26] and supported by Buldyrev et al. [20] Particularly in the second paper, where deviations from randomness of noncoding regions were discussed in general, the known process of "slippage" (repeat amplification caused by replication "mistakes") was suggested as a possible cause of low-complexity tracts expansion. This could be one of a series of processes which include those previously discussed in this article in relation to genome size increases, whose multiplicative nature by itself certainly may be the cause of nonrandom distribution of nucleotides and of genome sizes. It should be stressed, however, that a number of such processes at the molecular level is known to occur also in bacteria in which, for instance, mobile elements are present. However, in that case, most probably due to selection, the processes do not cause genome size increase or noncoding region expansion. It is, therefore, important to ascertain whether selection forces play a role in genome modeling, acting on noncoding sequences

TABLE 2   Repeat expansion and human diseases.

Sequence	Disease	Refernece
$(GCG)_2(GCA)_2 (GCG)_3$	Oculopharyngeal muscular dystrophy	[64]
$(CAG)_n$	Prostate cancer	[86]
$(CAG)_n$	Huntington's disease	[71]
$(CAG)_n$	Polycystic ovarian syndrome	[42]
$(CTG)_n$	Myotonic dystrophy 1	[77]
$(GAA)_n$	Friedreich's ataxia	[51]
$(CA)_n$	Wegener's granulomatosis	[97]
$(CAA)_n$, G-T mutation	Friedreich's ataxia	[62]
$(CGG)_n$	Fragile X	[14]
$(CAG)_n$	Cardiovascular diseases	[98]
$(AG/CTG)_n$	Huntington-like disease	[43]
$(CAG)_n$	Kennedy's disease	[27]
$(GT)_n$	Cardiovascular disease	[88]
$(CA)_n$	Breast cancer	[96]

TABLE 3   Some regulatory functions of low-complexity sequences.

Repeat	Function	Organism	Reference
dA:dT	Nucleosome formation	Yeast	[47]
AT rich	Transcriptional regulation	*Giardia*	[34]
$(GT)_n(GA)_n$	Zinc-dependent protein binding	Man	[61]
TG	Enhancer	Man	[40]
GT	Coenhancer	Man	[10]
TTTTA	Transcription control	Man	[61]
GC rich	Silencing	chimp BMV	[63]

nucleotide distributions. It's important to ascertain whether the lengthening of low-complexity tracts only in eukaryotes is in some way correlated with genome and organismal organization of this category of living beings. One strong suggestion of selection pressure against repeat expansion comes from the increasing evidence of the involvement of multiple repeats, mostly rich in GC, in many human diseases (table 2), both in coding and noncoding sequences. A functional role in transcription has also been suggested for several kinds of repeats (table 3), generally positive for AT-rich ones which, in many cases (table 4), also locally enhance recombination and mutation frequencies.

All these data induced us to develop two experimental systems with the aim of testing the possible roles of homogenous tracts.

The first was an in vitro evolving system of tomato plant cells of the same initial genotype on media containing different equilibria between two key classes of plant hormones (auxins and cytokinins). This system is fast evolving due to the known extremely high spontaneous mutation frequencies in in vitro cultures (over 10%) of plant cells, and due to the relatively short cell generation times

TABLE 4   Some low-complexity sequences influencing recombination.

Sequences	Features
Poly AT	Hypermutability [8]
AG py	Hot spot tripled in lg [76]
TA microsatellite	MICA hot spot [79]
(CTTT)(CTTTT)	HLA hot spot F [93]
Poly CT	Hot spot [60]
AT rich	Hot spot [32, 31]
TTAGGG	Hot spot [25]
GC rich	Recombination silencing [63]
AT rich	Class-switch recombination w[89]

(22 hours). It was, therefore, chosen for the monitoring of genome changes during a large number of cell generations in the presence of different physiological contexts. Screening of cell cultures grown for two years with a random molecular method (RAPD) showed the diversification of genotypes in cells grown in different media and resulting in specific combinations of variants of hypervariable (polymorphic) sequences in each physiological condition. Polymorphic DNA sequences isolated from the different, selected genotypes were found to be rich in low-complexity sequences and particularly in AT, suggesting that such sequences may be hypervariable [12] and possibly involved in the regulation of genes connected with the plant cell hormonal system. Further studies were then carried out on the same system; these analyzed the variability in low-complexity sequences located in noncoding sequences within specific genes known to be involved in phytohormone synthesis or regulatory action. It was thus shown that, indeed, the variation between cell clone genotypes was higher within and near the homogeneous tracts and that this led to the fixation of different lengths in different contexts [13]. Our attention was then directed to in vivo systems and particularly to the analysis of sequence variability between and within tomato species and in the evolution of primates. Therefore, a number of known tomato genotypes (20 cultivated varieties and species), and of eight primate species were chosen for the study and genes affecting key characters of the plant and tomato systems were selected, isolated, and sequenced. Sequence variation was analyzed in noncoding and coding sequences of the ACC-4 synthase gene in tomato involved in the synthesis of a key plant hormone, ethylene, and in EMX2 and OTX-1 genes, both controlling brain development in eight primate species. It was thus found that sequence variability was much higher in noncoding than in coding DNA, being limited in coding regions to a few single base changes that mostly result in synonymous codons and, therefore, do not affect protein sequences. Moreover, in all cases, changes in noncoding regions were more frequent and involved large

insertions and deletions in low-complexity sequences whose instability seemed to be extended to neighboring areas. Critical experiments were then carried out on tomatoes to ascertain whether the observed variability may have a functional role in the modulation of gene expression. For this purpose, ACC synthase expression was tested in tissue culture lines evolved in different physiological contexts and in cultivated varieties of the same genetic background differing in the length of a $(AT)_n$ repeated sequence. Gene transcription was found not only to happen at different levels, but also to respond differently to exogenous hormonal treatments. These results obtained by our group and others available in the literature suggest a clear functional role of low-complexity sequences at least in terms of transcription regulation. Increasing evidence, coming from physico-chemical studies of DNA structures, suggests that this role and, in general, the roles of specific fixed noncoding subsequences, may derive from their local conformational landscapes determined by nucleotide combinations. We introduce here the term "conformational landscape" of DNA as a comprehensive term underlining the presence of more than one conformation minimum for all sequences. Particularly, it is known that repeated regions may assume "abnormal" conformations like triple and tetra helices, cruciforms, hairpins, etc., each having a specific and different functional "meaning." Moreover, nucleotide composition determines the level of curvature, bendability, and the capacity of DNA to "twist" and "roll." This has, for instance, been proved by Gabrielian and Bolshoy [36], who found an inverse correlation between linguistic complexity and curvature. Now, chromosome organization and most DNA-based dynamic processes require specific DNA conformations and local "matching" between DNA and protein structures. Particularly in eukaryotes, a considerable part of sequences upstream of coding regions is bound to contain specific "motifs" with local conformations allowing recognition by complementary transcription factors which, combined with others, will form the "transcription complex." Furthermore, the formation of complexes between DNA and other molecules depends on bending, curving, twisting, and rolling of the nucleic acid. The amount, the site, and the time of gene expression are determined by the dynamics of complementary conformations of proteins and DNA in the transcription complex, and similar conditions are present in complexes involved in replication, splicing, rejoining of exons, etc.

The modulation of recognition is also at the base of chromosome packaging and opening, and at the base of the general structural organization of the cell components into compartments, which is largely based on molecular networks whose fractal nature is also partially due to formal constraints. It is worth noting that shape and functional correlations between DNA subsequences and proteins necessarily mean a correlational constraint in the noncoding upstream DNA of the expressed gene and in the coding DNA of the gene carrying the information for transcription factors. Therefore, for the optimal dynamics of the living system, concerted selection and evolution must have occurred, limiting the degrees of freedom (the entropy) of the connected sequences. However, this does not lead to low complexity of DNA subsequences in the case of coding ones, which have been selected by the functional and environmental lottery from the possible random nucleotide combinations and whose structural constraints are due to protein and not DNA functional rules. Noncoding sequences, on the other hand, are directly selected for optimal "matching" conformational landscapes and, probably, hypervariability. Although direct proofs of the possible long-term advantage of the presence of the recombination hot spots are lacking, indirect evidence of their possible role should be considered. Firstly, it has been known since the late 1970s that the efficiency of the immunoglobulin and histocompatibility systems efficiency is largely based on the presence of hypervariable sequences.

Moreover, the whole genetic homeostasis theory [54] and the balanced selection evolution theory [56] are based on evidence of the selective advantages of heterozygosity. The high-mutation frequencies in low-complexity sequences, the role of these sequences in gene regulation, and the dependence of quantitative characters on regulation suggest that high heterozygosity in intergenic sequences may increase the homeostatic capacity of organisms through the presence in the same organism of more than one modulation pattern in single genes. This may be at the base of evolution through changes in regulation more than through the "invention" of new genes, as it has been shown, for instance, to have occurred in primates. Within this framework, complexity reduction at the level of noncoding sequences may result in increased inter- and intra-individual entropy (variability) and consequent increased robustness of the higher organismal levels. In other words, the robustness (low entropy) at lower levels would increase stability at higher levels as supported by the known reduction in the phenotypic variability of heterozygotes found in many cases by Lerner and others and, recently, by the evidence of "molecular heterosis" due to the contemporary presence of two isozymes (see Nevo [65], for a recent review).

2   THE PROTEIN AND NETWORK LEVELS

2.1   MOLECULAR VERSATILITY GENERATORS

Low-complexity sequence fixation and the presence of other, specific, hypervariable regions suggest that genome flexibility, although maintained within functional limits, may be one of the processes needed both for living systems robustness and evolvability [52]. In the case of DNA, specifically, some low-complexity sequences, generated by an antonomous spontaneous extension process, have been allowed to expand particularly in eukaryotes, other homogeneous tracts being kept within lower dimension limits. The selected sequences are hypervariable but, at the same time, are endowed with specific conformational landscapes whose functional meaning mainly derives from interactions with proteins. Other sources of variations and flexibility versatility are present particularly in eukaryotes at the RNA and protein levels. Different initiation and termination of transcription and alternative splicing, all determined by the presence of multiple signals leading to DNA or RNA protein interactions, lead in eukaryotes to a previously unsuspected high level of gene ambiguity as shown in neurexins [75, 87], a group of more than a thousand proteins coded by only three genes. At the post translational level a source of versatility comes from the wide protein conformational and functional landscapes. For instance calmodulin, a very flexible and versatile inhibitor endowed with a variable "expansion joint" may assume different configurations, allowing the binding to different protein targets and, therefore, totally different functions [52]. Moreover, each protein is known to be made by the assemblage of different highly conserved domains which have been shuffled throughout evolution and which give rise to new functions. It has recently been shown that the occurrence of post-translational domain shuffling further increases the "positive" ambiguity of single genes.

All these processes, besides having a possible value in genome size parsimony, contribute to the differentiation of different cell types in multicellular organisms and to their global robustness during life cycles. In all cases, flexibility is achieved through and leads to recognition events. These, in turn, limit the range of flexibility to its "functional" component. In most cases, the variation conserved (functional) in the partners of recognition events, leads to complementary conformational landscapes and is therefore correlated. To put it another way, living systems seem to maintain a number of different "noise generators" needed to acquire the structural and functional variability necessary for robustness and adaptation to internal and external changes. According to this hypothesis, a part of the noise produced would be organized into "correlated information," whose flexibility would be used to constrain the disrupting damage derived from what M. Gell-Mann calls "ignorance" [37]. Living systems robustness seems to be, therefore, different from sheer stability and is strictly bound to the capacity to maintain communication between network components.

In the last five years, as reviewed recently by Strogatz [85], an entirely new approach has been applied to the problem of network robustness with results which seem to confirm our hypothesis. According to Albert et al. [2], networks can be divided into two major classes on the basis of the decay function of $P(K)$, the probability P of finding nodes with K connections. In one class $P(K)$ peaks at an average K and then decays exponentially. In the other, to which all biological networks studied so far belong, the connections are not random, and networks become inhomogeneous, showing clustering and a low number of nodes with high connectivity. In this case $P(K)$ decays with a power law $P(K) K^{-8}$. Metabolic [50, 95] and protein networks [50] are all coherent with the proposed scale-free models whose present features can be interpreted as being derived from growth through preferential attachment of new components to pre-existing nodes. Such a process has the advantage of allowing the average fast communication between nodes through the reduction of path dimension (d = average number of shortest path between two random nodes).

This behavior [50] is typical of living networks in which, at variance with nonliving ones, the dimension seems to have remained constant throughout evolution. Scale-free biological networks are particularly robust in the presence of random noise [2], but may be disrupted and subdivided into subnetworks when highly connected nodes are deleted. The capacity of biological networks to maintain low-average dimension is most probably due to, during evolution, the increase of the number of short connections to a relatively low and highly conserved number of highly connected nodes, with the consequent maintenance of a high homeorrhetic power. In Waddington's words, homeorrhesis is the dynamical version of homeostasis, in the same logical sense that the notion of flux applies to development in biological systems [94]. Both these features are proven by a good set of experimental data. It is well known, for instance, that not only a relevant number of genes is the same all through the "tree of life," but also that proteins share common domains and are derived from "bricolage" [48]. The hypothesis of network growth through preferential addition has also been suggested for ecosystems by Solé [81, 82]. These authors introduce the term "self-organized instability" as comprehensive of the introduction in an ecosystem of new immigrant species (noise generation), their rapid integration into the network whenever possible, and the resistance of the web to further immigrants not compatible with a connection threshold. Moreover, food webs' robustness has been shown to increase with "connectance," defined as the ratio between the number of links between species and the square of the number of species (the fraction of all possible trophic links including cannibalism) [30]. Diversity is considered as positive within this frame for ecosystem robustness, as long as it is integrated in the connection web, because of the so-called "insurance effect." This is given by redundancy (species replacing extinct ones), versatility (new functions useful under variable environmental conditions), and reduction of secondary extinction. Finally, "strong" connections (for instance, only one prey for a predator) are unanimously considered as negative for ecosystem endurance.

It is worth noting that very similar conditions are known to limit robustness at the metabolic and developmental levels. For instance Kirschner and Gerhart [52], in a comprehensive review on "evolvability" (organisms capacity to generate heritable phenotypic variation), consider versatility (low requirements for achieving different complex functional outcomes), weak linkages, and compartmentation as the main conditions for developmental robustness. Compartmentation is defined here as a division of the cell's total genomic potential into partially independent subsets of "expressed genes." It is, therefore, similar to the "modules" mentioned recently as network subsets induced or inhibited by signals [41]. It should be noted that compartments can be assimilated with groups of genes connected with the same node and that the robustness derived from compartmentation may be one reason for the structure typical of living networks.

3   CONCLUSIONS

We have tried to update the data from experimental biology and modeling from the point of view of constraint/randomness ratios at the different hierarchical levels of organization. The ongoing convergence of many disciplines in the study of living systems seems to be leading us to the verge of a new, unitarian vision of life. Living systems have been adopting very different adaptive strategies during evolution. A considerable number of them has maintained a relatively low level of complexity (the prokaryotes). Others have "chosen" the way of progressively increasing in complexity. However, the dimension of networks has remained constant and the increase in the number of components has been combined with an increase in constraints, due to an increase in the number of interactions. This is the case with DNA, where the formerly called "junk DNA" has been "filled" with constrained sequences. The process seems to have been characterized by the progressive addition of newly integrated elements in to pre-existing systems. Interactions and multiplicative increase have led to a "kingdom of power laws," describing the correlations within and between components and systems. Living systems always had inborn "noise generators" which, however, seem to be more important in eukaryotes than in prokaryotes, and in which part of the noise coming from internal and external sources (the environment), has been incorporated into the correlated, flexible systems. The correlated information is used to increase the robustness of the systems against noncorrelated noise as this information is the basis for flexibility, redundancy, and versatility—the "benevolent disorder of life."

ACKNOWLEDGMENTS

Thanks are due to Paolo Allegrini and Luigi Palatella for fruitful discussion on the dynamics and structure of DNA constraints and to Marco Buiatti for critical reading of the manuscript and suggestions on its basic concepts.

REFERENCES

[1] Acquisti, C., L. Fronzoni, G. Mersi, E. Caranese, L. Quera, and M. Buiatti. "Evolutionary Analysis of Homogeneous DNA Tracts in Coding and Noncoding Regions." Manuscript in preparation.

[2] Albert, R., H. Jeong, and A. Lazló Barabasi. "Error and Attack Tolerance of Complex Networks." *Nature* **406** (2000): 378–382.

[3] Allegrini, P., P. Grigolini, and B. J. West. "Dynamical Approach to Lévy Processes." *Phys. Rev. E.* **54** (1996): 4760–4765.

[4] Allegrini, P., M. Buiatti, P. Grigolini, and B. J. West. "Fractional Brownian Motion as a Non-stationary Process. Non-Gaussian Statistics of Anomalous

Diffusion: The DNA Sequences of Prokaryotes." *Phys. Rev. E.* **58** (1998): 3640–3647.

[5] Almirantis, Y. "A Standard Deviation-Based Quantification Differentiates Coding from Non-coding DNA Sequences and Gives Insight to Their Evolutionary History." *J. Theor. Biol.* **196** (1999): 297–308.

[6] Arneodo, A., Y. d'Aubenton-Carafa, E. Bacry, P. V. Graves, J. F. Muzy, and C. Thermes. "Wavelet-Based Fractal Analysis of DNA Sequences." *Physica D* **1328** (1995): 1–30.

[7] Audit, B., C. Thermes, C. Vaillant, Y. d'Aubenton-Carafa, J. F. Muzy, and A. Arneodo. "Long Range Correlations in Genomic DNA: A Signature of the Nucleosomal Structure." *Phys. Rev. Lett.* **86** (2001): 2471–2474.

[8] Bacon, A. L., S. M. Farrington, and M. G. Dunlop. "Mutation Frequency in Coding and Non-coding Repeat Sequences in Mismatch Repair Deficient Cells Derived from Normal Human Tissue." *Oncogene* **20** (2001): 7464–7471.

[9] Beaton, M. J., and P. D. N. Herbert. "Geographical Parthenogenesis and Polyploidy in *Daphia pulex*." *Amer. Natur.* **132** (1988): 157–163.

[10] Berg, D. T., J. D. Walls, A. E. Reifel-Miller, and B. W. Grinnel. "E1A-Induced Enhancer Activity of the poly(dG-dT), poly(dA-dC) Element (GT element) and Interactions with a GT-Specific Nuclear Factor." *Mol. Cell Biol.* **9** (1989): 5248–5253.

[11] Bernardi, G., B. Olofsson, and J. Filipski. "The Mosaic Genome of Warm-Blooded Vertebrates." *Science* **228** (1985): 953–958.

[12] Bogani, P., A. Simoni, P. Liò, A. Scialpi, and M. Buiatti. "Genome Flux in Tomato Cell Clones Cultured in vitro in Different Physiological Equilibria. II: A RAPD Analysis of Variability." *Genome* **39** (1996): 846–853.

[13] Bogani, P., A. Simoni, P. Liò, A. Germinario, and M. Buiatti. "Molecular Variation in Plant Cell Populations Evolving in vitro in Different Physiological Contexts." *Genome* **44** (2001): 1–10.

[14] Bontekoe, C. J., C. E. Bakker, I. M. Nievwenhuizen, H. Van der Linde, H. Lans, D. De Lange, M. C. Hirste, and B. A. Oostra. "Instability of a $(CGG)_{98}$ Repeat in the FMR 1 Promoter." *Human Mol. Genet.* **10** (2001): 1693–1699.

[15] Broschard, T. H., N. Koffel-Schwarts, and R. P. Fuchs. "Sequence-Dependent Modulation of Frameshift Mutagenesis at NarI-derived Mutation Hot Spots." *J. Mol. Biol.* **288** (1999): 191–199.

[16] Buiatti, M. *Lo stato vivente della materia.* Torino: Utet libreria, 2000.

[17] Buiatti, M., and P. Bogani. "Exploiting Genome Plasticity for the Detection of Hypervariable Sequences." In *Molecular Tools for Screening Biodiversity in Plants and Animals*, edited by Angela Karp, P. G. Isaac, and D. Ingram, 205–237. London: Chapman and Hall, 1998.

[18] Buiatti, M., and M. Buiatti. "The Living State of Matter." *Biol. Forum* **94** (2001): 59–82.

[19] Buiatti, M., C. Acquisti, G. Mersi, P. Bogani, and M. Buiatti. "The Biological Meaning of DNA Correlations." In *Fractals in Biology and Medicine*, edited by Gabriele A. Losa, D. Morlini, T. F Nonnnenmacher, and E. Weibel, 235–247. Stuttgart: Birkhäuser Press, 2002.

[20] Buldyrev, S. V., S. Harlin, and H. E. Stanley. "Distribution of Base Pair Repeats in Coding and Non-coding DNA Sequences." *Phys. Rev. Lett.* **79** (1997): 5178–5181.

[21] Buldyrev, S. V., A. L. Goldberger, S. Harlin, R. N. Mantegna, M. E. Masta, C. K. Peng, M. Simmons, and H. E. Stanley. "Long-Range Correlations Properties of Coding and Non-coding DNA Sequences-GenBank Analysis." *Phys. Rev. E* **51** (1995): 5084–5091.

[22] Cellini, E., P. Forleo, B. Nacmias, A. Tedde, S. Latorraca, S. Piacentini, L. Parnetti, V. Gallai, and S. Sorbi. "Clinical and Genetic Analysis of Hereditary and Sporadic Ataxia in Central Italy." *Brain Res. Bull.* **56** (2001): 363–366.

[23] D'Amato, F. *Nuclear Cytology in Relation to Development*. Cambridge: Cambridge University Press, 1977.

[24] Dawkins, R. *The Selfish Gene*. Oxford: Oxford University Press, 1976.

[25] Day, J. P., C. L. Limoli, and W. F. Morgan. "Recombination Involving Interstitial Telomere Repeat-like Sequences Promotes Chromosomal Instability in Chinese Hamster Cells." *Carcinogenesis* **19** (1998): 259–265.

[26] Dokholyan, N. V., S. V. Buldyrev, S. Havlin, and H. F. Stanley. "Distribution of Base Pair Repeats in Coding and Non-coding DNA Sequences." *Phys. Rev. Lett.* **79** (1977): 5178–5181.

[27] Domitrz, I., M. Jedrzejowska, M. Lipowska, T. Siddique, and H. Kwieczynski. "Kennedy's Disease: Expansion of the CAG Trinucleotide." *Neurol. Neurochir. Pol.* **35** (2001): 107–114.

[28] Doolittle, W. F., and C. Sapienza. "Selfish Genes, the Phenotype Paradigm and Genome Evolution." *Nature* **284** (1980): 601–603.

[29] Dover, G. A. "Ignorant DNA?" *Nature* **285** (1980): 618–620.

[30] Dunne, J. A., R. J. Williams, and N. D. Martinez. "Network Topology and Biodiversity Loss in Food Webs: Robustness Increases with Connectance." Working Paper 02-03-013, Santa Fe Institute, Santa Fe, NM, 2002.

[31] Edelmann, L., E. Spiteri, K. Koren, V. Pulijaal, M. G. Bialer, A. Shanske, R. Goldberg, and B. E. Morrow. "AT-rich Palindromes Mediate the Constitutional t(11,22) Translocation." *Am. J. Human Genet.* **68** (2001): 1–13.

[32] Edelmann, L., E. Spiteri, N. McCain, R. Goldberg, R. K. Pandita, S. Duong, J. Fox, D. Blumenthal, S. R. Lalani, L. G. Shaffer, and B. E. Morrow. "A Common Breakpoint on 11q23 in Carriers of the Constitutional t(11;22) Translocation." *Am. J. Human Genet.* **65** (1999): 1608–1616.

[33] Eigen, M., and P. Schuster. *The Hypercycle*. Berlin: Springer-Verlag, 1979.

[34] Elmendorf, H. G., S. M. Singer, J. Pierce, J. Cowan, and T. E. Nash. "Initiator and Upstream Elements in the 2-tubulin Promoter of *Giardia lamblia*." *Mol. Biol. Parasitol.* **113** (2001): 157–169.

[35] Enard, W., P. Khaitovich, J. Klose, S. Zöllner, F. Heissig, P. Giavalisca, K. Nieselt-Strune, E. Muchmore, A. Varki, R. Rand, G. M. Doxiadis, R. E. Bontrop, and S. Paabo. "Intra- and Interspecific Variation in Primate Gene Expression Patterns." *Science* **296** (2002): 340–343.

[36] Gabrielian, A., and A. Bolshoy. "Sequence Complexity and DNA Curvature." *Comp. Chem.* **23** (1999): 263–274.

[37] Gell-Mann, Murray. "Effective Complexity." This volume.

[38] Gregory, T. R., and P. D. N. Herbert. "The Modulation of DNA Content: Proximate Causes and Ultimate Consequences." *Genome Resh.* **9** (1999): 317–324.

[39] Grosse, I., H. Herzel, S. V. Buldyrev, and H. E. Stanley. "Species Independence of Mutual Information in Coding and Non-coding DNA." *Phys. Rev. E.* **61** (2000): 5624–5629.

[40] Hamada, K., S. L. Gleason, B. Z. Levi, S. Hirschfeld, E. Appella, and K. Ozato. "H-2RIIBP, a Member of the Nuclear Hormone Receptor Superfamily that Binds to Both the Regulatory Element of Major Histocompatibility Class I Genes and the Estrogen Response Element." *Proc. Natl. Acad. Sci. USA* **86** (1989): 8289–8293.

[41] Hartwell, L. H., J. J. Hopfield, S. Leibler, and A. W. Murray. "From Molecular to Cell Biology." *Nature* **402** (1999): c47–c50

[42] Hickey, T., A. Chandy, and R. J. Norman. "The Androgen (AG Repeat Polymorphism and X-Chromosome Inactivation in Australian Caucasian Women with Infertility Related to Polycystic Ovary Syndrome." *J. Clin. Endocrinol. Metab.* **87** (2002): 161–165.

[43] Holmes, S. E., E. O'Hearn, A. Rosenblatt, C. Callahan, G. Stevanin, A. Brice, N. T. Potter., C. A. Ross, and R. L. Marchis. "A Repeat Expansion in the Gene Encoding Juncophilin-3 is Associated with Huntington Disease-like 2." *Nat. Genet.* **29** (2002): 123.

[44] Holste, D., I. Grosse, and H. Herzel. "Statistical Analysis of the DNA Sequence of Human Chromosome 22." *Phys. Rev. E.* **64** (2001): 228–301.

[45] Hughes, A. L. *Adaptive Evolution of Genes and Genomes.* Oxford: Oxford University Press, 1999.

[46] Hughes, A. L., and M. K. Hughes. "Small Genomes for Better Fliers." *Nature* **377** (1995): 391

[47] Iyer, V., and K. Struhl. "Poly(dA:dT), a Ubiquitous Promoter that Stimulates Transcription via Its Intrinsic DNA Structure." *EMBO J.* **14** (1995): 2570–2579.

[48] Jacob, F. "Evolution and Tinkering." *Science* **196** (1977): 1161–1166.

[49] Jauert, P. A., S. N. Edminston, K. Conway, and D. T. Kirkpatrick. "RAD1 Controls the Meiotic Expansion of the Human HRAS1 Minisatellite in *Saccharomyces cerevisiae.*" *Mol. Cell Biol.* **22** (2002): 953–964.

[50] Jeong, H., B. Tombor, R. Albert, and Z. Oltvai. "The Large-Scale Organization of Metabolic Networks." *Nature* **407** (2001): 651–654.

[51] Juvonen, V., S. M. Kulmala, J. Ignatius, M. Penttinen, and M. L. Savontaus. "Dissecting the Epidemiology of a Trinucleotide Repeat Disease—Example of a FRDA in Finland." *Human Genet.* **110** (2002): 36–40.

[52] Kirschner, M., and J. Gerhart. "Evolvability." *Proc. Natl. Acad. Sci.* **95** (1988): 8420–8422.

[53] Kurahashi, H., and B. S. Emanuel. "Long AT-Rich Palindromes and the Constitutional t(11;22) Breakpoint." *Human Mol. Genet.* **1** (2001): 495–505.

[54] Lerner, I. M. *Genetic Homeostasis.* Edinburg: Oliver and Boyd, 1954.

[55] Lewis, W. H. *Polyploidy: Biological Relevance.* New York, NY: Plenum Press, 1980.

[56] Lewontin R. C. *The Genetic Basis of Evolutionary Change.* Columbia University Press, 1974.

[57] Liaofu, Lub, Neijang Lee, Lijun Jia, Fengmin Ji, and LuTsai. "Structural Correlation of Nucleotides in a DNA Sequences." *Phys. Rev. E.* **58** (1998): 861–871.

[58] Liò, P., A. Politi, M. Buiatti, and S. Ruffo. "High Statistics Block Entropy Measures of DNA Sequences." *J. Theor. Biol.* **180** (1996): 151–160.

[59] Liò, P., S. Ruffo, and M. Buiatti. "Third Codon G+C Periodicity as a Possible Signal for an Internal Selective Constraint." *J. Theor. Biol.* **171** (1994): 215–223.

[60] Majewski, J., and J. Ott. "GT Repeats are Associated with Recombination on Human Chromosome 22." *Genome Res.* **10** (2000): 1108–1114.

[61] Mäuler, W., G. Bassili, A. Rüdger, R. Renkawitz, and J. T. Epplen. "The $(GT)_n(GA)_n$ Containing Intron 2 of HLA-DRB Alleles Binds a Zinc-Dependent Protein and Forms Non-B-DNA Structures." *Gene* **226** (1999): 9–23.

[62] Mc Cabe, D. J., N. W. Wood, F. Ryan, M. G. Hanna, S. Connoly, D. P. Moore, A. Redmond, D. E. Barton, and R. P. Murphy. "Interfamilial Phenotypic Variability in Friedrich Ataxia Associated with AG 130 V Mutation in the FRDA Gene." *Arch. Neurol.* **59** (2002): 296–300.

[63] Nagy, P. D., and J. J. Bujarski. "Silencing Homologous RNA Recombination Hot Spot with GC-Rich Sequences in Brome Mosaic Virus." *J. Virol.* **72** (1998): 1122–1130.

[64] Nakamoto, M., S. Nakano, S. Kawashima, M. Ihara, Y. Nishimura, A. Shinde, and A. Kakizuka. "Unequal Crossing-Over in Unique PAB2 Mutations in Japanese Patients: A Possible Cause of Oculopharyngeal Muscular Dystrophy." *Arch. Neurol.* **59** (2002): 474–477.

[65] Nevo, E. "Evolution of Genome-phenome Diversity under Environmental Stress." *Proc. Natl. Acad. Sci. USA* **98** (2001): 6233–6240.

[66] Ohno, S. "So Much Junk DNA in Our Genome." In *Evolution of Genetic Systems*, 366–370. New York: Gordon and Breach, 1972.

[67] Orgel, L. E., and R. A. Johnstone. "Variation Across Species in the Size of the Nuclear Genome Supports the Junk DNA Explanation for the C-Value Paradox." *Proc. Roy. Soc. Lond. B Biol. Sci.* **249** (1992): 119–124.

[68] Parenti, R., E. Guillé, J. Grisvard, M. Durante, L. Giorgi, and M. Buiatti. "Transient DNA Satellite in Dedifferentiating Pith Tissue." *Nature New Biol.* **246** (1973): 237–239.

[69] Peng, C. K., S. V. Buldyrev, A. L. Goldberger, S. Harlin, F. Sciortino, M. Simon, and H. E. Stanley. "Long-Range Correlation's in Nucleotide Sequences." *Nature* **356** (1992): 168–170

[70] Peng, C. K., S. V. Buldyrev, J. M. Hausdorff, S. Havlin, J. E. Mietus, M. Simons, H. E. Stanley, and A. L. Goldberger. "Non-equilibrium Dynamics as an Indispensable Characteristics of a Healthy Biological System." *Integr. Physiol. Behav. Sci.* **29** (1994): 283–293.

[71] Petersen, A., K. E. Larsen, G. G. Behr, N. Romero, S. Przedborski, P. Brfundin, and D. Sulzer. "Expanded CAG Repeats in Exon 1 of the Huntington's Disease Gene Stimulate Dopamine-Mediated Striatal Neuron Autophagy and Degeneration." *Human Mol. Genet.* **10** (2001): 1243–1254.

[72] Petrov, D. A., T. A. Sangster, J. S. Johnston, and K. L. Shaw. "Evidence for DNA Loss as a Determinant of Genome Size." *Science* **287** (2000): 1060–1062.

[73] Provata, A., and Y. Almirantis. "Scaling Properties of Coding and Noncoding DNA Sequences." *Physica A.* **247** (1997): 482–496.

[74] Ritossa, F., and S. Spiegelman. "Localization of DNA Complementary to Ribosomal RNA in the Nucleolus Organizer Region of *Drosophyla melanogaster*." *Proc. Natl. Acad. Sci. USA* **53** (1965): 737–746.

[75] Rowen, L., J. Young, B. Birditt, A. Kaur, A. Madan, D. L. Philipps, S. Qin, P. Minx, R. K. Wilson, L. Hood, and B. R. Graveley. "Analysis of the Human Neurexin Genes: Alternative Splicing and the Generation of Protein Diversity." *Genomics* **79** (2002): 587–597.

[76] Saini, S. S., and A. Kaushik. "Extensive CDR3H Length Heterogeneity Exists in Bovine Foetal VDJ Rearrangements." *Scand. J. Immunol.* **55** (2002): 140–148.

[77] Savic, D., V. Rakolvic, D. Keckarevic, B. Culjkovic, O. Stojkovic, Mladenovic, S. Todorovic. S Apostolski, and S. Romac. "250 CT6 Repeats in DMPK is a Threshold for Correlation of Expansion Size and Age at Onset of Juvenile-Adult DM1." *Human Mutat.* **19** (2002): 131–139.

[78] Scafetta, N, V. Latora, and P. Grigolini. "Scaling Without Detrending: The Diffusion Entropy Method Applied to the DNA Sequences." *Phys. Rev. E* **66** (2002): 031906.

[79] Shiina T., G. Tamiya, A. Oka, T. Yamagata, N. Yamagata, E. Kikkawa, K. Goto, N. Mizuki, K. Watanabe, Y. Fukuzumi, S. Taguchi, C. Sugawara, A. Ono, L. Chen, M. Yamazaki, H. Tashiro, A. Ando, T. Ikemura, M. Kimura, and H. Inoko. "Nucleotide Sequencing Analysis of the 146-kilobase Segment

around the IkBL and MICA Genes at the Centromeric End of the HLA Class I Region." *Genomics* **47** (1998): 372–382.

[80] Shuter, B. J., J. L. Thomas, and N. D. Taylor. "Phenotypic Correlates of Genomic DNA Content in Unicellular Eukaryotes and Other Cells." *Amer. Natur.* **122** (1983): 26–44.

[81] Solé, R. V., and J. M. Montoya. "Complexity and Fragility in Ecological Networks." *Proc. Roy. Soc. Lond. B Biol. Sci.* **268** (2001): 2039–2045.

[82] Solé, R. V., D. Alonso, and A. McKane. "Self-Organized Instability in Complex Ecosystems." *Trans. Roy. Soc. Series B.* (2002): in press. Special issue: The Biosphere as a Complex Adaptive System.

[83] Sparrow, A. H., and A. F. Naumann. "Evolution of Genome Size by Doubling." *Science* **192** (1976): 524–527.

[84] Stanley, H. E., S. V. Buldyrev, A. L. Goldberger, Z. D. Goldberger, S. Havlin, R. N. Mantegna, S. M. Ossadnik, C. K. Peng, and M. Simons. "Statistical Mechanics in Biology: How Ubiquitous are Long-Range Correlations?" *Physica A* **205** (1994): 214–253.

[85] Strogatz, S. H. "Exploring Complex Networks." *Nature* **410** (2001): 268–272.

[86] Suzuki, H., K. Akakura, A. Komiya, T. Ueda, T. Imamoto, T. Furuya, T. Ichikawa, M. Watanabe, T. Shiraishi, and H. Ito. "CAG Polymorphic Repeat Lengths in Androgen Receptor Gene among Japanese Prostate Cancer Patients: Potential Predictor of Prognosis after Endocrine Therapy." *Prostate* **51** (2002): 219–224.

[87] Tabuchi, K., and T. C. Sudhof. "Structure and Evolution of Neurexin Genes: Insight into the Mechanism of Alternative Splicing." *Genomics* **79** (2002): 849–859.

[88] Tang, D. C., R. Prauner, W. Liv, K. H. Kim, R. P. Hirsch, M. C. Driscoll, and G. P. Rodgers. "Polymorphisms within the Angiotensin Gene (GT-repeat) and the Risk of Stroke in Pediatric Patients with Sickle Cell Disease: A Case-Control Study." *Am. J. Hematol.* **68** (2001): 164–169

[89] Tashiro, J., K. Kinoshita, and T. Honjo. "Palindromic but Not G-rich Sequences are Targets of Class Switch Recombination." *Intl. Immunol.* **13** (2001): 485–505

[90] Trifonov, E. N. "The Multiple Codes of Nucleotide Sequences." *Bull. Math. Biol.* **51** (1989): 417–432.

[91] Trifonov, E. N. "3-, 10.5-, 200-, 400-Base Periodicities in Genome Sequences." *Physica A.* **249** (1998): 511–516.

[92] Vinogradov, A. E. "Intron-Genome Size Relationship on a Large Evolutionary Scale." *J. Mol. Evol.* **49** (1999): 376–384.

[93] Vorechovsky, I., J. Kralovicova, M. D. Laycock, A. D. Webster, S. G. Marsh, A. Madrigal, and L. Hammarstrom. "Short Tandem Repeat (STR) Haplotypes in HLA: An Integrated 50-kb STR/linkage Disequilibrium/Gene Map between the RING3 and HLA-B Genes and Identification

of STR Haplotype Diversification in the Class III Region." *Eur. J. Human Genet.* **9** (2001): 590–598.

[94] Waddington, C. H. "A Catastrophe Theory of Evolution." *Ann. N.Y. Acad. Sci.* **231** (1974): 32–42.

[95] Wagner, A., and D. A. Fell. "The Small World Inside Large Metabolic Networks." *Proc. Roy. Soc. Lond. B Biol. Sci.* **268** (2001): 1803–1810.

[96] Yu, H., B. D. Li, M. Smith, R. Shi, H. J. Berkel, and I. Kato. "Polymorphic CA Repeats in the IGF-1 Gene and Breast Cancer." *Breast Cancer Res. Treat.* **70** (2001): 117–122.

[97] Zhou, Y., R. Giscombe, D. Huang, and A. K. Lefvert. "Novel Genetic Association of Wegner's Granulomatosis with the Interleukin 10 Gene." *J. Rheumatol.* **29** (2002): 317–320.

[98] Zitzmann, M., M. Brune, B. Kornman, J. Gromoll, S. Eckardstein, and E. Nieshlag. "The CAG Polymorphism in the AR Gene Affects High Density Lipoprotein Cholesterol and Arterial Vasoreactivity." *Clin. Endocrinol. Metab.* **86** (2001): 4867–4873.

[99] Zuckerhandl, E. "Polite DNA: Functional Density and Functional Compatibility in Genomes." *J. Mol. Evol.* **24** (1986): 12–27.

Plant Spread Dynamics and Spatial Patterns in Forest Ecology

Sergio A. Cannas
Diana E. Marco
Sergio A. Páez
Marcelo A. Montemurro

1 INTRODUCTION: THE BIOLOGICAL INVASION PROBLEM

Species in an ecosystem can be classified as *natives* or *exotics*. Native species are those that have coevolved in the ecosystem, while exotic ones have not. The introduction of exotic species into an ecosystem is usually associated with human influence, which can be intentional or accidental.

Some exotic species do not survive, at least not without artificial assistance. But some others do quite well on their own in a new environment. Exotic species may have no natural predators in the new environment or they may make better use of the natural resources than the natives, so they spread in the new territory and compete with some of the natives, who eventually become extinct. Exotic species that successfully establish and spread in an ecosystem are called *invaders*. The process by which an invader arrives and spreads into the new territory is called *biological invasion*.

It is worth mentioning that, although invaders are usually exotic species, sometimes native species may also behave like invaders. That is, if an ecosystem

Nonextensive Entropy—Interdisciplinary Applications
edited by Murray Gell-Mann and Constantino Tsallis, Oxford University Press

suffers a strong disturbance, like fire or heavy grazing, some native species whose populations were originally stable may start to grow, outcompeting other native species.

There are many examples of introduced species that became invaders, ranging from bacteria to cattle.

Accidental or intentional introductions by humans are responsible for most of the present biological invasions, threatening the structure and functioning of many ecosystems.

There are many effects associated with biological invasions, perhaps the most important one being the possible loss of biodiversity in the long term. But biological invasions may also introduce changes in different environmental traits, like climate, hydrology (invaders may consume more water than natives), and soil composition (for instance, some plants take up salt from soil and deposit it on the surface, making it unsuitable for some native species).

All these changes have strong economical impacts, considering their influences in agriculture, forestry, and public health [9]. Hence, it is of interest to understand this phenomenon in order to *predict the potential invasiveness* of a species before its introduction in an ecosystem, and to *develop strategies of control for invasive species that have already been introduced.*

Although the biological invasion has been recognized as potentially damaging for ecosystems functioning more than 40 years ago [5], there are still many important open questions about it.

Perhaps the most important one is: what makes a successful invader? That is, why does a given species behave as an invader and a similar one does not? Why does the same species behave as an invader in one ecosystem but not in another?

At present, specialists agree that this is an extremely complex problem and that the answer to this question depends on several factors. Among the most important factors for plants are:

- *Life history traits*, that is, reproductive and growth properties of individual trees, like seed-dispersal mechanisms, germination properties, and age of reproductive maturity. One point that is of particular interest is the possible existence of *long-range seed-dispersal distributions* (in a statistical sense). Trees disperse seeds through several mechanisms, which generate different types of spatial seed-dispersal distributions. There is evidence that some mechanisms may generate fat-tailed (that is, long-ranged) distributions. This is a very important factor in the description of plant migration processes, invasive or not, since this type of distribution may change the rates of spread of a species by orders of magnitude. We will discuss this topic in more detail later.
- *Demographic traits*, like the mortality rates at the different stages of growth of a tree.
- *Environmental conditions*, including disturbance regimes.

- *Interspecific interactions*, for instance, competition between the alien and native species, but also positive interactions (like symbiosis or facilitation).

However, the relative importance of all these factors is not clear and experimental results are frequently ambiguous and, sometimes, even contradictory. Hence, it is of interest to have accurate theoretical models to check all these hypotheses. In section 2 we present a recently proposed cellular automaton model [3] that describes the population dynamics of several interacting woody species (trees and shrubs) at an individual level. The model includes most of the features that are believed to be important in describing invasion processes, where the parametrization is made on the bases of measurable quantities. In section 3 we show a comparison between field data estimations and predictions of the model, both for invasion and noninvasion systems.

Another important question is, how can we characterize the invasion process? That is, how do we *measure* the invasiveness of a species or the resistance of the habitat to invasion? Usually the invasion process is characterized by quantifying the different rates of spread of the invading species, that is, the velocities at which it spreads in the new environment. Although these are good measures, sometimes the calculations involve a series of problems.

Experimentally, the rates of spread can be estimated, for example, from aerial photographs, by measuring the areas covered by the species at different times. However, this involves the usage of high-resolution photographs taken over long periods of time (typically several decades). Hence, the available data are scarce and not very accurate, and it is important to find alternative ways of characterizing the invasion process.

Sections 4 and 5 are devoted to the study of different scale-invariant (i.e., fractal) spatial patterns that may appear during invasion processes. The results presented in those sections suggest that the fractal dimensions of the particular fractal patterns are good (and measurable) quantities for characterizing the invasion processes.

A general discussion is presented in section 6.

2   THE MODEL

We first present a cellular automaton for the description of a single, noninteracting species. This model will be generalized later to the case of several interacting species.

2.1   SINGLE SPECIES CELLULAR AUTOMATON

The model [3] is defined on a square lattice of $N = L_x \times L_y$ sites or cells with open boundary conditions. Each cell can be occupied at most by one adult. This rule fixes the spatial scale, in the sense that the area of a unit cell of the lattice will correspond to the average area covered by the canopy of the trees under consideration. For instance, in the case of the species considered in section 3, the average area is $25\,\mathrm{m}^2$ [10], that is, the lattice unit corresponds to $5\,\mathrm{m}$.

To each cell we associate an integer variable $a_i(t)$ that represents the age of the individual located there at time t, where the time unit is chosen to be one year. This time scale appears naturally, since we consider annual rates of reproduction and mortality. We use a parallel dynamics.

Reproductive traits are described by the following parameters:

- Every mature individual produces n seeds in a seed crop. We assume a self-breeding species; that is, every individual produces seeds.
- The dispersal of those seeds is described by a density function $f(r)$, defined as the fraction of seeds dispersed by one individual to a distance r per unit area. This density is normalized in the plane.
- Every seed germinates with a probability f_g.
- Reproductive or mature individuals are those whose age is greater than t_m.
- Individuals produce seed crops every t_s years.
- An important factor for invasion is the existence of *juvenile banks*. Saplings of some species can survive under the shade of the parent, but they grow only to a certain height and then stop. They do not become reproductive while living under the shade of the parent. When the parent dies, they can resume their growth; one of them can replace the parent and become reproductive. The collection of saplings below the shade of the parent is called the juvenile bank. For species with juvenile banks we include another parameter, the average age t_J of the saplings.

Mortality in trees is much higher during the first stages of growth (usually during the first and second years of life) than the later stages. Hence, we consider two probabilities of survival:

- q is the annual adult survival probability.
- P_s is the juvenile (< 2 years) survival probability.

The dynamics of the model are as follows. Let's first consider an occupied cell i at time t, that is, a cell for which $a_i(t) \neq 0$. If the species does not have a juvenile bank, then the individual can survive and grow with probability q, and, therefore, $a_i(t+1) = a_i(t)+1$, or it dies with probability $1-q$, leaving the cell empty $a_i = 0$. If the species has a juvenile bank, the rule is similar, except that when it dies, instead of leaving the cell empty, it is replaced by another individual with age t_J.

Let's now consider an empty cell i at time t, that is, $a_i(t) = 0$.

This cell can be colonized at time $t + 1$ with probability $p_i(t)$, that depends on the seeds received from other cells; $p_i(t)$ is the probability that *at least* one of the seeds received at time t germinates *and* that the corresponding juvenile survives until the adult stage. Assuming these two events as independent, p_i is given by

$$p_i(t) = 1 - (1 - P_s f_g)^{s_i(t)} \qquad (1)$$

where $s_i(t)$ is the total number of seeds received by the cell at time t. How do we calculate s_i? Suppose that the species produces crops every year. Then, the number of seeds received by the cell i, coming from a mature tree located at a cell j is $n f(r_{ij})$, where r_{ij} is the distance between cells. Hence, $s_i(t)$ is obtained by summing this quantity over all cells containing mature trees. If the interval between seed crops t_s is greater than one, then cells containing mature trees only contribute to these sums every t_s years.

Now, what is the appropriate choice for the seed-dispersal function $f(r)$? For many tree species, seed dispersal has a limited range; that is, seeds are dispersed in the neighborhood of the tree. In this case $f(r)$ will be *short ranged*, that is, without long tails, and the global spread properties are not expected to depend on the specific form of $f(r)$. However, as we mentioned before, one open question about plant spread dynamics is the possible existence of long-range seed-dispersal mechanisms. This question was first addressed when people tried to explain the postglacial migration of different species of trees, like oaks. After the glaciers retreated, they left an open area that the woods recolonized at a velocity of about 1 km/year [4]. Short-range dispersal models with realistic parametrizations, like the reaction-diffusion model, when applied to this problem predicted velocities ranging between 1 and 50 m/year [13]. This difference of about two orders of magnitude suggests the presence of long-range seed-dispersal distributions [8].

But then the question is, which types of mechanisms can lead to the occurrence of rare events of seeds dispersed over very long distances? Trees use one or more different mechanisms of seed dispersal as a result of evolutionary adaptations. The simplest one is gravity; that is, seeds just fall down near the canopy. This is clearly a short-range mechanism.

Some plant species use mutualistic relationships with animals, like birds or mammals, as dispersal strategies. They produce fruits that have attractive colors and flavors. Then the animals eat the fruits together with the enclosed seeds, which pass unharmed through their digestive tracts. So, the animals move and defecate the seeds in another place. Whether or not this type of mechanism could lead to long-range dispersal effects depends on several factors, like the effective mobility and the habits of the animals. Experimental evidence is not conclusive at all.

Finally, the best candidates are wind dispersal mechanisms. Some seeds have a wing with the form of an helix (gyroscopic seeds). In the presence of strong winds these seeds leave the canopy, rotate, and fly, so they can be driven by the wind over long distances. However, the experimental evidence is not conclusive in this case either. Unfortunately, it is very difficult to measure seed-dispersal distribution functions. Field measurements of seed distributions have large error bars and limited ranges. Therefore, attempts to fit data with short- and long-range distributions cannot give definite answers. Hence, it is interesting to look for other ways of detecting long-range seed-dispersal distributions.

To analyze this problem we used two different types of models for the seed-dispersal function. For simulating short-range dispersal mechanisms we proposed an exponential function:

$$f(r) = \frac{2}{\pi d^2} e^{-2r/d} \tag{2}$$

where d is the mean dispersal distance.

For simulating long-range dispersal mechanisms we proposed a power law:

$$f(r) = \begin{cases} \frac{A}{r^\alpha} & \text{if } r \geq 1/2\,; \\ 0 & \text{if } 0 < r < 1/2\,. \end{cases} \tag{3}$$

where A is a normalization constant and $\alpha > 2$ (otherwise the density function f cannot be normalized). We expect three different behaviors, according to the values of α. For $\alpha > 4$ this function has finite first and second moments. Therefore, the central limit theorem holds and short-range behavior is expected. When $3 < \alpha \leq 4$, the first moment remains finite but the second moments becomes infinite. The mean dispersal distance is given by $d \equiv \langle r \rangle = (\alpha - 2)/2(\alpha - 3)$.

Finally, when $2 < \alpha \leq 3$, both the first and second moment are infinite.

2.2   INTERACTIVE MULTIPLE CELLULAR AUTOMATA MODEL

Let's now consider the case of several interacting species. In this case we assigned to each species one single cellular automaton of the type described in the previous subsection, all of these cellular automata are defined over the same lattice. In this way each cell of the lattice is associated with several dynamical variables. Interactions are introduced by coupling them using new dynamical rules. We called this an interactive multiple cellular automaton model, or IMCA [3].

The type of interaction we are interested in is competition, which in a broader sense is competition for space and involves competition for resources. This means that only one of the different dynamical variables associated with a given cell can be different from zero at any given time.

Concerning the colonization of an empty cell, we count the seeds received by the cell from each one of the different species. Then we calculate the different colonization probabilities and compare them with independent random numbers.

If only one species succeeds in the colonization, the cell follows the dynamics of a single species.

If more than one species succeeds we sort the winner with some probability that may depend on environmental conditions. If the different species use resources in a similar way, the winner is sorted with equal probability from the different species present in the cell.

If different species use resources in different ways the probability depends on the environmental conditions, which are described by a new set of variables. Those environmental variables may be dynamical or fixed. In the last case we can consider them as external conditions. An example of colonization rules that depend on environmental variables will be presented in the next section.

3   SIMULATIONS VS. FIELD DATA VALUES

To compare the predictions of our model with real situations, we considered four different tree species from the mountain forest in Córdoba, Argentina, two of them exotics (*Gleditsia triacanthos* and *Ligustrum lucidum*) and two natives (*Lithraea ternifolia* and *Fagara coco*). *L. ternifolia* was once the dominant tree species in this region. The native woodlands are being invaded by several exotic species. *F. coco* coexists in some parts of the mountains with the dominant *L. ternifolia*. Seed dispersal in all these cases is short ranged; another remarkable fact is that both invaders have juvenile banks while the natives do not [10].

We started our simulations by considering an initial dense forest of the dominant native (*L. ternifolia*) with random ages, located in a rectangular area L_x sites of width. At the bottom of the area $y = 0$ we put a row of individuals of one of the two exotic species, with random ages between zero and the age of first reproduction t_m. This configuration simulates a very typical situation, since these plants usually spread into the forests from the sides of roads or rivers.

In figure 1 we see the typical pattern of invasion at two different times, where gray points represent the natives, black ones are the invaders (*L. lucidum* in this case), and white points are empty sites. We see that the invaders form a widespread band that moves in the y direction, leaving a dense invaded forest behind it. The presence of juvenile banks turns the invasion inevitable and over a long period of time the invaders cover the whole area.

To characterize the invasion process we defined the *invasion front* $h(x)$ as the coordinate y of the farthest occupied site corresponding to the x position along the bottom line (see fig. 1).

We then calculated the mean front position

$$\overline{h} = \frac{1}{L_x} \sum_{x=1}^{L_x} h(x) \tag{4}$$

as a function of time, and we averaged this quantity over different initial conditions and over different sets of random numbers.

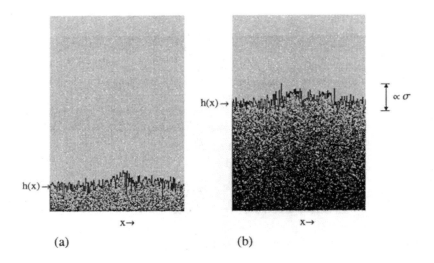

(a)   (b)

FIGURE 1   *L. lucidum* invasion (black cells) in a dense *L. ternifolia* forest (gray cells), for a simulation area of 80×160 cells; white cells correspond to empty cells. (a) $t = 50$ years; (b) $t = 150$ years.

In the case of invasion by *G. triacanthos*, it is necessary to introduce environmental variables. *L. ternifolia* has a special ability to establish in shallow soils and rock crevices. *G. triacanthos* does not present this ability; that is, it grows much more slowly in shallow soils than in deep soils. In this case, we introduced a set of soil parameters $\{c_i\}$ that can take two values: $c_i = 0$ represents a shallow soil cell, while $c_i = 1$ represents a deep soil cell. These parameters are sorted with some distribution at the beginning of the simulation and kept fixed through it. The colonization rule of empty cells is the following: if $c_i = 0$ the native species always win with probability 1; if $c_i = 1$ the winner is sorted with equal probability.

In figure 2 we see the behavior of the mean front position in this system, for the cases of homogeneous deep and rocky soil. We see that the effect of the soil type is just to slow down the process. Both curves show the same qualitative behavior; that is, after some transient period, the front increases linearly with time, showing a well-defined velocity V. We also performed several simulations with random mixtures of deep and rocky sites in different proportions. The behavior is the same, with velocities between these two values. Hence, we can consider these values as *upper and lower bounds for the velocity prediction* of the model in this particular case.

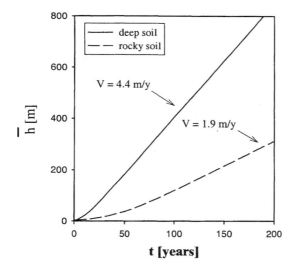

FIGURE 2   Mean front position as a function of time for *G. triacanthos* invasion in a dense forest of *L. ternifolia*.

TABLE 1   Comparison between IMCA model predictions of the invasion velocitiy with field data estimations.

Invader	Field Estimation [m/y]	Model Prediction [m/y]
G. triacanthos	2.5 − 4	1.9 − 4.4
L. lucidum	11 − 12.5	13.6

We also performed similar simulations, but considering the other invader *L. lucidum*. In this case we did not introduce soil variables, because this invader grows as well as the native in any type of soil. The results are qualitatively the same as in the previous case.

In table 1 we show a comparison between the predictions of the IMCA model with field data estimations made by using aerial photographs of the region [11].

In order to perform another comparison with field data values, we simulated the dynamics of both native species *L. ternifolia* and *F. coco*. In this case, as long as *F. coco* spreads into the area, cells behind the spreading front can be reoccupied by *L. ternifolia*. This leads, after some time, to a stationary situation of a mixed forest with a distribution of patches of both species. The stationary values of the population densities of both species are independent of the initial conditions, showing a proportion of 2.3 between both densities that is consistent with the observed value 2 ± 1 [3].

4   FRACTAL GROWTH OF INVASION FRONTS

As mentioned in the introduction, it is interesting to look beyond the rate of spread for alternative ways of characterizing the invasion process. If we look at the invasion front, we see that it has some structure (see fig. 1). To analyze this structure we studied the behavior of the average width of the front, which is proportional to the standard deviation:

$$\sigma(t) = \sqrt{\frac{1}{L_x} \sum_{x=1}^{L_x} \left(h(x,t) - \overline{h}(t) \right)^2}. \tag{5}$$

We calculated σ as a function of t and averaged this quantity over different initial conditions and different sequences of the random noise.

In figure 3(a) we show the temporal behavior of σ for different widths of the simulation area in a log-log plot. We see that, for any value of L_x, σ presents three distinct regimes: a transient period that is independent of L_x, a power-law regime, and a saturation regime, where it becomes constant. For large values of L_x the power-law regime also becomes independent of L_x, and it is characterized by a single exponent β. Both the crossover time τ from the power law to the saturation regimes and the saturation value depend on L_x. Moreover, it can be seen that for large values of L_x they present a power-law dependency, characterized by two exponents α and z.

All this phenomenology is characteristic of what is known as a *roughening process*.

Roughening refers to some nonequilibrium phenomena associated with the growth of certain types of interfaces between two different media. Growing interfaces appear in a variety of phenomena in nature, like the fluid motion in a porous media. If you put a drop of ink on a sheet of paper, you see that the patch grows with a well-defined interface between the wet and dry parts. Also, if you burn one edge of the paper, you see a propagating interface between the burned and the unburned parts. If you look at any one of these interfaces under a microscope, you see that it is not smooth, but rather irregular. Moreover, as the interface propagates, the sizes of the irregularities increase. It is said that the interface *roughens* [1].

There is not a complete theory based on first principles that explains dynamic roughening. However, most of the basic properties of these types of processes are well described by a general *phenomenological scaling approach*, which can be summarized in the following scaling relation [1]:

$$\sigma(L_x, t) \sim L_x^\alpha F\left(\frac{t}{L_x^z} \right) \tag{6}$$

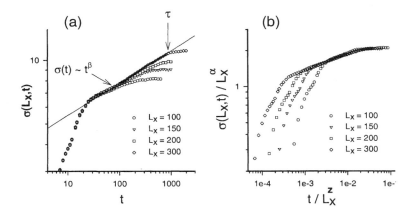

FIGURE 3   (a) Standard deviation of the invasion front as a function of t and the system width L_x. (b) Data collapse of the curves shown in (a) for $\alpha = 0.294$ and $z = 1.93$.

where the scaling function F can be represented by

$$F(x) \sim \begin{cases} x^\beta & \text{if } x \ll 1, \\ \text{constant} & \text{if } x \gg 1; \end{cases} \qquad (7)$$

and $z = \alpha/\beta$. This relation implies a crossover time $\tau \sim L_x^z$ such that $\sigma(L_x, t) \sim t^\beta$ for $t \ll \tau$ and $\sigma(L_x, t) \sim L_x^\alpha$ for $t \gg \tau$. In the last regime, the interface develops a *self-affine* structure; i.e., for a fixed time the profile $h(x)$ satisfies (in a statistical sense) the property $h(x) \sim b^{-\alpha} h(bx)$, for arbitrary values of the scale factor b [1]. From this property it follows that the local fractal or box dimension D of the profile is $D = 2 - \alpha$ for short-length scales (for long-length scales D always equals one) [15]. The data collapse displayed for long times in figure 3(b) shows that the scaling assumption (eq. (6)) holds for the invasion fronts with nontrivial values of the exponents α, β, and $z = \alpha/\beta$.

We then analyzed how variations in the life-history parameters influence the roughening process of the invasion front. It can be shown that the invasion velocity is mainly determined by two parameters: the mean dispersal distance d and the age of first reproduction t_m [3]. So we calculated the two independent exponents of the roughening process α and z for different combinations of values of d and t_m. We found that α is sensitive to variations in d and t_m, while z is

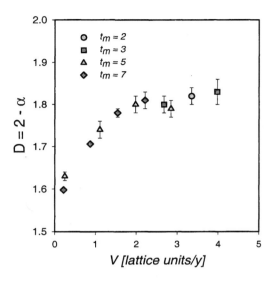

FIGURE 4 Parametric plot of the local fractal dimension *vs.* the invasion velocity.

almost constant, with a value near two for all the combinations of d and t_m that we checked.

In figure 4 we show a parametric plot of the local fractal dimension $D = 2 - \alpha$ *vs.* the invasion velocity V; every point in this graph corresponds to the calculated values of (α, V) for a particular pair of values of (d, t_m). We see that all points appear to fall into a *single curve* that saturates at a constant value around $D \approx 1.8$ for large values of V, showing that large invasion velocities can be expected for very rough fronts. This result also suggests that the local fractal dimension of the invasion front is a *single-valued, monotonic increasing function of the velocity*. But the local fractal dimension is something that, in principle, we could *measure* from a high-resolution aerial photograph. Hence, this result suggests that we would not need photographs taken at different times to estimate the invasion velocity; we could do that with just one photograph. In other words, we could make *predictions about the long-time dynamical behavior* from *pure geometrical properties* of a spatial pattern at a fixed time, at least in the case of an invader with short-range dispersal and in the presence of a strong native competitor.

5   LONG-RANGE SEED-DISPERSAL AND SPATIAL-PATTERN FORMATION

To analyze the influence of long-range seed dispersal, we simulated the spread of a species with long-range dispersal from a single focus; that is, we started the simulations with a single mature individual located at the center of a square area. To discriminate between the interaction and seed-dispersal effects we *neglected competition*; that is, we considered the spread of a single species in a clean area. We chose a reasonable set of life-history parameters (that is, a set of values inside the ranges of values for the different species considered), and we varied the exponent of the power-law seed-dispersal function.

In figure 5 we can see the difference between the typical spatial patterns generated by short- and long-range seed-dispersal distributions, where the parameter values are the same in both cases (figs. 5(a) and 5(b)). The short-range case is characterized by a single compact cluster with an almost circular shape, surrounded by a few isolated trees.

The long-range case presents a much more complex pattern. In the first years there is again a unique large cluster, which is more irregular and is surrounded by a few small clusters and a broad distribution of isolated trees. After a characteristic period of time that depends mainly on t_m, we observe the sudden appearance of a distribution of clusters of several sizes, including some large ones. This occurs because some of the trees located far away from the initial focus started to reproduce and formed new secondary focuses. As time goes on, the main cluster continues growing and absorbs neighboring clusters. This effect generates a very complex border structure that becomes self-similar (i.e., fractal) at large times, while new clusters of different sizes are being created continuously. The predicted pattern agrees qualitatively with those observed in *Cryptostegia grandiflora* (a shrub with seed dispersal by wind) in northeast Australia [7].

We calculated the fractal dimension D of the main cluster border using a box-counting procedure. In figure 6 we see D as a function of time for different values of the exponent α, half of them corresponding to distributions with infinite first moment and the other half with finite first moment. We see in all the cases that, after some transient period, D saturates into a constant value, which seems to be independent of α with a value around $D = 1.72 \pm 0.03$ for distributions with infinite first moment. The fact that physical properties of the system become independent of α, when α is such that the first moment diverges, is characteristic of systems with interactions that decay as $1/r^\alpha$ [2, 6].

Finally, in figure 7(a) we can see the typical time evolution of the normalized frequency $P(s)$ of clusters with area s, for a particular value of α. We also see that this quantity reaches a stationary state at long times. An approximation of the stationary distribution is displayed in figure 7(b). We see that the long-range quality of the basic interactions generates a complex distribution with crossovers between different power-law regimes, which are also expected to contain infor-

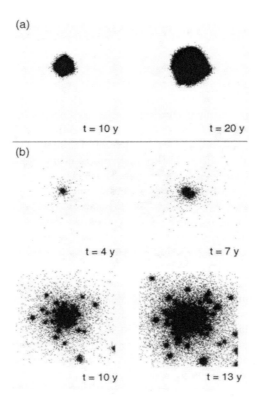

FIGURE 5   Comparison between spatial patterns generated by short- and long-range seed-dispersal strategies, in a simulation area of 200 × 200 cells and parameters values of *G. triacanthos*. (a) Exponential distribution function (eq. (2)). (a) Power-law distribution function (eq. (3)).

mation about the invasion process. This point deserves further investigation and some related work is in progress.

6   CONCLUSIONS

We have shown that the study of spatial pattern formation is indeed a valuable tool for the analysis and detection of several features associated with biological invasion processes. That is, we saw how stationary fractal patterns may be developed during different invasion processes and that the corresponding fractal dimensions contain information about them. In particular, we showed that the competition between exotic and native species leads to the appearance of a

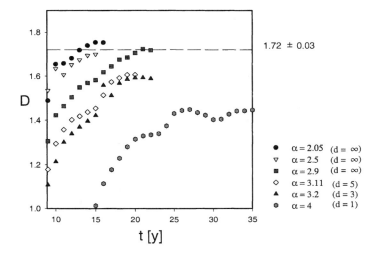

FIGURE 6   Fractal dimension of the main cluster border for different values of α and and $L_x = 1024$.

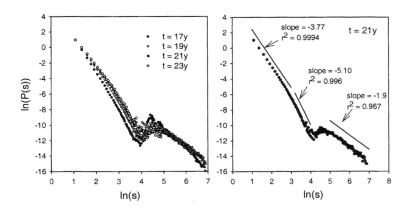

FIGURE 7   Normalized frequency of clusters with area s for $\alpha = 2.05$ and $L_x = 512$.

self-affine invasion front (as a result of a roughening process), whose local fractal dimension is proportional to the invasion velocity. This result suggests a possible technique for estimating invasion velocities through the analysis of spatial patterns in aerial photographs. But the suggestion of a constant value of the dynamical exponent z also opens the possibility of estimating the *long-time* behavior (i.e., the asymptotic velocity through the local fractal dimension D) by monitoring the *short-time* behavior of the growth exponent β of the standard deviation. Works along these lines are in progress.

Concerning the effect of long-range seed dispersal, we showed that it generates a very particular stationary pattern that may serve as an indirect way of detecting its presence. In this case, the fractal dimension D of the main cluster border increases as $\alpha \to 3^+$ and it seems to be independent of α when $2 < \alpha \le 3$. Of course, as α decreases, the different rates of spread increase (population growth is nonlinear in this case). Hence, we see again what could be a general rule; that is, large fractal dimensions appear to be associated with fast invasion processes.

Finally, the presence of several power-law regimes (and the corresponding crossovers between them) in the stationary cluster size distribution of a species with long-range seed dispersal suggests a nonextensive statistical formalism [14] (see also Montemurro [12]), as the appropriated framework for the analysis of this problem. Work along this line is also in progress.

ACKNOWLEDGMENTS

This research was supported by grants from Secyt—Universidad Nacional de Córdoba, Agencia Córdoba Ciencia, and CONICET (Argentina).

REFERENCES

[1] Barabási, A. L., and H. E. Stanley. *Fractal Concepts in Surface Growth.* Cambridge: Cambridge University Press, 1995.

[2] Cannas, S. A. "Phase Diagram of a Stochastic Cellular Automaton with Long-Range Interactions." *Physica A* **258** (1998): 32–44.

[3] Cannas, S. A., D. E. Marco, and S. A. Páez. "Modelling Biological Invasions: Species Traits, Species Interactions, and Habitat Heterogeneity." *Math. Biosci.* (2003): in press.

[4] Delcourt, D. A., and H. R. Delcourt. *Long Term Forest Dynamics of the Temperate Zone.* New York: Springer-Verlag, 1987.

[5] Elton, D. E. *The Ecology of Invasions by Animals and Plants.* London: Methuen and Company, 1958.

[6] Gleiser, P. M., S. A. Tamarit, and S. A. Cannas. "Self-Organized Criticality in a Model of Biological Evolution with Long-Range Interactions." *Physica A* **275** (2000): 272–280.

[7] Grice, A. C., I. J. Radford, and B. N. Abbot. "Regional and Landscape Patterns of Shrub Invasion in Tropical Savannas." *Biol. Invasions* **2** (2000): 187–205.

[8] Higgins, S. I., and D. M. Richardson. "Predicting Plant Migration Rates in a Changing World: The Role of Long-Distance Dispersal." *Amer. Natur.* **153** (1999): 464–475.

[9] Mack, R. N., D. Simberloff, W. M. Lonsdale, H. Evans, M. Clout, and F. A. Bazzaz. "Biotic Invasions: Causes, Epidemiology, Global Consequences and Control." *Ecol. Appl.* **10** (2000): 689–710.

[10] Marco, D. E., and S. A. Páez. "Invasion of *Gleditsia triacanthos* in *Lithraea ternifolia* Montane Forests of Central Argentina." *Envir. Mgmt.* **26** (2000): 409–419.

[11] Marco, D. E., S. A. Páez, and S. A. Cannas. "Species Invasiveness in Biological Invasions: A Modelling Approach." *Biol. Invasions* **4** (2002): 193–205.

[12] Montemurro, M. "A Generalization of the Zipf-Mandelbrot Law in Linguistics." This volume.

[13] Skellam, J. G., "Random Dispersal in Theoretical Populations." *Biometrika* **38** (1951): 196–218.

[14] Tsallis, C, G, Bemski, and R. S. Mendes. "Is Re-association in Folded Proteins a Case of Nonextensivity?" *Phys. Lett. A* **257** (1999): 93–98.

[15] Vicsek, T. *Fractal Growth Phenomena*. 2d ed. Singapore: World Scientific, 1992.

[16] Vitousek, P. M. "Biological Invasions and Ecosystem Processes: Toward an Integration of Population Biology and Ecosystem Studies." *Oikos* **57** (1990): 7–13.

Generalized Information Measures and the Analysis of Brain Electrical Signals

A. Plastino
M. T. Martin
O. Rosso

The traditional way of analyzing brain electrical activity, on the basis of electroencephalogram (EEG) records, relies mainly on visual inspection and years of training. Although it is quite useful, of course, one has to acknowledge its subjective nature that hardly allows for a systematic protocol. In order to overcome this undesirable feature, a quantitative EEG analysis has been developed over the years that introduces objective measures. These reflect not only characteristics of the brain activity itself, but also clues concerning the underlying associated neural dynamics. The processing of information by the brain is reflected in dynamical changes of the electrical activity in (i) time, (ii) frequency, and (iii) space. Therefore, the concomitant studies require methods capable of describing the qualitative variation of the signal in both time and frequency. In the present work we introduce new information tools based on the wavelet transform for the assessment of EEG data. In particular, different complexity measures are utilized.

Nonextensive Entropy—Interdisciplinary Applications
edited by Murray Gell-Mann and Constantino Tsallis, Oxford University Press

1   INTRODUCTION

The traditional electroencephalogram (EEG) tracing is now interpreted in much the same way as it was 50 years ago. More channels are used now and much more is known about clinical implication of the waves, but the basic EEG display and quantification of it are quite similar to those of its predecessors. The clinical interpretation of EEG records is made by a complex process of visual pattern recognition and the association with external and evident characteristics of clinical symptomatology. Analysis of EEG signals always involves the queries of quantification, i.e., the ability to state objective data in numerical and/or graphic form that simplify the analysis of long EEG time series. Without such measures, EEG appraisal remains subjective and can hardly lead to logical systematization [36].

Spectral decomposition of the EEG by computing the Fourier transform has been used since the very early days of electroencephalography. The rhythmic nature of many EEG activities lends itself naturally to this analysis. Fourier transform allows separation of various rhythms and estimation of their frequencies independently of each other, a difficult task to perform visually if several rhythmic activities occur simultaneously. Spectral analysis can also quantify the amount of activity in a frequency band. The results of EEG spectral analysis are often grouped into the traditional frequency bands, i.e., delta $0.5 - 4$ Hz, theta $4 - 7$ Hz, alpha $7 - 14$ Hz, beta $14 - 30$ Hz, and gamma $30 - 70$ Hz. There is much physiological and statistical evidence for independence of several of these bands, but their boundaries can vary a little according to the particular experiment being considered, and they can be adjusted as required [36].

Recently, oscillatory EEG activity has been discussed in relation to functional neuronal mechanisms. In this regard, it is of major interest to investigate how brain electric oscillations become synchronized in pathological or physiological brain states (e.g., epileptic seizures, sleep-wake stages), or by external and internal stimulation (event-related potentials [ERP] or evoked potentials [EP]). This issue can be addressed by applying system analysis methods to the EEG signals, because changes in EEG activity occur in temporal relation to triggering events, and could be thought of as transitions from disordered to ordered states (or vice versa).

The EEG can be regarded as reflecting the activity of ensembles of generators producing oscillations in several frequency ranges, which are active in a very complex manner [4, 5]. Upon stimulation, a resonance phenomenon occurs, and the generators begin to act together in a coherent way. This transition from a disordered to an ordered state gives rise to superimposed event-related oscillations in several frequency ranges [4, 5]. Although different methods have provided indirect evidence for synchronization in EEG processes [1], an adequate tool for a *quantitative evaluation* of both the complex EEG signal synchronization and its temporal dynamics is still lacking.

The application of the nonlinear dynamic techniques to the analysis of EEG time series can be justified, taking into account the characteristic behavior of the neuronal activity. The neurons are the constitutive elements of the brain. There are about 10^{11} elements and their interconnections are about 10^{14}. Because the neurons are highly nonlinear elements, we can expect that the EEG activity, which can be thought of as the result of the temporal and spatial average of the postsynaptic potentials, present similar nonlinear characteristics or sometimes chaotic behavior. The treatment of EEG series under the approach of nonlinear dynamic systems has opened new possibilities for understanding the brain dynamic. However, the aims are not limited to this, but include finding new ways to quantify differences in the EEG series that have some kind of clinical application.

In the characterization of the time evolution of the complex EEG dynamics, quantifiers based on nonlinear dynamics have been applied. They analyze (i) the temporal evolution of the EEG signal's complexity (associated with the measurement of the correlation dimension D_2 [15, 29]) and (ii) the degree of chaoticity (in terms of the largest Lyapunov exponent, Λ_{max} [26, 27]). One of the results of such analysis is that a transition from a rather complex behavior (of the neural network) to a simpler one can be detected at or even before epileptic seizure onset [26, 27, 29, 37]. However, despite the obvious physiological relevance of such findings, a basic requirement for using nonlinear dynamic metric tools (chaos theory) in conjunction with experimental data is that the relevant time series be stationary, which suggests that the time series is representative of a unique and stable attractor. Unfortunately, this is not the case with EEGs. To make the situation even worse, for the evaluation of D_2 and Λ_{max} (defined as asymptotic properties of the attractor), long time recordings are required. Moreover, these metric invariants require the computation of primary parameters that rapidly degrade with additive noise (for a review see Abarbanel [1], Başar [4, 5], and Elbert et al. [20]). Thus, the applicability of deterministic chaos ideas to EEG data remains highly controversial. The concomitant interpretation of data is still under discussion [30].

2 THE TOOLS

2.1 SPECTRAL ENTROPY

Alternatively, a natural approach to quantify the degree of order of a complex signal is to consider its spectral entropy, as defined from the Fourier power spectrum [41]. The spectral entropy is a measure of how concentrated or widespread the Fourier power spectrum of a signal is. An ordered activity represented by a sinusoidal signal appears as a narrow peak in the frequency domain. This concentration of the frequency spectrum in one single peak corresponds to a low-entropy value. On the other extreme, a disordered activity (e.g., the one generated by pure noise or by a deterministic chaotic system) will have a wide band response in the frequency domain that is reflected in higher entropies. However, the Fourier transform (FT) requires stationarity of the signal as well, and EEGs are highly nonstationary. Furthermore, the FT does not yield the time evolution of the pertinent frequency patterns. Consequently, the spectral entropy does not get defined as a function of time.

The disadvantages of the spectral entropy defined from the FT can be partially overcome by using a short-time Fourier transform (STFT). Powell and Percival [41] defined a time-evolving entropy from the STFT by using a Hanning window. With this approach, the FT is applied to time-evolving windows of a few seconds of data refined with an appropriate function, so that the time evolution of the frequencies can be followed. The stationarity requirement is partially satisfied by considering the signals as quasi-stationary for a few seconds. Due to the uncertainty principle [2, 18, 32, 50], one critical limitation appears when windowing data: if the window is too narrow, the frequency resolution will be poor. Conversely, if the window is too wide, the time localization will be less precise.

All of the above-mentioned difficulties can be overcome by appeal to the wavelet transform [2, 18, 32, 50], an efficient time-frequency decomposition method. In particular, the orthogonal discrete wavelet transform (ODWT) makes no assumptions about a record's stationarity. The only input needed is the time series itself. If the entropy is computed via the wavelet transform, the time evolution of frequency patterns can be followed with an optimal time-frequency resolution [10, 13, 14, 21, 44, 46, 49, 45]. The ensuing Shannon entropy form, based on the wavelet transform, is called the "Shannon wavelet entropy" (SWS). It reflects the degree of order/disorder of the signal. The wavelet entropy appears thus as a natural measure of order for EEG signals, more specifically of the synchrony of the group of cells involved in the different neural responses [10, 45, 49].

2.2   BEYOND SHANNON: GENERALIZED INFORMATION MEASURES

Some twelve years ago, Tsallis proposed a generalization of the celebrated Boltzmann-Gibbs entropic measure [39, 56, 57]. The new entropy functional introduced by Tsallis [56], along with its associated generalized thermostatistics, is nowadays being hailed as the possible basis of a theoretical framework appropriate to deal with nonextensive settings. This entropy has the form

$$S_T^{(q)}[\,p\,] \equiv S_q = \frac{1}{q-1} \sum_j [\,p_j - (p_j)^q\,], \tag{1}$$

where p is a discrete probability distribution and the entropic index q is any real number. This entropy recovers the standard Boltzmann-Gibbs-Shannon entropy (the subindex stands for Shannon)

$$S_S[\,p\,] = -\sum_j p_j \ln p_j \tag{2}$$

in the limit $q \to 1$. The entropy S_q is *nonextensive*, for example,

$$S_q(\ A + B\ )\ =\ S_q(A)\ +\ S_q(B)\ +\ (1 - q) \cdot S_q(A) \cdot S_q(B)\ , \tag{3}$$

where A and B are two systems, independent in the sense that $p(A + B) = p(A) \cdot p(B)$. It is clear that q measures the degree of nonextensivity.

Many relevant mathematical properties of the standard thermostatistics are preserved by Tsallis' formalism or admit natural generalizations. Tsallis' proposal was shown to be consistent both with Jaynes' information theory formulation of statistical mechanics [40], and with the dynamical thermostatting approach to statistical ensembles. The application of Tsallis' theory to an increasing number of physical problems provides a picture of the kinds of scenarios where the new formalism is useful. Tsallis' bold attempt to develop a complete thermostatistical formalism on the basis of a nonlogarithmic entropy function has raised many interesting issues related both to the mathematical structure and physical implications of general thermostatistical formalisms [38].

2.3   ESCORT DISTRIBUTIONS

Tsallis' pioneering work has stimulated the exploration of the properties of other generalized or alternative information measures. Moreover, it has been recently realized that some important features are shared by extended families of thermostatistical formalisms [38]. In Martin et al. [33] it was shown that a new information measure of the Tsallis kind, originally introduced by Di Sisto [19], seems to be more sensitive to a variation of relevant parameters of complex systems than the original measure in eq. (1). The new measure is indeed Tsallis' original one, but expressed in terms of escort distributions of the order q, as discussed below.

Let us briefly review the useful concept of escort probabilities (see Beck and Schlögl [7] and references therein). To such an end, one introduces the following transformation between an original, normalized probability distribution r and a new, related probability distribution R, namely, $r \to R$ with

$$R_i\ =\ \frac{(\ r_i\ )^q}{\sum_j\ (\ r_j\ )^q}\ , \tag{4}$$

q being any real parameter (here we identify it with Tsallis' index). We reiterate: r is the *original* probability distribution of concern. For $q = 1$ we have $R \equiv r$ and, obviously, R is normalized to unity. General global quantities formed with escort distributions of different order q, such as the different types of information or mean values, will often give more revealing information than those formed with the original distribution only. Changing q is quite a useful device for scanning the structure of the original distribution [7]. Starting now with Tsallis' information measure constructed with some probability distribution r, one may think of computing the associated Tsallis measure that results from replacing r by R.

This was investigated in Di Sisto et al. [19]. If, in terms of r, one casts eq. (1) in the fashion

$$S_q = \frac{1}{q-1} \sum_i [\, r_i - r_i^q \,],$$ (5)

then the associated escort Tsallis measure, as a function of the *original* measure, reads

$$S_q^{\text{esc}} = \frac{1}{q-1} \left\{ 1 - \left[\sum_i r_i^{1/q} \right]^{-q} \right\}.$$ (6)

We will here adapt for our EEG purposes the measure introduced by Di Sisto et al. [19], with reference to escort distributions. In our present notation the escort probability distribution of order q reads

$$P_i = \frac{(\, p_i \,)^q}{\sum_j (\, p_j \,)^q},$$ (7)

while the associated escort Tsallis information measure acquires the appearance (the subindex G stands for "generalized")

$$S_G^{(q)}[\, P \,] = \frac{1}{q-1} \left\{ 1 - \left(\sum_i P_i^{1/q} \right)^{-q} \right\}.$$ (8)

The entropic functional (8) is a legitimate measure in its own right, whose properties can be studied without reference to the original probability set $\{p_i\}$. It reduces to the standard Boltzmann-Gibbs entropy for $q \to 1$. Note also that $S_G^{(q)}[P] = S_T^{(q)}[p]$. The entropy functional (8) verifies Tsallis' q-generalized additivity law. As stated above, the generalized escort Tsallis entropy measure $S_G^{(q)}$ has proved to be useful on the analysis of some aspects of EEG signals [33]. We must stress here that the scenario of Martin et al. [33] is quite different from the one to be addressed here. As for the methodology, no spectral entropies were used. In particular, the probability distribution is evaluated from the signal amplitudes in Martin et al. [33]. The ensuing entropic quantifier is useful for studying and detecting morphological changes (such as, spikes) [33]. Here, instead, we are interested in the tonic-clonic transition and in the characterization of the dynamic associated with the epileptic seizure, topics that were not even mentioned in Martin et al. [33].

2.4   SHINER-DAVISON-LANDSBERG COMPLEXITY MEASURE

Complexity is a measure of off-equilibrium "order." It refers to nonequilibrium structures that arise spontaneously in certain situations. This type of "order" is not the one associated, for instance, with crystal structures, for which the entropy is very small. Biological life is a typical example of the kind of "new" order one has in mind here, associated with relatively large entropic values. Current definitions of *complexity measures* can be divided into three categories [53]. The measure can either (a) grow with increasing disorder (decrease as order increases), (b) be quite small for large amounts of the degree of either order or disorder, with a maximum at some intermediate stage, or (c) grow with increasing order (decrease as disorder increases).

Shiner, Davison, and Landsberg (SDL) have recently proposed a *measure of complexity*, based on appropriately defined notions of order and disorder, which has a considerable degree of flexibility in its dependence on these concepts [53]. We will choose the parameters of the SDL complexity measure so that it belongs to the second category mentioned above. Their measure is easy to calculate and behaves like an intensive thermodynamical quantity. The possible functional dependencies of the SDL measure encompass those of many earlier definitions of complexity [53]. SDL define, for a given probability distribution $\{p_i\}$ and for its associated information measure I, an amount of "disorder" H in the fashion $H = I/I_{\max}$, where $I_{\max} = I\{$uniform probability distribution$\}$. Obviously, the associated definition of "order" reads $\Omega = 1 - H$. Both Ω and H lie between 0 and 1. Among the members of the SDL complexity measures (characterized by two parameters α, β) [53] we choose here to work with the $\alpha = \beta = 1$ instance. One has

$$C^{(\mathrm{SDL})} = 4\,H\cdot(1 - H) = 4\,\Omega\cdot(1 - \Omega)\,. \tag{9}$$

Note that the SDL complexity measure is calculated from the normalized information measure, or entropy. In our case, it is calculated from the normalized total wavelet entropy, an entropy of the distribution over different scales (see below) [47]. One could raise the objection that $C^{(\mathrm{SDL})}$ is just a simple function of the entropy. As a consequence, it might not contain new information vis-à-vis the measure of order. Such an objection is discussed at length in Binder and Perry [8], Crutchfield et al. [16], and Shiner et al. [52].

2.5 LÓPEZ-RUIZ-MANCINI-CALBET COMPLEXITY MEASURE

López-Ruiz, Mancini, and Calbet (LMC) have recently proposed a measure of complexity based on the notion of "disequilibrium" [12, 31]. The LMC measure is easy to calculate, as it is evaluated in terms of common concepts of statistical mechanics. The definition of the LMC complexity measure reads

$$C^{(\mathrm{LMC})} = Q\cdot H\,, \tag{10}$$

where Q stands for the so-called "disequilibrium" [31] and H has been defined above ($0 \leq H \leq 1$). Following LMC [31] we define, in addition to H, the disequilibrium Q. This is a "distance" in probability space. It measures "how far" $\{p_i\}$ is located, in this space, from the uniform distribution p_e that characterizes equilibrium in Gibbs' statistical mechanics: $p_e = 1/N$. One has, for the disequilibrium

Q,

$$Q = \sum_j [\, p_j - p_e \,]^2 \,. \qquad (11)$$

Note that, since $0 \leq H \leq 1$ and $0 \leq Q \leq (N-1)/N$, the complexity $C^{(\mathrm{LMC})}$ is normalized as well; that is, $0 \leq C^{(\mathrm{LMC})} \leq 1$. It should be noticed that the LMC complexity *is not a trivial function of the entropy*, in the sense that, for a given H value, there exists a range of complexities between a minimal value $C_{\min}$ and a maximal value $C_{\max}$ [3, 12]. Thus, evaluating the complexity provides one with important additional information regarding the peculiarities of a probability distribution.

Landsberg and co-workers have shown that their definition of complexity encompasses the LMC one [53]. Note that the LMC complex definition [31] involves the disequilibrium Q. Alternatively, instead of using the quadratic Euclidean distance between the probability distributions $\{p_i\}$ and $\{p_e\}$, one could use the fact that Q is proportional to $S(p|p_e)$, the Kullback-Leibler cross information (relative entropy) [28]. The Kullback-Leibler relative entropy for the two probability distributions $\{p_i\}$ and $\{q_i\}$ is given by

$$S(p|q) = \sum_j p_j \cdot \ln\left(\frac{p_j}{q_j}\right) \,, \qquad (12)$$

and ascertains to what extent both probability distributions are alike. Taking $Q \sim S(p|p_e) = (1-H) \cdot \ln N$, it is easy to see that, starting from the LMC definition, we arrive at the SDL one. Summing up, both definitions of complexity are strongly linked.

In statistical mechanics, one is usually interested in isolated systems characterized by an initial, arbitrary, and discrete probability distribution. The evolution toward equilibrium will be described. At equilibrium, the distribution is the equiprobability one. To study the time evolution of the complexity, a diagram of C versus time t should then be used. But, as we know, the second law of thermodynamics states that entropy grows monotonically with time ($dH/dt \geq 0$). This implies that an equivalent manner of studying the temporal evolution of the complexity can be obtained by plotting $C^{(\mathrm{LMC})}$ versus H. In this way, the normalized entropy substitutes for the time. LMC have shown that, for an isolated system evolving in time, the complexity measure cannot attain any arbitrary value in a $C^{(\mathrm{LMC})}$ versus H map. The value $C^{(\mathrm{LMC})}$ must always stay within the bounds $C_{\min}$ and $C_{\max}$ [3, 12]. These are the maximum and minimum possible values of $C^{(\mathrm{LMC})}$, given H. A procedure for the evaluation of these two curves, for a given value of N, is given in Calbet and López-Ruiz [12].

2.6   OUR GOAL

It is our objective here to undertake an entropic-complexity analysis of EEGs by using the three measures, (i) Shannon, (ii) Tsallis, and (iii) generalized escort

Tsallis for the information I. It is our hope that such a mixture of measures and tools will be useful. The motivation for developing it is rather obvious: we confront highly nonlinear, long-range (for interneuronal distances) phenomena. The preconditions for the success of a nonextensive treatment are all present. Interesting results will ensue, in particular with regards to transitions in tonic-clonic epileptic seizures.

3 CLINICAL DATA AND EXPERIMENTAL SETUP

A scalp EEG signal is, essentially, a nonstationary time series that presents artifacts due to electrooculograms (EOG), electromyograms (EMG), and electrocardiograms (ECG), among others [36]. Artifacts related to muscle contractions are especially troublesome in the case of tonic-clonic epileptic seizures, where they reach very high amplitudes that contaminate the whole seizure recording. Sometimes artifacts are present during a just a few seconds and can be obviated because they obscure only a small portion of the EEG. In other cases, almost the total signal is obscured by them, and very little information about the underlying brain activity can be extracted. An example of this kind of scalp EEG signal is one that corresponds to an epileptic tonic-clonic seizure [36]. A tonic-clonic (TC) seizure is characterized by violent muscle contractions. Initial massive tonic spasms are replaced seconds later by the clonic phase with violent flexor movements and characteristic rhythmic spasms toward the ending of the seizure. In these seizures, artifacts related to muscle contractions are especially troublesome because they reach very high amplitudes [36]. In fact, not only do they limit the traditional visual analysis to the pre- and post-ictal periods, but they also restrict the application of some mathematical methods. Analysis of the brain activity during this seizure has been previously performed only in special circumstances, such as in patients treated with curare (an inhibitor of the muscle responses) [22, 23] or by eliminating the high-frequency muscle activity with the use of traditional filters [24]. Gastaut and Broughton [22] described a frequency pattern during a tonic-clonic epileptic seizure from patients with muscle relaxation from curarization and artificial respiration. After a short period (which may be as short as 1 to 3 seconds) characterized by phase desynchronization, they found an "epileptic recruiting rhythm" at about 10 Hz [23] with a rapidly increasing amplitude dominating the EEG; later, as the seizure ends, there is a progressive increase of the lower frequencies associated with the clonic phase. About 10 seconds after the seizure onset, lower frequencies of delta and theta (0.5 to 3.5 Hz) are observed gradually diminish their activity. The clonic activity is accompanied by generalized polyspike bursts at each myoclonic jerk. Very slow irregular delta activity then dominates the EEG, accompanied by a gradual frequency increase of the theta (3.5 to 7.5 Hz) and alpha bands (7.5 to 12.5 Hz), indicative of the end of the seizure.

In figure 1 we present a scalp EEG signal, corresponding to a tonic-clonic epileptic seizure recorded in a central right location ($C4$ channel). We chose this electrode, after visual inspection of the pertinent EEG records, as the one with the minimum amount of artifacts. The signal was digitized at 409.6 Hz through a 12-bit A/D converter and filtered with an antialiasing, eight-pole, lowpass Bessel filter with a cutoff frequency of 50 Hz. Afterward, the signal was digitally filtered with a Butterworth filter with a bandwidth of 1 to 50 Hz and stored, after decimation, at 102.4 Hz on a PC hard drive. Recordings were performed under video control in order to have an accurate determination of the different stages of the seizure. The different stages of the EEG signals were determined by a team of physicians. The epileptic seizure starts at 80 seconds, with a "discharge" of slow waves superposed by fast ones with lower amplitude. This discharge lasts approximately 8 seconds and has a mean amplitude of 100 μV. Afterward, the seizure spreads, making the analysis of the EEG more complicated due to muscle artifacts; however, it is possible to establish the beginning of the clonic phase at around 125 seconds, and the end of the seizure at 155 seconds, where there is an abrupt decay of the signal's amplitude.

4   WAVELET ANALYSIS

4.1   WAVELET TRANSFORM

Wavelet analysis is a method which relies on the introduction of an appropriate basis and a characterization of the signal by the distribution of amplitude in this basis. If the basis is required to be a proper orthogonal basis, any arbitrary function can be uniquely decomposed and the decomposition can be inverted [2, 18, 32, 50]. Wavelet analysis is a suitable tool for detecting and characterizing specific phenomena in time and frequency planes. The *wavelet* is a smooth and quickly vanishing oscillating function with good localization in both frequency and time. A *wavelet family* $\psi_{a,b}$ is the set of elementary functions generated by dilations and translations of a unique admissible *mother wavelet* $\psi(t)$,

$$\psi_{a,b}(t) \; = \; |a|^{-1/2} \psi \left(\frac{t-b}{a} \right) \; , \qquad (13)$$

where $a, b \in \mathcal{R}$, $a \neq 0$ are the scale and translation parameters respectively, and t is the time. As a increases, the wavelet becomes narrower. Thus, one has a unique analytic pattern and can see its replications at different scales, with variable time localization.

The *continuous wavelet transform* (CWT) of a signal $\mathcal{S}(t) \in L^2(\mathcal{R})$ (the space of real square summable functions) is defined as the correlation between the function $\mathcal{S}(t)$ with the family wavelet $\psi_{a,b}$ for each a and b:

$$(W_\psi \mathcal{S})\,(a,b) \; = \; |a|^{-1/2} \int_{-\infty}^{\infty} \mathcal{S}(t)\,\psi^* \left(\frac{t-b}{a} \right) \, dt \; = \; \langle\, \mathcal{S},\, \psi_{a,b}\, \rangle. \qquad (14)$$

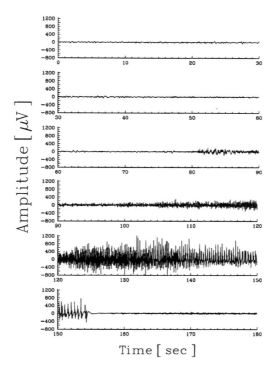

FIGURE 1   Scalp EEG signal for an epileptic tonic-clonic seizure, recorded at central right location ($C4$). The seizure starts at 80 seconds and the clonic phase at 125 seconds. The seizure ends at 155 seconds.

For special election of the mother wavelet function $\psi(t)$ and for the discrete set of parameters $a_j = 2^{-j}$ and $b_{j,k} = 2^{-j}k$, with $j, k \in \mathcal{Z}$ (the set of integers), the family

$$\psi_{j,k}(t) \;=\; 2^{j/2}\, \psi(\, 2^j\, t \,-\, k\,), \qquad j,\, k \,\in\, \mathcal{Z}\,, \tag{15}$$

constitutes an orthonormal basis of the Hilbert space $L^2(\mathcal{R})$ consisting of finite-energy signals.

The wavelet transform (DWT) provides a nonredundant representation of the signal, and the values $\langle\, \mathcal{S},\ \psi_{a,b}\, \rangle$ constitute the coefficients in a wavelet series. These wavelet coefficients provide (i) relevant information in a simple way and (ii) a direct estimation of local energies at the different scales. Moreover, the information can be organized in a hierarchical scheme of nested subspaces called multiresolution analysis in $L^2(\mathcal{R})$. In the present work, we employ orthogonal cubic spline functions as mother wavelets. Among several alternatives, cubic spline functions are symmetric and combine smoothness in suitable proportions to numerical advantages (for a complete and very good discussion of these topics, see Thévenaz et al. [55] and Unser [58]). These functions have become a useful tool for representing natural signals.

In what follows, the signal is assumed to be given by the sampled values $\mathcal{S} = \{s_0(n), n = 1, \cdots, M\}$, corresponding to an uniform time grid with sampling time t_s. For simplicity, the sampling rate is taken as $t_s = 1$. If the decomposition is carried out over all resolution levels, the wavelet expansion will read ($N = \log_2(M)$)

$$\mathcal{S}(t) = \sum_{j}^{-1} \sum_{k} C_j(k)\, \psi_{j,k}(t) = \sum_{j}^{-1} r_j(t) \,, \tag{16}$$

where the wavelet coefficients $C_j(k)$ can be interpreted as the local residual errors between successive signal approximations at scales j and $j + 1$, and $r_j(t)$ is the *residual signal* at scale j. It contains the information of the signal $\mathcal{S}(t)$ corresponding to the frequencies $2^{j-1}\omega_s \leq |\omega| \leq 2^j\omega_s$.

4.2 WAVELET TRANSFORM AND SIGNAL SEPARATION

Natural phenomena produce time series that, usually, contain, in addition to the desired "clean" signal, (i) effects of contamination by the environment through which the signal passes on its way from the source to the measurement device or (ii) contamination by properties of the measurement process itself. In many applications one wishes for a separation between signal and noise since, otherwise, when nonlinear invariants or other quantifiers are evaluated, the contaminating noise can yield spurious results, and the associated values will either underestimate or overestimate the real values of the system's quantifiers under study.

The elimination of the high-frequency muscle activity (in the frequency range given by frequency bands B_1 and B_2, see section 5) with the use of traditional filters has some disadvantages [34, 35, 54]: (a) filtering frequencies related to muscle artifacts also affect the morphology of the remaining ones and, (b) the filtering process alters the nonlinear metric invariants. To overcome the above limitations, a signal separation based on orthonormal wavelets was used. When noise (signal contamination) is present only in specific frequency bands, a filtering process or signal separation based on an orthonormal wavelet transform can be efficiently implemented. Supposing, as in the case of tonic-clonic seizures, that we are interested in eliminating high-frequency noise. Using eq. (16) at a j_0-scale, we can obtain a smoothed version of the original signal by recourse to

$$\widetilde{\mathcal{S}}(t) = \sum_{j=-N}^{-j_0} r_j(t) \,. \tag{17}$$

This smoothed signal has a lesser number of high frequencies as compared to the $(j_0 + 1)$ level, but it possesses just half the amount of data as that of the previous level. The original sample ratio can be obtained by a cubic spline interpolation.

One of the main reasons for using a signal separation method based on orthogonal wavelets lies in the fact that it enables one to analyze the "de-noised" signal without seriously impairing the interpretation of the associated dynamics. Mallat [32] proved that, for a one-dimensional signal denoising process (assuming additive noise), an orthogonal wavelet-based method is better than a method based on Fourier transforms, because wavelets do not change the original signal. The main advantage of signal separation, with orthonormal wavelets, lies in the fact that the morphology of the nonfiltered frequencies is not affected, and, therefore, the dynamics associated with these nonfiltered frequencies do not change either.

4.3   RELATIVE WAVELET ENERGY

Since the family $\{\psi_{j,k}(t)\}$ is an *orthonormal* basis for $L^2(\mathcal{R})$, the concept of energy is linked with the usual notions derived from Fourier's theory. The wavelet coefficients are given by $C_j(k) = \langle S, \psi_{j,k} \rangle$ and the energy, at each resolution level $j = -1, \ldots, -N$, will be the energy of the detail signal

$$E_j = \|r_j\|^2 = \sum_k |C_j(k)|^2 . \tag{18}$$

The total energy can be obtained in the fashion

$$E_{\text{tot}} = \|S\|^2 = \sum_j \sum_k |C_j(k)|^2 = \sum_j E_j . \tag{19}$$

Finally, we define the normalized p_j-values, which represent the *relative wavelet energy*

$$p_j = \frac{E_j}{E_{\text{tot}}} \tag{20}$$

for the resolution levels $j = -1, -2, \ldots, -N$. The p_j yield, at different scales, the probability distribution for the energy. Clearly, $\sum_j p_j = 1$ and the distribution $\{p_j\}$ can be considered as a time-scale density that constitutes a suitable tool for detecting and characterizing specific phenomena in both the time and the frequency planes.

4.4   WAVELET ENTROPY

The logarithmic Shannon entropy (2) [51] gives a useful criterion for analyzing and comparing probability distributions. It provides a measure of the information contained in any distribution. We define the Shannon wavelet entropy (SWS) [10, 46, 49] as

$$S_S = -\sum_j p_j \cdot \ln[\, p_j \,] . \tag{21}$$

The SWS appears as a measure of the degree of order/disorder of the signal. It provides useful information about the underlying dynamical process associated with the signal. Indeed, a very ordered process can be represented by a periodic monofrequency signal (a signal with a narrow band spectrum). A wavelet representation of such a signal will be resolved at one unique wavelet resolution level, i.e., all relative wavelet energies will be (almost) zero except at the wavelet resolution level, which includes the representative signal frequency. For this special level the relative wavelet energy will (in our chosen energy units) almost equal unity. As a consequence, the Shannon wavelet entropy will acquire a very small, vanishing value. A signal generated by a totally random process can be taken as representative of a very disordered behavior. This kind of signal will have a wavelet representation with significant contributions coming from all frequency bands. Moreover, one could expect that all contributions will be of the same order. Consequently, the relative wavelet energy will be almost equal at all resolution levels, and the Shannon wavelet entropy will acquire its maximum possible value.

We now need the nonextensive counterparts of Shannon's wavelet entropy. For the Tsallis measure we obviously have a Tsallis wavelet entropy (TWS) [46] given by

$$S_T^{(q)} = \frac{1}{(q-1)} \sum_j [\, p_j - (\, p_j \,)^q \,] \,, \tag{22}$$

while, for the generalized escort Tsallis information measure, the pertinent generalized escort Tsallis wavelet entropy (GWS) [46] is

$$S_G^{(q)} = \frac{1}{(q-1)} \left\{ 1 - \left[\sum_j (\, p_j \,)^{1/q} \right]^{-q} \right\} . \tag{23}$$

4.5   WAVELET COMPLEXITY

Adapting the language of SDL and LMC [31, 53] to wavelet parlance, we define, for the above wavelet energy probability distribution $\{p_i\}$, an amount of "disorder" $H_\kappa^{(q)}$ (*for a normalized "κ" wavelet entropy*) [47] in the fashion

$$H_\kappa^{(q)} = \frac{S_\kappa^{(q)}}{S_\kappa^{\max}} \tag{24}$$

where the subindex κ denotes one of the three different functional forms described above for the evaluation of the information measure (S for Shannon, T for Tsallis, and G for generalized escort Tsallis measure). $S_\kappa^{\max}$ is the maximum possible entropy value. The number of wavelet resolution levels included in such an evaluation is called N_J. For Shannon's functional form we have $q = 1$ and $S_S^{\max} = \ln N_J$. For both the Tsallis and the generalized escort-Tsallis functional forms, one has $S_T^{\max} = S_G^{\max} = [1 - (N_J)^{(1-q)}]/(q-1)$.

Obviously, the associated definition of "order" reads $\Omega_\kappa^{(q)} = 1 - H_\kappa^{(q)}$. Both $\Omega_\kappa^{(q)}$ and $H_\kappa^{(q)}$ lie between 0 and 1. Our SDL *normalized κ wavelet complexity* thus reads [47]

$$C_\kappa^{(\text{SDL})}(q) \;=\; 4\,H_\kappa^{(q)} \cdot \left(1 \,-\, H_\kappa^{(q)} \right) \;=\; 4\,\Omega_\kappa^{(q)} \cdot \left(1 \,-\, \Omega_\kappa^{(q)} \right) \tag{25}$$

while the LMC *normalized κ wavelet complexity* [47] is

$$C_\kappa^{(\text{LMC})}(q) \;=\; Q \cdot H_\kappa^{(q)} \tag{26}$$

where, again, the index κ denotes one of three options, namely, S for normalized Shannon wavelet complexity (NSWC), T for normalized Tsallis wavelet complexity (NTWC), and G for normalized generalized escort Tsallis complexity (NGWC).

4.6   WAVELET QUANTIFIERS OF TIME EVOLUTION

In the case of a diadic wavelet decomposition, the number of wavelet coefficients at resolution level j is two times smaller than at the previous, $j-1$, one. In order to follow the temporal evolution of the above-defined quantifiers (SWS, TWS, GWS, NSWC, NTWC, NGWC), the wavelet coefficient series of the analyzed signal are divided into nonoverlapping temporal windows of length L and, for each interval i ($i = 1, \cdots, N_T$, with $N_T = M/L$), appropriate quantifier signal-values are assigned to the central point of the time window. The minimum length of the temporal window will, therefore, include at least one wavelet coefficient at each level.

The wavelet energy at resolution level j for the time window i is given by

$$E_j^{(i)} \;=\; \sum_{k=(i-1)\cdot L+1}^{i\cdot L} |C_j(k)|^2 \qquad \text{with } i = 1, \cdots, N_T \,, \tag{27}$$

while the total energy in this time window will be

$$E_{\text{tot}}^{(i)} \;=\; \sum_{j<0} E_j^{(i)} \,. \tag{28}$$

As for the temporal behavior we have:

1. for the relative wavelet energy

$$p_j^{(i)} \;=\; \frac{E_j^{(i)}}{E_{\text{tot}}^{(i)}} \,, \tag{29}$$

TABLE 1   Frequency boundaries (in Hz) associated with the different resolution wavelet levels j, and associated time resolution ΔT (in s), with sample-frequency $\omega_s = 102.4$ Hz. The traditional EEG frequency bands correspond to the following frequencies: δ (0.5 to 3.5 Hz); θ (3.5 to 7.5 Hz); α (7.5 to 12.5 Hz); β (12.5 to 30 Hz); and γ (greater than 30 Hz).

Notation	Wavelet Band				EEG Band
	$\omega_{\min}$	$\omega_{\max}$	j	ΔT	
B_1	25.6	51.2	-1	0.0195	$\beta,\ \gamma$
B_2	12.8	25.6	-2	0.0391	β
B_3	6.4	12.8	-3	0.0781	$\theta,\ \alpha$
B_4	3.2	6.4	-4	0.1562	θ
B_5	1.6	3.2	-5	0.3125	δ
B_6	0.8	1.6	-6	0.6250	δ

2. for the Shannon wavelet entropy

$$S_S(i) = -\sum_j p_j^{(i)} \cdot \ln\left[p_j^{(i)} \right] , \tag{30}$$

3. for the Tsallis wavelet entropy

$$S_T^{(q)}(i) = \frac{1}{(q-1)} \sum_j \left[p_j^{(i)} - \left(p_j^{(i)} \right)^q \right] , \tag{31}$$

4. for the generalized escort Tsallis wavelet entropy

$$S_G^{(q)}(i) = \frac{1}{(q-1)} \left[1 - \sum_j \left(p_j^{(i)} \right)^{1/q} \right]^{-q} , \tag{32}$$

and,

5. for the normalized κ wavelet complexity

$$C_\kappa^{(\mathrm{SDL})}(q,i) = 4\, H_\kappa^{(q)}(i) \cdot \left(1 - H_\kappa^{(q)}(i) \right) , \tag{33}$$

$$C_\kappa^{(\mathrm{LCD})}(q,i) = Q(i) \cdot H_\kappa^{(q)}(i) , \tag{34}$$

with $\kappa = S,\ T,\ G$ and

$$Q(i) = \sum_j \left[p_j^{(i)} - p_e \right]^2 . \tag{35}$$

5 RESULTS AND DISCUSSION

The stationarity of the experimental time series data is the basic requirement for using nonlinear dynamics metric tools (chaos theory): the time series should be representative of a unique and stable attractor. Also, for the evaluation of both the correlation dimension D_2 and the maximum Lyapunov exponent $\Lambda_{\max}$, long time recordings are required, because they are defined as asymptotic properties of the attractor. With the help of D_2 and $\Lambda_{\max}$, different brain states can be characterized. One must perform static measurements of these quantities for selected portions of the brain (EEG time series) such that all the necessary mathematical requirements are fulfilled [9, 11, 45]. Unfortunately, for long recordings of neural mass activity we cannot assume a stable attractor and, therefore, we cannot straightforwardly arrive at significant conclusions concerning the characterization of the dynamics. Moreover, one is often interested in different state transitions. The times at which these transitions take place, however, are not sharply defined. In these cases one can relax the stationary requirement of the time series and introduce new quantifiers: the *dimensionality* and the *chaoticity*, formally are defined as D'_2 [15, 29] and $\Lambda'_{\max}$ [26, 27], respectively. They are not equivalent, because of the violation of the basic mathematical hypothesis of stationarity. Anyway, computing their values using sliding time-windows (with or without overlapping), the EEG-associated dynamics can be accessed and valuable information may be obtained for comparative purposes (this can also be used in order to characterize state transitions).

EEG spectral analysis is traditionally performed by studying different frequency bands with well-defined boundaries. Some small variations can be found, according to the particular experiment under consideration. Absolute and relative intensities of these bands are usually analyzed and correlated with different pathologies. In this work we define six frequency bands for an appropriate wavelet analysis within the multiresolution scheme to be used. We denote these band-resolution levels by B_j ($|j| = 1, \cdots, 6$). Their frequency limits, time resolution, as well as their correspondence with traditional EEG frequency bands, are given in table 1. Note that the coefficients were nonoverlapping for each scale or frequency band. In order to make a behavior's "quantification" we divided the total signal into time-window intervals. The minimum width of the time window would be one which includes at least one wavelet coefficient of the lowest frequency band. In the present study the time-window width employed contained 256 data = 2.5 seconds.

If the frequency bands B_1 and B_2, at wavelet resolution levels $j = -1$ and -2, contain high-frequency artifacts related to muscular activity that blur the EEG, then their contributions in the evaluation of EEG quantifiers will be considered null. Although high-frequency brain activity is thereby also eliminated, its contributions during the ictal stage are not as important as it is in the middle and low frequencies. This has been conclusively demonstrated [43, 48]. Once the high-frequency artifacts are eliminated, we can analyze the time evolution of the chaoticity and quantifiers based on ODWT [45, 46, 47].

5.1 CHAOTICITY

The dynamical evolution associated with a time series can be studied by evaluating, in a continuous fashion, just one relevant dynamical variable, like the largest Lyapunov exponent. One divides the time series into time portions of L *data*-length and, for each interval, evaluates the largest Lyapunov exponent [26, 27] and assigns the resulting value to the "beginning" of the time window. Of course, we must ensure that in each time window the time series is stationary. It remains unclear just how long the brain remains in a specific state that would produce a stationary signal governed by a unique attractor. The EEG time series is not able to remain stationary over intervals long enough to yield sufficient data for a reliable estimation of $\Lambda_{\max}$. Thus, in order to characterize the dynamical evolution of a nonstationary signal like the EEG, one can introduce a new parameter called *chaoticity*, which is formally defined in the same manner as the largest Lyapunov, save for the fact that the stationarity constraints are removed. With reference to the present case of scalp EEG tonic-clonic records, evaluation of quantifiers based on nonlinear metrics tools requires that muscular activity be previously removed, as shown in Rosso and Mairal [45]. The chaoticity for the "clean" EEG signal (signal without contributions of wavelet frequency bands B_1 and B_2) was evaluated for sliding time windows of length L=2560 data = 25 seconds with an overlapping of $\delta = 256$ data = 2.5 seconds [45]. The parameters for attractor reconstruction were $\tau = 3 \cdot t_s$ and $D_e = 10$. They were taken to be equal for all time windows, and were chosen in such a way as to obtain the best possible pre-ictal and ictal attractor reconstruction [45]. The time evolution of the chaoticity is displayed in figure 2, in which we see that the chaoticity indicator seems to detect relevant changes in the signal.

Note that fluctuations in the chaoticity constitute advance signatures of morphological changes in the signal. They can be regarded as indicators of changes in the associated dynamics.

After inspection of figure 2, we can assert that the chaoticity's mean values are lower for the ictal than for the pre-ictal stages. This agrees with previous results reported in the literature [4, 5, 20], and also with those obtained by the present authors for the quantities D_2 and $\Lambda_{\max}$, evaluated in static fashion for selected portions representative of different stages of the process (we used the nonlinear, dynamical metrics protocol described in Blanco et al. [9, 11] and Rosso and Mairal [45]). We notice also in figure 2 that the chaoticity exhibits a decreasing trend, with oscillating values between 80 seconds and 125 seconds. Starting around 125 seconds, the chaoticity increases but, in the interval (80 to 155 seconds), its mean value is smaller than the ones prevailing at the pre-ictal stage. Note that, as we mention above, the physician's team identifies the 80 seconds and 125 seconds as the beginning of (i) the epileptic seizure and

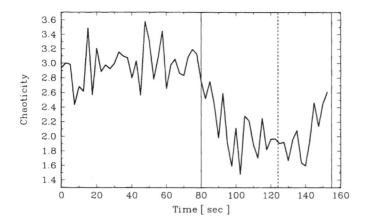

FIGURE 2   Time evolution of the chaoticity corresponding to EEG noise-free signal. The vertical solid lines represent the start and end of the epileptic seizure. The vertical dashed line represents the tonic to clonic phase transition.

(ii) the clonic stage, respectively. Finally, it is important to stress that, in static and in dynamic analysis, the evaluation of both the largest Lyapunov exponent and the chaoticity yields, for the EEG signal, values greater than zero. This fact is indicative of a chaotic behavior that could be associated with the whole of the EEG signal displayed in figure 1. The chaos quantifier becomes smaller at the recruiting phase (90 to 145 seconds), though.

6   RELATIVE WAVELET ENERGY

Figure 3(a) displays the RWE without electromyographic contributions (only bands B_3 to B_6 contribute) [43, 48, 45, 46]. We see that the pre-ictal phase is characterized by a dominance of low rhythms (pre-ictal: $[B_5 + B_6] \sim 50\%$). The seizure starts at 80 seconds with a discharge of slow waves superimposed to low-voltage fast activity. This discharge lasts approximately 8 seconds and produces a marked "activity-rise" in the frequency bands B_5 and B_6 (delta band), which reaches 80% of the RWE. Starting at 90 seconds, the low-frequency activity, represented in our analysis by B_5 and B_6, abruptly decreases to relative values lower than 10%, while the other frequency bands (theta and alpha bands) become more important. We also observe in figure 3(a) that the start of the clonic phase is correlated with increased activity in the B_4 frequency band. After 140 seconds, when clonic discharges become intermittent, the B_5 activity increases again until the end of the seizure, when the B_6 frequency activity also increases in very rapid fashion and both frequency bands become clearly dominant. The B_5 and

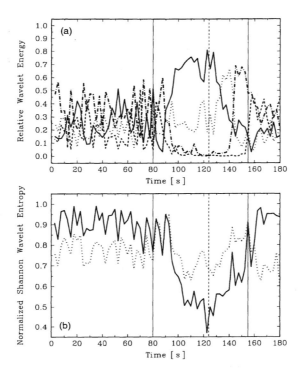

FIGURE 3   Time evolution of the relative wavelet energy corresponding to EEG noise-free signal (fig. 1(b)), for the frequency bands B_3 (solid line), B_4 (dotted line), B_5 (dot-dashed line) and B_6 (dashed line). The vertical solid lines represent the start and end of the epileptic seizure. The vertical dashed lines represents the tonic to clonic phase transition.

B_6 (delta band) frequency bands maintain this predominance throughout the postictal phase. We conclude from this example that the seizure was dominated by the middle frequency bands B_3 and B_4 (alpha and theta rhythms, 12.8 − 3.2 Hz), with a corresponding abrupt activity decrease in the low-frequency bands B_5 and B_6 (delta rhythm, 3.2 − 0.8 Hz) [43, 45, 46, 48]. Clearly, this behavior can be associated with the putative "epileptic recruiting rhythm" of Gastaut and Broughton [22] and Gastaut and Fischer-Williams [23].

A more precise characterization of the evolution of those frequencies associated with the "epileptic recruiting rhythm" (EPR) was made on the basis of a "time-scale-frequency" description that employs trigonometric wavelet packets [10]. From that work one infers that the central frequency associated with the EPR decreases with the time (see figs. 3 and 4 from Blanco et al. [10], and also [42]). One can thus reasonably conjecture that the tonic phase's spasms are answers to brain oscillations generated with such high frequencies that, because muscles can not contract in such a rapid fashion, muscle activity becomes restricted to tonic contraction until the frequency of brain oscillations becomes slow enough for the muscle to be capable of oscillating in resonance with them [42]. One important point to emphasize is that our results were obtained with scalp recordings without the use of either curare or any filtering method. Since intracranial recordings are nearly free of artifacts, the fact that the same pattern [17] is seen in both situations reinforces the idea that the results obtained with scalp electrodes were not spurious effects of muscle activity.

6.1   NORMALIZED WAVELET ENTROPY

For the data presented above, the ensuing normalized Shannon, Tsallis, and generalized escort Tsallis wavelet entropies, as a function of time, are depicted in figures 3(b) and 4, respectively [46]. We point out that the temporal evolution for the three functional forms is quite similar.

In figure 3(b) the dashed line represents the time evolution of the normalized Shannon wavelet entropy (all frequency bands are included), while the continuous line corresponds to results which ignore contributions due to high-frequency bands (B_1 and B_2). It is interesting to observe the behavior of the normalized SWS during the first 10 seconds following the seizure onset. We see that in this time interval the normalized SWS exhibits increasing values if all wavelet frequency bands are included. Comparison is to be made with normalized SWS values in the pre-ictal stage. If the wavelet frequency bands B_1 and B_2 (bands that mainly reflect muscular activity) are *not* included, the largest normalized SWS value is lower than that for the ictal onset. Thus, the behavior of the normalized SWS after the onset of seizure is compatible with an increase in the degree of disorder of the system, induced by a high-frequency activity. Superimposed low- and medium-frequency activities, however, are responsible for the "remaining signal's" more ordered behavior. The normalized SWS behavior after 90 seconds (in both cases with and without the inclusion of high-frequency bands) is indicative of the fact that the system exhibits a tendency to be more "ordered." This tendency is better appreciated without muscle activity. Moreover, note that the normalized SWS in the last case adopts a minimum value around 125 seconds, in coincidence with the beginning of the clonic phase. The peak observed in the normalized SWS at ~ 145 seconds could be associated with the disappearance of the epileptic recruitment rhythm. After this point, the normalized SWS displays increasing values until 155 seconds, which is defined as the seizure's ending time. We see that the normalized SWS for the post-ictal stage displays almost constant values, comparable to those obtained for the pre-ictal stage.

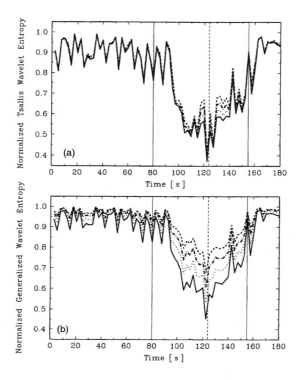

FIGURE 4 Normalized Tsallis wavelet entropy's time evolution for the EEG signal (Fig. 1.b) free of muscle activity (without the contribution of the frequency bands B_1 and B_2) for $q = 1.25$ (solid line), $q = 1.5$ (dotted line), $q = 1.75$ (dot-dashed line), and $q = 2.0$ (dashed line). The vertical solid lines represent the start and ending of the epileptic seizure. The vertical dashed lines represents the tonic to clonic phase transition.

Figures 4(a) and 4(b) depict the time evolution of both the normalized Tsallis and the generalized escort Tsallis wavelet entropies, for $q = 1.25, 1.5, 1.75,$ and 2.0, respectively. Comparing results depicted in these figures with those for the normalized Shannon wavelet ones, a great similitude is detected for $q = 1.25$. As expected, no differences can be appreciated if $q \to 1$. For larger q-values, the generalized escort Tsallis functional form is the one exhibiting the largest discrepancies in comparison to the Shannon results. Tsallis' results (fig. 4(a)) do not change very much q during the pre-ictal period. Small q-variations are seen during the temporal phase associated with the emergence of the epileptic recruiting rhythm. NTWS numerical results (independent of the q-value) during the ictal stage are significantly smaller than those of the pre-ictal stage. On the contrary, in the case of the wavelet entropy associated with generalized escort Tsallis distribution, numerical results for distinct q-values are different (see fig. 4(b)) in the whole temporal range. They are significantly different for the ictal stage. During the pre-ictal stage, NGWS values exhibit an almost constant behavior, with a dispersion that diminishes as q grows. For all q, NGWS values during the ictal stage are much smaller than for the pre-ictal stage.

In our three instances (figs. 3(b) and 4), the minimum absolute value is to be found in the vicinity of ~ 125 seconds, in agreement with the medical diagnosis: in that neighborhood one encounters the tonic-clonic "phase transition." The minimum NGWS value markedly depends upon q. For the three wavelet measures, two relative maxima are observed at ~ 145 seconds and ~ 155 seconds. As stated above, these times are associated with (i) the end of the epileptic recruiting rhythm and (ii) the epileptic seizure end, respectively. Changes in the EEG series around 125v (transition from tonic to clonic stage) are the result of a mechanism entirely different from the one that produces variations at 145 and 155 seconds (neuronal "fatigue," decrement in neuronal firing, and preponderance of inhibitory mechanisms are critical factors in the process underlying seizure ending). Summing up, one can associate a more robust degree of order with the EEG activity during the ictal phase than during the pre- and post-ictal stages, which is compatible with a dynamic process of synchronization in the brain activity. This behavior may be thought of as induced by an hypothetical epileptic focus, which generates the observed epileptic recruitment rhythm. Comparing figures 2 and 3(b) (fig. 4), it is clear that changes in the time evolution of the NSWS (NTWS, NGWS) quantifier are more synchronized with the changes in the signal (see fig. 1) than changes in the chaoticity. This is an important result: the NSWS (NTWS, NGWS) seems to be a better quantifier than chaoticity. Both quantifiers exhibit similar behaviors and do reflect morphological changes in the signal. They can indeed be taken as representative of changes in the associated dynamics.

One interesting point to underline is that, although the grouping in frequency bands implies a loss of frequency resolution, this procedure can be more useful than a study of single frequencies or peaks, due to the relation between frequency bands and functions or sources in the brain. In this context, the relative wavelet energy allows for an easy interpretation of several minutes of frequency variations in a single display, something that is sometimes difficult to achieve with traditional scalp EEGs. Being independent of the amplitude or of the energy of the signal, the wavelet entropy yields new information about EEG signals in comparison with that obtained by using frequency analysis or other standard methods. The normalized wavelet entropy (NSWS, NTWS and NGWS) has the following advantages.

1. In contrast to the spectral entropy, the total wavelet entropy is capable of detecting changes in a nonstationary signal due to the localization characteristics of the wavelet transform.

2. In comparison with dimensional analysis and Lyapunov exponents (which are only defined for stationary behaviors), or with dimensionality and chaoticity measures (stationary constraints removed), the computational time required for total wavelet entropy studies is significantly shorter. The algorithm for total wavelet entropy evaluation involves just the use of the wavelet transform in a multiresolution framework.
3. Contaminating noises' contributions (if they are basically concentrated in some frequency bands) can be easily eliminated.
4. An associated complexity can be evaluated.
5. Last but not least, the wavelet entropy is parameter free.

6.2 WAVELET COMPLEXITY

We pass now to a consideration of the complexity quantifier. Complexity results are displayed in figures 5 to 7 [47].

The temporal evolution of both the SDL and the LMC Shannon wavelet complexity (NSWC) are depicted in figure 5 for the EEG original signal and for the signal without the contribution of the muscular activity. The region of "epileptic recruiting rhythm" (90 to 145 seconds) looks quite different according to whether all frequency bands are included or not. We conclude that muscular activity destroys the underlying neuronal complexity. This is to be expected, of course, but it shows just how reliable our complexity tools are. The new feature that we are emphasizing here is that the zone of 90 to 145 seconds reveals a great degree of order (low entropy) *together* with higher complexity values (in particular $C^{(\mathrm{SDL})} \sim 1$). Order and complexity are seen to *coexist*. The complexity drop at 145 seconds is indicative of the end of the recruiting phase. The more pronounced drop at 155 seconds signals the seizure ending. In figures 6 and 7 the results of SDL and LMC Tsallis and the generalized escort Tsallis (normalized) wavelet complexities, as a function of time for $q = 1.25$, 1.5, 1.75, and 2.0 are shown. Note again, that in the case of $q = 1.25$, results resemble the Shannon ones. The normalized wavelet complexity behavior during the epileptic seizure markedly differs from that of the preseizure region. At the epileptic recruiting rhythm stage, normalized wavelet complexity values are almost constant with "valleys" at ~ 145 seconds and ~ 155 seconds, corresponding, respectively, to the disappearance of the epileptic recruiting rhythm and to the ending of the epileptic crisis.

The normalized Tsallis complexity does not strongly depend upon q (fig. 6). On the other hand, normalized generalized escort Tsallis complexity does markedly depend on q (fig. 7).

One appreciates here the fact that, during the epileptic recruiting rhythm interval, the SDL normalized wavelet complexity exhibits almost constant values, save for those in the vicinity of 125 seconds, where a maximum is exhibited (clearly an "outsider" point). Notice that these maxima become more noticeable as q becomes larger. We are speaking here precisely of the tonic-clonic phase

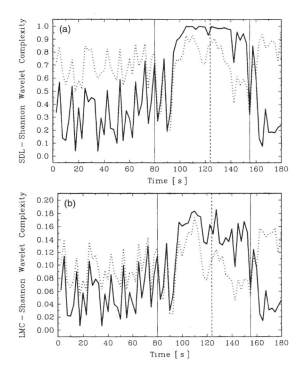

FIGURE 5   (a) Time evolution of the SDL—Shannon wavelet complexity. The dashed and the solid lines represent, respectively, the entropy's time evolution for the signal depicted in figure 1(a) and for that plotted in figure 1(b), that excludes contributions from the frequency bands B_1 and B_2. (b) Same for time evolution of the LMC—Shannon wavelet complexity. The vertical solid lines represent the start and ending of the epileptic seizure. The vertical dashed line represents the tonic to clonic phase transition.

transition time (125 seconds). The abrupt change of behavior of the NGWC (the "outsider" peak) is a signature of a muscular activity that begins to be able to respond to the brain oscillations (see the discussion in connection in section 5).

Figure 8(a) shows the time evolution of the EEG series without B_1 and B_2 contributions in the (H, C) phase space $(C \equiv C^{(\mathrm{LMC})})$. The maximum and minimum possible amounts of complexity $C_{\max}$ and $C_{\min}$ are drawn as a function of the entropy H, for a number of wavelet resolution levels (included in the evaluation of NSWS) $N = 4$. Note that the values corresponding to the pre- and postseizure stages are clearly concentrated in the zone of both relatively high entropy ($H > 0.8$) and intermediate complexity (at the "eastern" part of the

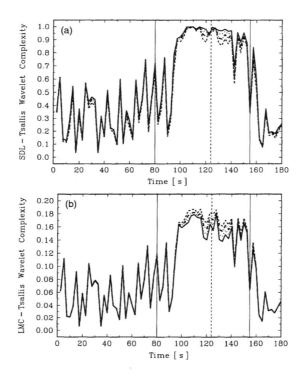

FIGURE 6 (a) SDL—Tsallis wavelet complexity's time evolution for the EEG signal (fig. 1(b)) free of muscle activity (without the contribution of the frequency bands B_1 and B_2) for $q = 1.25$ (solid line), $q = 1.5$ (dotted line), $q = 1.75$ (dot-dashed line), and $q = 2.0$ (dashed line). (b) Same for LMC—Tsallis wavelet complexity's time evolution. The vertical solid lines represent the start and end of the epileptic seizure. The vertical dashed line represents the tonic to clonic phase transition.

diagram), while a very different behavior can be associated with the ictal stage (triangles). Moreover, as can be observed in figure 8(b), it is possible to clearly distinguish between tonic (up triangles) and clonic (down triangles) stages. This tonic-clonic distinction is one of the main contributions of the present work. Figure 8(b) depicts data corresponding to the interval 90 to 145 seconds, associated with the epileptic recruiting rhythm. We see that the LMC complexity is higher in the tonic stage of 90 to 125 seconds than in the clonic one of 125 to 140 seconds. Moreover, it can be nitidly appreciated that the EEG "trajectory" remains close to $C_{\max}$ during the tonic stage, while, after a short transition time of 120 to 125 seconds, it tends to approach the minimum complexity path $C_{\min}$ as the

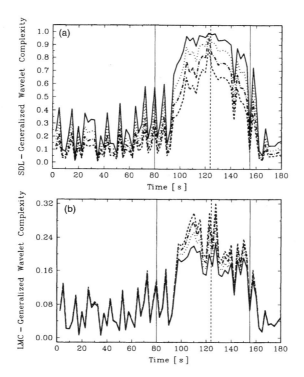

FIGURE 7 (a) SDL—generalized escort Tsallis wavelet entropy's time evolution for the EEG signal (fig. 1(b)) free of muscle activity (without the contribution of the frequency bands B_1 and B_2) for $q = 1.25$ (solid line), $q = 1.5$ (dotted line), $q = 1.75$ (dot-dashed line), and $q = 2.0$ (dashed line). (b) Same for LMC—generalized escort Tsallis wavelet entropy's time evolution. The vertical solid lines represent the start and end of the epileptic seizure. The vertical dashed lines represent the tonic to clonic phase transition.

trajectory evolves toward the end of the recruiting rhythm (145 seconds). Similar behavior is observed when complexities based on the Tsallis and the escort Tsallis information measures are used.

One of our goals here was to introduce the notion of *complexity* in the wavelet scenario so as to gather new insight into the background activity leading to (i) epileptic seizures and (ii) tonic-to-clonic phase transitions. With the help of both the normalized wavelet entropy and the wavelet complexity notion (SDL and LMC), we detect the presence of a rather special brain-state characterized by *both* order and large complexity.

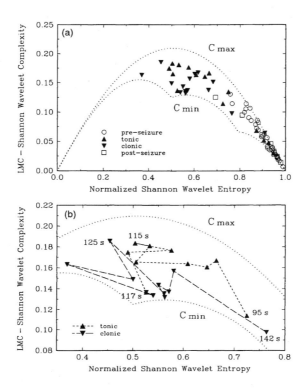

FIGURE 8   (a) The time evolution of the noise-free EEG is represented in a (C, H) (Shannon) phase space. The maximum complexity $C_{\max}$ and minimum complexity $C_{\min}$ (dotted line) are drawn as a function of the normalized Shannon wavelet entropy. (b) Tonic (black triangles) trajectory and clonic (white triangles) one in a (C, H) phase space.

Similar results to those presented above for the quantifiers based on ODWT were obtained in 20 tonic-clonic epileptic EEG time series recorded at $C4$ position, corresponding to eight different patients [43]. One critical aspect of our results is the possible distortion due to the spatial propagation of the seizure, since we analyzed data gathered by $C4$ electrodes, while the sources of the seizures are to be found mostly in temporal locations (see table 1 in Quian Quiroga et al. [43]). In order to overcome this difficulty we used data collected by $T3$ and $T4$ electrodes as well, obtaining similar results to the ones reported for $C4$ electrodes.

7 CONCLUSIONS

It is well established that an EEG is directly proportional to the local field potential recorded by electrodes on the brain's surface. Furthermore, one single EEG electrode placed on the scalp records the aggregate electrical activity from up to 6 cm^2 of the brain surface, and hence from many millions of neurons. With such large numbers, it seems quite natural to model the neocortex as a continuous sheet of neurons (neuronal matter) whose activity varies with time. The present work describes the use of quantitative parameters derived from the orthogonal discrete wavelet transform as applied to the analysis of brain electrical signals. The relative wavelet energy provides information about the relative energy associated with different frequency bands present in the EEG and enables one to ascertain their corresponding degrees of importance. The normalized wavelet entropy carries information about the degree of order/disorder associated with a multifrequency signal response, and allows for the evaluation of the complexity behavior associated to the signal by recourse to the normalized wavelet complexity. In addition, the time evolution of these quantifiers yields information about the dynamics associated with the EEG records.

In particular, we have shown that the epileptic recruitment rhythm behavior reported by Gastaut and Broughton [23] for epileptic tonic-clonic seizures is accurately described by the relative wavelet energy concept. Moreover, our studies do not require the use of curare or of digital filtering. In addition, a significant decrease in the total wavelet entropy was observed in the recruitment epoch, indicating a more rhythmic and ordered behavior of the EEG signal, compatible with a dynamical process of synchronization in the brain activity. We have also shown, via the evaluation of the chaoticity (a quantifier associated with the biggest Lyapunov exponent) as a function of time, that chaotic behavior could be associated to the whole of the tonic-clonic epileptic EEG signal, although it becomes smaller in the recruiting phase. We have also shown here that, this fact notwithstanding, this phase also exhibits maximum complexity. Thus, some of the brain's neurons in such a state would acquire spatial, or spatio-temporal, structures by means of internal processes, without specific interference from the outside.

As pointed out by many authors (see, for instance Başar et al. [6] and Haken [25]), the coexistence of chaos with complexity is a manifestation of self-organization. We have thus shown, on the basis of experimental EEG data, that an epileptic focus triggers a self-organized brain state characterized by both order and maximal complexity.

Summing up this work, we have reviewed extensive and nonextensive tools associated with wavelet entropies and with complexity measures for the analysis and characterization of EEG time series. Our main goal was to be in a position to detect the transition between tonic and clonic phases. One of our new tools clearly detects such a phase transition, the so-called generalized escort Tsallis wavelet complexity measure, associated with the generalized escort Tsallis entropy. The use of the quantifiers here advanced, based on time-frequency methods (like ODWT), can contribute to the analysis of brain electrical responses and may also lead to a better understanding of the underlying neural dynamics. Certainly, the use of these quantifiers is not intended to replace conventional EEG analyses, but to provide further insights into the intricacies of brain mechanisms.

REFERENCES

[1] Abarbanel, H. D. I. *Analysis of Observed Chaotic Data.* New York: Springer, 1996.

[2] Aldroubi A., and M. Unser. *Wavelets in Medicine and Biology.* Boca Raton: CRC Press, 1996.

[3] Anteneodo, C., and A. R. Plastino. "Some Features of the López-Ruiz-Mancini-Calbet (LMC) Statistical Measure of Complexity." *Phys. Lett. A* **223** (1997): 348–354.

[4] Başar, E. *Brain Function and Oscillations (I): Brain Oscillations, Principles and Approaches.* Berlin: Springer, 1998.

[5] Başar, E. *Brain Function and Oscillations (II): Integrative Brain Function, Neurophysiology and Cognitive Processes.* Berlin: Springer, 1999.

[6] Başar, E., H. Flohr, H. Haken, and A. J. Mandell. *Synergetics of the Brain.* Berlin: Springer–Verlag, 1983.

[7] Beck, C., and F. Schlögl. *Thermodynamics of Chaotic Systems.* Cambridge: Cambridge University Press, 1993.

[8] Binder, P. M., and N. Perry. "Comment II on Simple Measure for Complexity." *Phys. Rev. E* **62** (2000): 2998–2999.

[9] Blanco, S., S. Kochen, R. Quian Quiroga, L. Riquelme, O. A. Rosso, and P. Salgado. "Characterization of Epileptic EEG Time Series (I): Gabor Transform and Nonlinear Dynamics Methods." In *Wavelet Theory and Harmonic Analysis in Applied Sciences*, edited by C. E. D'Attellis and E. M. Fernandez-Berdaguer, 179–226. Boston: Birkhäuser Publishers, 1997.

[10] Blanco S., A. Figliola, R. Quian Quiroga, O. A. Rosso, and E. Serrano. "Time-Frequency Analysis of Electroencephalogram Series (III): Wavelet Packets and Information Cost Function." *Phys. Rev. E* **57** (1998): 932–940.

[11] Blanco, S., A. Figliola, S. Kochen, and O. A. Rosso. "Using Nonlinear Dynamics Metrics Tools for Characterizing Brain Structures." *IEEE Eng. Med. and Biol. Mag.* **16** (1997): 83–92.

[12] Calbet, X., and R. López-Ruiz. "Tendency towards Maximum Complexity in a Nonequilibrium Isolated System." *Phys. Rev. E* **63** (2001): 066116-1–066116-9.

[13] Capurro, A., L. Diambra, D. Lorenzo, O. Macadar, M. T. Martin, C. Mostaccio, A. Plastino, J. Perez, E. Rofman, M. E. Torres, and J. Velluti. "Human Brain Dynamics: The Analysis of EEG Signals with Tsallis Information Measure." *Physica A* **265** (1999): 235–254.

[14] Capurro, A., L. Diambra, D. Lorenzo, O. Macadar, M. T. Martin, C. Mostaccio, A. Plastino, E. Rofman, M. E. Torres, and J. Velluti. "Tsallis Entropy and Cortical Dynamics: The Analysis of EEG Signals." *Physica A* **257** (1998): 149–155.

[15] Casdagli, M. C., L. D. Iasemedis, R. S. Savit, R. L. Gilmore, S. N. Roper, and J. C. Sackellares. "Non-linearity in Invasive EEG Recordings from Patients with Temporal Lobe Epilepsy." *Electroenceph. Clin. Neurophysiol.* **102** (1997): 98–105.

[16] Crutchfield, J. P., D. P. Feldman, and C. R. Shalizi. "Comment I on Simple Measure for Complexity." *Phys. Rev. E* **62** (2000): 2996–2997.

[17] Darcey, T. M., and P. D. Willamson. "Spatio-temporal EEG Measures and Their Applications to Human Intracranially Recorded Epileptic Seizures." *Electroenceph. Clin. Neurophysiol.* **61** (1985): 573–587.

[18] Daubechies, I. *Ten Lectures on Wavelets*. Philadelphia: SIAM, 1992.

[19] Di Sisto, R. P., S. Martinez, R. B. Orellana, A. R. Plastino, and A. Plastino. "General Thermostatistical Formalisms, Invariance under Uniform Spectrum Translations, and Tsallis q-Additivity." *Physica A* **265** (1999): 590–613.

[20] Elbert, T., W. J. Ray, Z. J. Kowalik, J. E. Skinner, K. E. Graf, and N. Birbaumer. "Chaos and Physiology: Deterministic Chaos in Excitable Cell Assemblies." *Physiol. Rev.* **74** (1994): 1–47.

[21] Gamero, L. G., A. Plastino, and M. E. Torres. "Wavelet Analysis and Nonlinear Dynamics in a Nonextensive Setting." *Physica A* **246** (1997): 487–495.

[22] Gastaut, H., and R. Broughton. *Epileptic Seizures*. Springfield: Thomas, 1972.

[23] Gastaut, H., and M. Fischer-Williams. "The Physiopathology of Epileptic Seizures." In *Handbook of Physiology*, edited by J. Field, H. W. Magoun and V. E. Hall, vol. 1, 329–364. Baltimore: Williams and Wilkins, 1959.

[24] Gotman, J., J. R. Ives and P. Gloor. "Frequency Content of EEG and EMG at Seizure Onset: Possibility of Removal of EMG Artifact by Digital Filtering." *Electroenceph. Clin. Neurophysiol.* **52** (1981): 626–639.

[25] Haken, H. *Information and Self-Organization: A Macroscopic Approach to Complex Systems*. Berlin: Springer-Verlag, 2000.

[26] Iasemedis, L. D., and J. C. Sackellares. "The Evolution with Time of Spatial Distribution of the Largest Lyapunov Exponent on the Human Epileptic Cortex." In *Measuring Chaos in the Human Brain*, edited by D. Duke and W. Pritchards, 49–82. Singapore: World Scientific, 1991.

[27] Iasemedis, L. D., J. C. Sackellares, H. P. Zaveri, and W. J. Williams. "Phase Space Topography and Lyapunov Exponent of Electrocorticograms in Partial Seizures." *Brain Topography* **2** (1990): 187–201.

[28] Kullback, S. *Information Theory and Statistics*. New York: Dover, 1997.

[29] Lehnertz, K., and C. E. Elger. "Can Epileptic Seizures be Predicted? Evidence from Nonlinear Time Series Analysis of Brain Electrical Activity." *Phys. Rev. Lett.* **80** (1998): 5019–5022.

[30] Lehnertz, K., C. E. Elger, J. Arnholds, and P. Grassberger. *Chaos in Brain?* Singapore: World Scientific, 2000.

[31] López-Ruiz, R., H. L. Mancini, and X. Calbet. "A Statistical Measure of Complexity." *Phys. Lett. A* **209** (1995): 321–326.

[32] Mallat, S. *A Wavelet Tour of Signal Processing.* San Diego: Academic Press, 1999.

[33] Martin, M. T., A. R. Plastino, and A. Plastino. "Tsallis-Like Information Measures and the Analysis of Complex Signals." *Physica A* **275** (2000): 262–271.

[34] Mitschake, F. "Acausal Filters for Chaotic Signals." *Phys. Rev. A* **41** (1990): 1169–1171.

[35] Mitschake, F., M. Möller, and W. Large. "Measuring Filtered Chaotic Signals." *Phys. Rev. A* **37** (1988): 4518–4582.

[36] Niedermeyer, E., and F. H. Lopes da Silva. *Electroencephalography, Basic Principles, Clinical Applications, and Related Fields.* Baltimore: Urban and Schwarzenberg, 1987.

[37] Pijn, J. P., J. Van Neerven, A. Noestt, and F. H. Lopes da Silva. "Chaos or Noise in EEG Signals: Dependence on State and Brain Site." *Electroenceph. Clin. Neurophysiol.* **79** (1991): 371–381.

[38] Plastino, A., and A. R. Plastino. "Tsallis Entropy and Jaynes' Information Theory Formalism." *Braz. J. Phys.* **29** (1999) 50–60.

[39] Plastino, A. R. "Tsallis Theory, the Maximum Entropy Principle and Evolution Equations." In *Nonextensive Statistical Mechanics and Its Applications*, edited by S. Abe and Y. Okamoto, 157–192. Lecture Notes in Physics. Berlin: Springer-Verlag, 2001.

[40] Plastino A. R., and A. Plastino. "Tsallis Entropy, Ehrenfest Theorem and Information Theory." *Phys. Lett. A* **177** (1993): 177–182.

[41] Powell, G. E., and I. C. Percival. "A Spectral Entropy Method for Distinguishing Regular and Irregular Motion of Hamiltonian Systems." *J. Phys. A: Math. Gen.* **12** (1979): 2053–2071.

[42] Quian Quiroga, R. "Quantitative Analysis of EEG Signals: Time-Frequency Methods and Chaos Theory." Ph.D. Thesis, Institute of Physiology and Institute of Signal Processing, Medical University of Lübeck, Lübeck, Germany, 1998.

[43] Quian Quiroga, R., S. Blanco, O. A. Rosso, H. García, and A. Rabinowicz. "Searching for Hidden Information with Gabor Transform in Generalized Tonic-Clonic Seizures." *Electroenceph. Clin. Neurophysiol.* **103** (1997): 434–439.

[44] Quian Quiroga, R., O. A. Rosso, E. Başar, and M. Schürmann. "Wavelet Entropy in Event-Related Potentials: A New Method Shows Ordering of EEG Oscillations." *Biol. Cyber.* **84** (2001): 291–299.

[45] Rosso, O. A., and M. L. Mairal. "Characterization of Time Dynamical Evolution of Electroencephalographic Epileptic Records." *Physica A* (2002): in press.

[46] Rosso, O. A., M. T. Martin, and A. Plastino. "Brain Electrical Activity Analysis using Wavelet-Based Informational Tools." *Physica A* (2002): in press.

[47] Rosso, O. A., M. T. Martin, and A. Plastino. "Brain Electrical Activity Analysis using Wavelet-Based Informational Tools (II): Complexity Measures." *Physica A* (2002): submitted.

[48] Rosso, O. A., S. Blanco, and A. Rabinowicz. "Wavelet Analysis of Tonic-Clonic Epileptic Seizures." *Signal Processing* (2001): submited.

[49] Rosso, O. A., S. Blanco, J. Yordanova, V. Kolev, A. Figliola, M. Schürmann, and E. Başar. "Wavelet Entropy: A New Tool for Analysis of Short Duration Brain Electrical Signals." *J. Neurosci. Meth.* **105** (2001): 65–75.

[50] Samar, V., A. Bopardikar, R. Rao, and K. Swartz. "Wavelet Analysis of Neuroelectric Waveforms: A Conceptual Tutorial." *Brain and Language* **66** (1999): 7–60.

[51] Shannon, C. E. "A Mathematical Theory of Communication." *Bell System Technol. J.* **27** (1948): 379–423; 623–656

[52] Shiner, J. S. , M. Davison, and P. T. Landsberg. "Replay to Comments on Simple Measure for Complexity." *Phys. Rev. E* **62** (2000): 3000–3003.

[53] Shiner, J. S., M. Davison, and P. T. Landsberg. "Simple Measure for Complexity." *Phys. Rev. E* **59** (1999): 1459–1464.

[54] Theiler, J., and S. Eubank, "Don't Bleach Chaotic Data." *Chaos* **3** (1993): 771–782.

[55] Thévenaz, P., T. Blue, and M. Unser. "Interpolation Revisited." *IEEE Trans. Med. Imag.* **19** (2000): 739–758.

[56] Tsallis, C. "Nonextensive Statistics: Theoretical, Experimental and Computational Evidences and Connections." *Braz. J. Phys.* **29** (1999): 1–35.

[57] Tsallis, C. "Nonextensive Statistical Mechanics and Thermodynamics: Historical Background and Present Status." In *Nonextensive Statistical Mechanics and Its Applications*, edited by S. Abe and Y. Okamoto, 3–98. Lecture Notes in Physics. Berlin: Springer-Verlag, 2001.

[58] Unser, M. "Spline: A Perfect Fit for Signal and Image Processing." *IEEE Sig. Proc. Mag.* **16** (1999) 22–38.

Nonextensive Diffusion Entropy Analysis and Teen Birth Phenomena

N. Scafetta
P. Grigolini
P. Hamilton
B. J. West

A complex process is often a balance between nonscaling and scaling components. We show how the nonextensive Tsallis q-entropy indicator may be interpreted as a measure of the nonscaling condition in time series. This is done by applying the nonextensive entropy formalism to the diffusion entropy analysis (DEA). We apply the analysis to the study of the teen birth phenomenon. We find that the number of unmarried teen births is strongly influenced by social processes that induce an anomalous memory in the data. This memory is related to the strength of the nonscaling component of the signal and is more intense than that in the married teen birth time series. By using a wavelet multiresolution analysis, we attempt to provide a social interpretation of this effect.

1 INTRODUCTION

One of the most exciting and rapidly developing areas of modern research is the quantitative study of "complexity." Complexity has special interdisciplinary

Nonextensive Entropy—Interdisciplinary Applications
edited by Murray Gell-Mann and Constantino Tsallis, Oxford University Press

impacts in the fields of physics, mathematics, information science, biology, sociology, and medicine. No definition of a *complex system* has been universally embraced, so here we adopt the working definition, "an arrangement of parts so intricate as to be hard to understand or deal with." Therefore, the main goal of the science of complexity is to develop mathematical methods in order to discriminate among the fundamental microscopic and macroscopic constituents of a complex system and to describe their interrelations in a concise way.

Experiments usually yield results in the form of time series for physical observables. Typically, these time series contain both a slow regular variation, usually called a "signal," and a rapid erratic fluctuation, usually called "noise." Historically, the techniques applied to processing such time series have been based on equilibrium statistical mechanics and, therefore, they are not applicable to phenomena far from equilibrium. Among the fluctuating phenomena, a particularly important place is occupied by those phenomena characterized by some type of self-similar or scaling-fractal structures [4].

In this chapter we show that the nonextensive Tsallis q-entropy indicator may be interpreted as a measure of the strength of the nonscaling component of a time series. This is done by applying the nonextensive entropy formalism to the diffusion entropy analysis (DEA). DEA is a recent and very efficient method developed to detect the scaling of a stationary complex process. The scaling of the probability density function (pdf) of the diffusion process generated by time series is imagined as a physical source of fluctuations (see Allegrini et al. [1], Grigolini et al. [3], Scafetta et al. [8, 9, 10, 11], and Scafetta and Grigolini [7]).

We apply the above analysis to the study of the teen birth phenomenon. The daily birth data cover the number of births to married and unmarried teens in Texas during the period 1994 to 1998. Time series analysis in the social sciences is traditionally done using linear models, such as analyses of variance and linear regression. Underlying these techniques is the assumption that the phenomena of interest, such as adolescent sexuality, pregnancy, and other developmental processes, are stationary [6]. However, this is not a comprehensive approach, because the births by teenagers may be characterized by a complicated annual cycle that is the source of a deterministic nonstationarity. We find that unmarried teens seem to be more strongly influenced by social processes than are the married teens. Finally, by using the wavelet multiresolution analysis, we attempt to give a social interpretation of this effect.

2   NONEXTENSIVE DIFFUSION ENTROPY ANALYSIS AND ITS NONSCALING MEANING

Diffusion entropy analysis detects the scaling of a stationary process through the study of the time evolution of the Shannon entropy of the pdf in terms of the diffusion process generated by the time series. A time series $\{\xi_i\}$ may be interpreted as diffusion fluctuations. As in a random walk, we define diffusion trajectories by

$$x^{(z)}(t) = \sum_{i=1}^{t} \xi_{i+z} \, , \tag{1}$$

where $z = 0, 1, 2, \ldots$. These trajectories generate a diffusionlike process that is described by a pdf $p(x,t)$, where x denotes the variable collecting the fluctuations and t is the diffusion time. The pdf of a diffusion process may be expected to have the fundamental scaling property

$$p(x,t) = \frac{1}{t^{\delta}} \, F\left(\frac{x}{t^{\delta}}\right) \, , \tag{2}$$

where the coefficient δ is the scaling exponent. For example, fractional Gaussian noise with a Hurst exponent H [4] generates a diffusion process that fulfills eq. (2) and it is expected $\delta = H$.

The Shannon entropy is defined by

$$S(t) = - \int_{-\infty}^{+\infty} dx \; p(x,t) \ln[p(x,t)] \, . \tag{3}$$

Using the scaling condition of eq. (2) we obtain

$$S(t) = A + \delta \; \ln(t) \, , \tag{4}$$

with

$$A \equiv - \int_{-\infty}^{\infty} dy \; F(y) \; \ln[F(y)] \, , \tag{5}$$

where $y = x/t^{\delta}$. Equation (4) indicates that in the case of a diffusion process with a scaling pdf, its entropy $S(t)$ increases linearly with $\ln(t)$. Numerically, the scaling exponent δ can be evaluated by using fitting curves with the function of the form $f_S(t) = K + \delta \ln(t)$ that, when graphed on linear-log graph paper, yields straight lines.

The breakdown of the scaling condition may be simulated by assuming that the scaling exponent δ of eq. (2) changes with time. This can be implemented by assuming eq. (2) has the general form

$$p(x,t) = \frac{1}{t^{\delta(t)}} \, F\left(\frac{x}{t^{\delta(t)}}\right) \, . \tag{6}$$

If we assume that

$$\delta(t) = \delta_0 + \eta \ln(t), \tag{7}$$

where δ_0 and η are two constants, we notice that, in the new general formalism, the traditional entropy (3) yields:

$$S(t) = A + \delta_0 \ln(t) + \eta \, [\ln(t)]^2 \, . \tag{8}$$

The quadratic form of eq. (8) suggests that the choice of $\delta(t)$ given by eq. (7) has the mathematical meaning of the quadratic term in the Taylor expansion of the diffusion entropy (3). As a consequence, we should expect that, in general, $\delta(t)$ always assumes the form of eq. (7), at least for small values of $\ln(t)$.

Let us see how all this may be related to the nonextensive Tsallis q-indicator [12]. The Tsallis nonextensive entropy reads

$$S_q(t) = \frac{1 - \int_{-\infty}^{+\infty} dx\, p(x,t)^q}{q - 1}.$$ (9)

It is straightforward to prove that this entropic indicator coincides with that of eq. (3) in the limit where the entropic index $q \to 1$. Let us make the assumption that in the diffusion regime, the departure from this traditional value is weak, and assume $\epsilon \equiv q - 1 \ll 1$. This allows us to use the following approximate expression for the nonextensive entropy

$$S_q(t) = -\int_{-\infty}^{+\infty} dx\, p(x,t)\ln[p(x,t)] - \frac{\epsilon}{2}\int_{-\infty}^{+\infty} dx\, p(x,t)\ln^2[p(x,t)].$$ (10)

In the specific case where the nonscaling condition of eq. (6) applies, this entropy yields the form

$$S_q = A - \epsilon B + (1 - \epsilon A)\delta(t)\ln(t) - \frac{\epsilon}{2}\left[\delta(t)\ln(t)\right]^2,$$ (11)

where A and B are two constants related to $F(y)$ of eq. (6). The regime of linear increase in $\ln(t)$ is recovered when ϵ is assigned the value

$$\epsilon = q - 1 = \frac{\eta}{(\delta_0^2/2) + \eta A}.$$ (12)

These theoretical remarks demonstrate that the nonextensive approach to the diffusion entropy makes it possible to detect the entropic strength of the deviation from the scaling condition. In fact, eq. (12) establishes that $\epsilon = q - 1 = 0$ implies $\eta = 0$ that, according to eq. (7), indicates the scaling condition. The conclusion of this section is that the breakdown of the scaling property of eq. (2) can be revealed by the DEA under the form of an entropic index q departing from the condition of ordinary statistical mechanics, namely $q = 1$.

Figure 1 shows the effect of the nonextensive Tsallis q-entropy indicator as a function of time t applied to the binomial distribution generated by a simple random walk: $q = 1$ (solid line), $q = 1.2$ (dotted line), and $q = 0.8$ (dashed line). The figure shows clearly the relation between q and the bending shape of the entropic curve, typical of the nonscaling condition expressed by eq. (8). Of course, in the case of a random walk, the linear increase in $\ln(t)$ of the entropy $S_q(t)$ is recovered when $q = 1$ and the scaling exponent is $\delta = 0.5$. This is the value of the scaling exponent of the Gaussian distribution to which the binomial distribution converges after few diffusion steps.

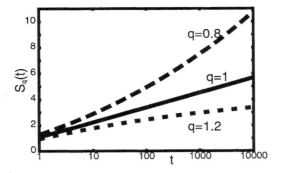

FIGURE 1 The nonextensive Tsallis entropy as a function of time t applied to the binomial distribution generated by the random walk: $q = 1$ (solid line), $q = 1.2$ (dotted line), and $q = 0.8$ (dashed line).

A curiosity: What happens if we adopt the Rényi entropy [2] instead of the Tsallis entropy? It is easy to prove that the Rényi q-entropy indicator has a simple parallel shifting effect instead of a bending effect upon the diffusion entropy and, therefore, it is not useful for our goal.

3 THE TEEN BIRTH PHENOMENON ANALYSIS

Texas is second only to California in the number of births to teens in the United States. Rates of birth to teens of all ages and racial/ethnic groups have been dropping in the United States since 1990 [13]. However, the size of the problem in Texas remains significant. In 1996, in Texas there were 80,490 pregnancies and 52,273 births to girls 15–19 years old [14]. The U.S. rate of pregnancy among young women 15 to 19 years old was 97 per 1000 girls of that age; the rate in Texas was 113 per 1000. The mean age of teens giving birth was 17.62 years in Texas. Approximately 66% of births to teenagers in Texas were out of wedlock and 24% of births to teens were to girls who had given birth at least once previously. Data for the study reported here were abstracted from birth certificates obtained from the Texas Department of Health. The original time series was constructed from the daily count of births from January 1, 1994, through December 31, 1998. Every recorded birth to a woman under the age of 20 was included. Data on the marital status of the mother allowed us to analyze married and unmarried births separately. Reliable and valid birth certificate information regarding marital status did not become available in Texas until January 1, 1994. More information about the data may be found in Scafetta et al. [11].

The DEA of the data is preceded by a preliminary detrending to free the data from easily understandable linear and cyclical deterministic trends. In fact,

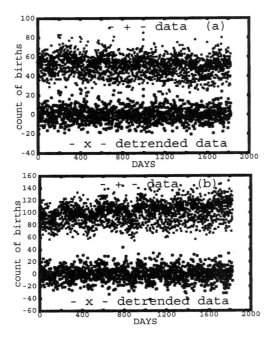

FIGURE 2   Number of births ("+" symbol) to (a) married and (b) unmarried teenagers from January 1, 1994, through December 31, 1998. The "x" symbol indicates the detrended data of the nonworking days and of the linear and of the seasonal trends.

the data show slight linear decrease (for married teens, fig. 2(a)) and increase (for unmarried teens, fig. 2(b)) trends, and two strong periodicities: the annual trend due to the seasonal cycle and the weekly cycle due to social organization of the week into workdays and weekends. The weekends consistently record a lower rate of births. We eliminate the data associated with weekends and holidays, and detrend the linear ramp and annual frequency through the fitting curve:

$$\Xi(t) = A + Bt + C\cos(\omega t) + D\sin(\omega t). \tag{13}$$

In the case of the unmarried teens, the fit gives $A = 97.5$, $B = 0.00893$, $C = 1.29$, $D = -6.30$, and $\omega = 2\pi/365.25$. In the case of the married teens, we set $A = 57.8$, $B = -0.00353$, $C = -0.277$, $D = -4.14$, and $\omega = 2\pi/365.25$. Figures 2(a) and 2(b) show the original data as well as the detrended data for the two groups.

Before applying the diffusion entropy algorithm to the two detrended datasets, for simplicity, we dichotomize the two signals, and associate the positive values to $+1$ and the negative values to -1. In this way, an easy confrontation with the random walk theory is possible.In fact, if the new dichotomous series of $+1$ and -1 is random, the diffusion produced by its walks gives the binomial distribution of the random walk that corresponds to the scaling condition with $q = 1$ and $\delta = 0.5$. On the other hand, if the new dichotomous series is not completely random, but modulated by some type of anomalous memory lacking fractal properties (for example, an anomalous cycle), the correspondent diffusion process shows some type of nonscaling behavior and we expect $q \neq 1$.

Let us apply the nonextensive DEA to the two dichotomous detrended datasets. First, we build the diffusion trajectories according to the prescription of eq. (1). Second, as done in the random walk model, we calculate the probabilities/frequencies $p_i(t)$ that a trajectory occupies the ith position at the diffusion time t. Finally, we evaluate the discrete nonextensive Tsallis entropy

$$S_q(t) = \frac{1 - \sum_i p_i(t)^q}{q - 1} \tag{14}$$

and we look for the *magic* $q = Q$ that makes $S_q(t)$ increase linearly in $\ln(t)$, at least in the first decades of the diffusion steps. Figures 3(a) and 3(b) show the diffusion entropy curves for $q = 1$ and for $q = Q$. For married teens we get a magic Q close to 1; this means that the corresponding dichotomous series is random. However, for unmarried teens we get $Q = 1.257$ that reveals a nonscaling diffusion process and, therefore, a nonfractal memory component in the signal.

To investigate the nature of the memory in the unmarried teen data, left after the detrending of the linear and cyclical annual trends, we use the wavelet analysis [5]. This is a powerful method of analysis that localizes a signal simultaneously in time and frequency. With a judicious use of the multiresolution wavelet transform, as explained in detail in Scafetta et al. [8], we can obtain an approximated distribution of conceptions which result in births during the year to both married and unmarried teenagers. The estimated errors are ± 2 births against ± 2 weeks. Moreover, we point out that identifying conception distribution from delivery dates among teens may be imprecise because of the high number of miscarriages or abortions, almost 49% of conceptions, and a sensitive seasonal dependency preterm delivery in teens. Figure 4 shows our estimation of the daily number of conceptions, relative to the annual mean value, for married ("+") and unmarried ("x") teenagers. The standard error is two births. The conception rate in married teens changes regularly following the annual seasonal temperature cycle, with a higher rate during the cold months and a lower rate during the hot months. Instead, the conception rate in unmarried teens seems more strongly influenced by the school-holiday yearly calendar. For example, there is a sharp drop in conceptions during the summer, due probably to the fact that the schools are closed and the interactions with other teenagers are greatly reduced. It is this complex social component of the births to unmarried teens that is detected as a nonscaling component by the DEA.

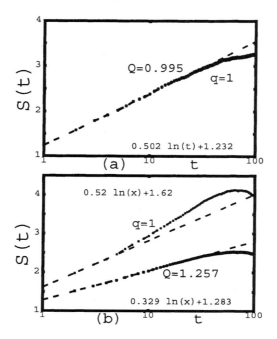

FIGURE 3   Nonextensive DEA for (a) married and (b) unmarried teenagers. The symbol "+" indicate the curves for $q = 1$. The married teens $Q = 0.995$ and the unmarried teens $Q = 1.257$ curves are indicated by the symbol "x." The fitting straight lines are shown as well.

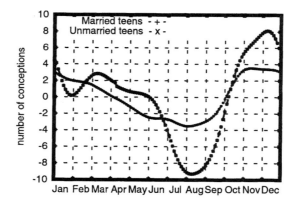

FIGURE 4   Approximated daily number of conceptions which result in birth relative to the annual mean value in married (symbol "+") and unmarried (symbol "x") teenagers. The standard errors are two births against two weeks.

4   CONCLUSION

The new technique of analysis, based on the entropy of diffusion process, and consequently called DEA, has been proved by Allegrini et al. [1], Grigolini et al. [3], Scafetta et al. [8, 9, 10, 11], and Scafetta and Grigolini [7] to be a very efficient method of scaling detection, which ensures the possibility of measuring the correct scaling coefficient δ when the property of eq. (2) applies. This chapter shows that the adoption of the Tsallis entropy rather than the Shannon entropy, as an entropy measure of the diffusion process, allows us to interpret the deviation of the Tsallis index q from the ordinary value $q = 1$, as an indicator of the strength of a nonfractal memory.

REFERENCES

[1] Allegrini, P., V. Benci, P. Grigolini, P. Hamilton, M. Ignaccolo, G. Menconi, L. Palatella, G. Raffaelli, N. Scafetta, M. Virgilio, and J. Jang. "Compression and Diffusion: A Joint Approach to Detect Complexity." *Chaos, Solitons & Fractals* **15(3)** (2003): 517–535.

[2] Beck, C., and F. Schlögl. *Thermodynamics of Chaotic Systems*. Cambridge: Cambridge University Press, 1993.

[3] Grigolini, P., D. Leddon, and N. Scafetta. "The Diffusion Entropy and Waiting Time Statistics of Hard X-Ray Solar Flares." *Phys. Rev. E* **65** (2002): 046203.

[4] Mandelbrot, B. B. *The Fractal Geometry of Nature*. New York: Freeman, 1983.

[5] Percival, D. B., and A. T. Walden. *Wavelet Methods for Time Series Analysis*. Cambridge: Cambridge University Press, 2000.

[6] Rodgers, J. L., and D. C. Rowe. "Social Contagion, Adolescent Sexual Behavior, and Pregnancy: A Nonlinear Dynamic EMOSA Model." *Dev. Psychol.* **34** (1998): 1096–1113.

[7] Scafetta, N., and P. Grigolini. "Scaling Detection in Time Series: Diffusion Entropy Analysis." *Phys. Rev. E* **66** (2002): 036130.

[8] Scafetta, N., P. Grigolini, P. Hamilton, and B. J. West. "Complexity, Multiresolution, Nonstationarity and Entropic Scaling: Teen Birth Thermodynamics." (2002): unpublished.

[9] Scafetta, N, V. Latora, and P. Grigolini. "Lévy Statistics in Coding and Noncoding Nucleotide Sequences." *Phys. Lett. A* **299 (5-6)** (2002): 565–570.

[10] Scafetta, N, V. Latora, and P. Grigolini. "Scaling Without Detrending: The Diffusion Entropy Method Applied to the DNA Sequences." *Phys. Rev. E* **66** (2002): 031906.

[11] Scafetta, N., P. Hamilton, and P. Grigolini. "The Thermodynamics of Social Process: The Teen Birth Phenomenon." *Fractals* **9** (2001): 193.

[12] Tsallis, C. "Possible Generalization of Boltzmann-Gibbs Statistics." *J. Stat. Phys.* **52** (1988): 479.

[13] Ventura, S. J., W. D. Mosher, S. C. Curtin, and J. C. Abma. "Highlights of Trends in Pregnancies and Pregnancy Rates by Outcome: Estimates for the United States, 1976–1996." *Natl. Vital Stat. Rep.* **47(26)** (1999): 1–12.

[14] *The National Campaign to Prevent Teen Pregnancy.* 1999. ⟨http://www.teenpregnancy.org⟩

The Pricing of Stock Options

Lisa Borland

We describe how a stock price model based on nonextensive statistics
can be used to derive a generalized theory for pricing stock options. A
review of theoretical and empirical results is presented.

1 INTRODUCTION

In 1973, Black and Scholes [1] and Merton [12] published their seminal papers
which developed a theory of the fair price of *options*. Scholes and Merton were
later to receive the 1997 Nobel prize for this famous work (Fisher Black had
unfortunately passed away two years earlier). Options are important financial
instruments which are traded in a huge volume all around the world on a variety
of exchanges. There are options on underlying assets ranging from orange juice
to gold, stocks to currency. In principle, an option is simply the right—but not
the obligation—to execute some previously agreed upon action, for example, the
right to buy or sell the underlying asset at some predetermined price, called the
strike. It is not difficult to understand that the existence of such instruments

could be extremely useful—for example, the right to buy an asset at a certain price protects against unforeseen events which could lead to huge price rises and thereby losses to someone who knows that they will need the asset at some time in the future. Similarly, the right to sell the asset at a certain price will protect against unforeseen drops in its value. These examples illustrate the use of options to *hedge* oneself against possible future events. Another use is more speculative: If a trader believes that the price of a stock will rise above a certain price at some date in the future, then it is in his interest to secure an option to buy the stock at some fixed lower price. Then, if the price of the stock does rise above that price, the trader can execute his option, just to turn around and resell the stock again at the higher market price.

Clearly then, options can potentially yield monetary value to the holder, depending on the future price of the underlying asset and the terms of the agreement and so on. But they can also prove to be worthless if there is never the need to exercise them. Given these scenarios, what then is the fair price of an option? The Black-Scholes model gives the answer in a very elegant manner, and the resulting Black-Scholes option prices have been used as a proxy by traders for the past 30 years or so. One interesting feature of the Black-Scholes model is that it yields a pricing formula which does not depend on the actual rate of return of the stock, but rather only on its volatility. In addition, the formula depends on the risk-free rate of return (i.e., the return on government T bills), the time to expiration of the option, the current stock price, and the strike price.

There is only one problem with the option prices of the Black-Scholes model, namely that they do not quite match observed, empirically traded ones. The deviations appear in a rather systematic way. In particular, in order to make the Black-Scholes prices match actual ones, a different value of the volatility must be used for each value of the strike price. A plot of the necessary values of the volatility versus the strike is typically a convex function, looking a bit like a smile or a smirk. For this reason, the feature has been dubbed the *volatility smile* in the financial literature. Obviously, this must be an artifact of false assumptions in the Black-Scholes approach. If not, then the smile would imply that the actual volatility of the underlying stock must vary depending on how much profit is potentially to be made in a particular deal. But the volatility should really be a well-defined inherent property of the price movements of the stock, so that a plot of volatility versus strike should ideally result in a flat line.

One of the assumptions upon which the Black and Scholes model is built is that log stock returns are normally distributed, with log stock prices modeled by a simple Langevin equation with Gaussian driving noise. Based on this model, they were able to set the foundations for much of modern mathematical finance. However, it is quite clear from empirical observations [13, 17] that real log returns are not normally distributed, as the Black-Scholes model implies. This model misspecification is certainly one source of the observed discrepancy between the theoretical option prices and the market prices. Over the years many significant modifications to the standard model have been made, in order to make it more realistic (cf. Bouchaud and Potters [5], Eberlein et al. [9], and Hull [11]). These approaches have typically been rather ad hoc or very complicated, and do not result in manageable closed form solutions which is one the main advantages of the Black-Scholes results.

In this chapter we follow an alternative approach. We generalize the standard stock price model to the framework of generalized thermostatistics, recently introduced by Tsallis [6, 7, 8, 20]. Then we proceed to derive an option pricing paradigm utilizing some of the concepts of the standard theories of mathematical finance. Furthermore, we obtain closed form solutions for certain important types of options. The Black-Scholes scenario is included as a special case. This work is for most part a review of the papers [3, 4], but some original empirical results have been added.

2 THE STOCK PRICE MODEL

Let us now briefly review the standard stock price model, which we proceed to generalize to the Tsallis framework. One assumes that the stock price S at time $\tau + t$ is given by

$$S(\tau + t) = S(\tau)e^{Y(t)},\tag{1}$$

where Y follows the stochastic process

$$dY = \mu dt + \sigma d\omega.\tag{2}$$

The drift μ is the mean rate of return and σ^2 is the variance of the stock logarithmic return. The driving noise ω is a Brownian motion defined with respect to a probability measure F. It represents a Wiener process and has the property

$$E^F[d\omega(t)d\omega(t')] = dtdt'\delta(t - t')\tag{3}$$

where the notation $E^F[\ ]$ means the expectation value with respect to the measure F. This model gives rise to a Gaussian distribution for the variable $Y(t) = \ln S(\tau + t)/S(\tau)$ resulting in a normal distribution for the log stock price changes

$$P\left(\ln \frac{S(t + \tau)}{S(t)}\right) = \frac{1}{\sqrt{2\pi}\sigma t}\exp\left\{-\frac{[\ln \frac{S(\tau+t)}{S(\tau)} - \mu t]^2}{2\sigma^2 t}\right\}.\tag{4}$$

Based on the stock price model equation (2), Black and Scholes were able to establish a pricing model to obtain the fair value of options on the underlying stock S.

In this chapter we shall show that the same goal can be accomplished for a new class of stochastic processes, motivated by those recently introduced [2] within the framework of the generalized thermostatistics of Tsallis [6, 7, 8, 20].

In this setting, we assume that the log price changes follow the process (see also our discussion in Osorio et al. [17])

$$dY = \mu dt + \sigma d\Omega \tag{5}$$

where Ω is an anomalous Wiener process that evolves according to a Tsallis distribution rather than a Gaussian. Such a process can be described by the statistical feedback process [2]

$$d\Omega = P(\Omega)^{\frac{1-q}{2}} d\omega . \tag{6}$$

The probability distribution P satisfies the nonlinear Fokker-Planck equation

$$\frac{\partial}{\partial t} P(\Omega, t \mid \Omega', t') = \frac{1}{2} \frac{\partial}{\partial \Omega^2} P^{2-q}(\Omega, t \mid \Omega', t') . \tag{7}$$

Explicit time-dependent solutions for P can be found [21], and are given by Tsallis distributions (or q-Gaussians)

$$P_q(\Omega, t \mid \Omega(0), 0) = \frac{1}{Z(t)} \left\{ 1 - \beta(t)(1-q)[\Omega(t) - \Omega(0)]^2 \right\}^{\frac{1}{1-q}} . \tag{8}$$

It can be verified that choosing $\beta(t) = c^{(1-q)/(3-q)}((2-q)(3-q)t)^{-2/(3-q)}$ and $Z(t) = ((2-q)(3-q)ct)^{1/(3-q)}$ ensures that the initial condition $P_q = \delta(\Omega(t) - \Omega(0))$ is satisfied. The index q is known as the entropic index of the generalized Tsallis entropy. We find that the q-dependent constant c is given by $c = \beta Z^2$ with $Z = \int_{-\infty}^{\infty} (1 - (1-q)\beta\Omega^2)^{1/(1-q)} d\Omega$ for any β. By choosing $\Omega(0) = 0$, we obtain a generalized Wiener process, distributed according to a zero-mean Tsallis distribution In the limit $q \to 1$ the standard theory is recovered, and P_q becomes a Gaussian. In that case, the standard Gaussian driving noise of eq. (2) is also recovered. Distributions for $q < 1$ exhibit cutoff thresholds beyond which $P_q = 0$. In this chapter we are concerned with the range $1 \le q < 5/3$ in which positive tails and finite variances are found [22].

For the variable $Y(t) = \ln S(\tau + t)/S(\tau)$, the following distribution is obtained

$$P_q \left(\ln \frac{S(\tau + t)}{S(\tau)} \right) = \frac{1}{Z(t)} \left\{ 1 - \tilde{\beta}(t)(1-q) \left[\ln \frac{S(\tau + t)}{S(\tau)} - \mu t \right]^2 \right\}^{\frac{1}{1-q}} \tag{9}$$

with $\tilde{\beta} = \beta(t)/\sigma^2$. This implies that log-returns $\ln[S(\tau+t)/S(\tau)]$ over the interval t follow a Tsallis distribution.

This result is consistent with empirical evidence [13, 17], with $q \approx 1.5$. It was shown that the distribution of returns over the timescale t evolved according to a Tsallis distribution. Note that from this viewpoint, the stochastic equation (5) should not be interpreted as corresponding to the evolution of log prices themselves but rather to the evolution of log price changes over the interval t. In the standard case where the distributions are Gaussian, the two interpretations coincide.

This anomalous diffusion across time scales can be seen as the result of the fact that subsequent price changes are not independent—traders react to previous price changes. This is captured phenomenologically by the statistical feedback into the system, from the macroscopic level characterized by P, to the microscopic level characterized by the dynamics of Ω, and thereby ultimately by changes of $\ln S$. This yields a nonhomogenous reaction to the price changes: depending on the value of q, rare events (i.e., extreme returns) will be accompanied by large reactions. The net effect of this is that extreme returns will tend to be followed by large returns in either direction. In fact, this type of "volatility clustering" is a well-known property of real stock returns. On the other hand, if the price takes on less extreme values, then the size of the noise acting upon it is more moderate.

3   DERIVATIVES

A *derivative* $f(S)$ is a financial asset which depends on the price movements of the underlying asset S. Futures, forwards, swaps, and options are all examples of derivative financial assets. To obtain an expression for the price movements of f in our generalized framework, let us first write down the dynamics of S. Using the rules of Ito stochastic calculus [10, 18], the evolution of S on the timescale t follows from eq. (5) as

$$dS = \tilde{\mu} S dt + \sigma S d\Omega \tag{10}$$

where

$$\tilde{\mu} = \mu + \frac{\sigma^2}{2} P_q^{1-q}. \tag{11}$$

The term $(\sigma^2/2)P_q^{1-q}$ is a result of the noise-induced drift. Remember that P_q is a function of Ω with

$$\Omega(t) = \frac{\ln \frac{S(\tau+t)}{S(\tau)} - \mu t}{\sigma}. \tag{12}$$

With $q = 1$ the standard result is recovered. Price movements $f(S)$ can now be obtained again through application of Ito calculus, namely

$$df = \frac{\partial f}{\partial S} dS + \frac{\partial f}{\partial t} dt + \frac{1}{2} \frac{\partial^2 f}{\partial S^2} (\sigma^2 P_q^{1-q}) dt \tag{13}$$

Here, dS which is given by eq. (10). Through this, df also depends on the noise $d\Omega$.

Just as in the standard case (cf. Hull [11] and Neftci [15]), the noise term driving the price of the shares S is the same as that driving the price f of the derivative. As a result, it should be possible to invest one's wealth in a portfolio of shares and derivatives in such a way that the noise terms cancel each other, yielding the so-called *risk-free portfolio*, the return on which is the risk-free rate r. This argumentation is a consequence of a famous principle in finance, known as the *arbitrage theorem*. In essence, it simply means that market dynamics will keep prices in equilibrium such that excess profits cannot be made without additional risk. If not, there would be a profitable risk-less trading strategy, i.e., an arbitrage opportunity. If one such opportunity were to arise, traders would quickly take advantage of it and it would disappear.

Explicitly, the risk-free portfolio is formed as

$$\Pi = -f + \frac{\partial f}{\partial S} S. \tag{14}$$

A small change in this portfolio is given by

$$d\Pi = -df + \frac{\partial f}{\partial S} dS \tag{15}$$

with a risk-free return

$$d\Pi = r\Pi. \tag{16}$$

After insertion of eq. (13) and eq. (10), this results in a partial differential equation of the form

$$\frac{\partial f}{\partial t} + rS\frac{\partial f}{\partial S} + \frac{1}{2}\frac{\partial^2 f}{\partial S^2}\sigma^2 S^2 P_q^{1-q} = rf \tag{17}$$

where $P_q(\Omega(t))$ evolves according to eq. (7). In the limit $q \to 1$, we recover the standard Black-Scholes PDE. Equation (17) depends explicitly only on the risk-free rate and the variance, not on μ. However, there is an implicit dependency on μ for $q \neq 1$ through the dependency on $P_q(\Omega)$, with Ω given by eq. (12).

This is an important point, because the independence of μ is an essential ingredient for many concepts in mathematical finance which rely upon *risk-free* asset pricing theory. To be consistent with risk-free pricing theory, we should first transform our original stochastic equation for S into a *martingale* before we apply the above analysis. Under the new martingale measure there are no arbitrage opportunities, and the rate of return on all assets is equal to the risk-free rate. Therefore, our results will not be affected other than that $\tilde{\mu}$ will be replaced by the risk-free rate r, ultimately eliminating the dependency on μ. In the following we describe how this can be done.

4   MARTINGALES AND RISK-FREE ASSET PRICING

A martingale is a stochastic process that displays no discernible trends. It is entirely random with an expectation equal to zero. This implies that a process is a martingale if its future movements are entirely unpredictable given knowledge of the past. Financial assets such as stocks and bonds are typically not martingales. Over any small time interval they usually drift in proportion to some instantaneous rate of return. However, it turns out that they can be easily transformed into martingales, either by appropriately subtracting the drift term or by transforming the probability measure under which the driving noise is defined. The former method is known as the Doob-Meyer decomposition and is not usually used, although it seems more intuitive. Instead, the latter approach which introduces *synthetic* probabilities has proven to be an extremely powerful tool and is widely used.

One may wonder why it would be of interest to transform these assets into martingales. The answer is that in a representation where the discounted stock price is a martingale, there will be no arbitrage opportunities. Therefore, the fair price of an option can be obtained simply by taking expectations with respect to the risk-neutral equivalent martingale probability measure. We shall illustrate how these concepts apply to our current model, but first we review the standard case. The discounted stock price

$$\tilde{S} = e^{-rt}S(\tau + t) \tag{18}$$

follows

$$d\tilde{S} = (\mu - r)\tilde{S}dt + \sigma\tilde{S}d\omega \tag{19}$$

where ω is a Brownian noise term associated with a probability measure F. It is not a martingale because of the drift term $(\mu - r)\tilde{S}dt$. According to the Girsanov theorem [16], one can however find an equivalent measure Q corresponding to an alternative noise term dz, such that the process is transformed into a martingale, by rewriting it as

$$d\tilde{S} = \sigma\tilde{S}\left(\frac{\mu - r}{\sigma}dt + d\omega\right) \tag{20}$$

$$= \sigma\tilde{S}dz. \tag{21}$$

The new driving noise term z is related to ω through

$$z = \int_0^t u\,ds + \omega \tag{22}$$

with

$$u = \frac{\mu - r}{\sigma}. \tag{23}$$

The noise term z is defined with respect to the equivalent martingale measure Q which is related to F through the Radon-Nikodym derivative [16]

$$\zeta(t) = \frac{dQ}{dF} = \exp\left(-\int_0^t u\,d\omega - \frac{1}{2}\int_0^t u^2\,ds\right). \tag{24}$$

Under the measure F, the original random variable ω follows a zero-mean process with variance equal to t. Under that same measure, the new noise term $z(t)$ is normal with non zero mean equal to $\int_0^t u\,ds$ and variance t. However, with respect to the equivalent probability measure Q one can easily verify that $z(t)$ is normal with 0 mean and variance t. This follows because the relationship

$$E^Q[Y] = E^F[\varsigma Y] \tag{25}$$

holds. In the above discussion, u, μ and σ may all depend on the variable $S(t)$ as well. The only criterion that must be satisfied for the Girsanov theorem to be valid is that

$$\exp\left(-\frac{1}{2}\int_0^t u^2 ds\right) < \infty \tag{26}$$

which implies that ς is a square integral martingale. (For details see Oksendal [16].) Note also that a comparison between eqs. (19) and (21) shows that the effect of the martingale transformation has been to set $\mu = r$. This reflects the idea that in the risk-neutral representation, the return on all assets is r.

Now we would like to formulate similar equivalent martingale measures for the present class of probability-dependent stochastic processes. The discounted stock price follows the dynamics

$$d\tilde{S} = (\tilde{\mu} - r)\tilde{S}dt + \sigma\tilde{S}d\Omega \tag{27}$$

over the timescale t, where $d\Omega$ follows eq. (6), which in turn is defined with respect to a probability measure F. For there to be no arbitrage opportunities, risk-free asset pricing theory requires that this process be a martingale, but this is not the case due to the drift term. However, we can define an alternative driving noise z associated with an equivalent probability measure Q so that, with respect to the new noise measure, the discounted stock price has zero drift and is thereby a martingale. Explicitly

$$d\tilde{S} = (\tilde{\mu} - r)\tilde{S}dt + \sigma\tilde{S}P^{\frac{1-q}{2}}d\omega. \tag{28}$$

Here, P is a nonvanishing bounded function of Ω. With respect to the initial Brownian noise ω, Ω relates to S via eq. (12). That is why for all means and purposes, P in eq. (28) is simply a function of S (or $\tilde{S}$), and the stochastic process can be seen as a standard state-dependent Brownian one. As a consequence, both the Girsanov theorem (which specifies the conditions under which we can transform from the measure F to Q) and the Radon-Nikodym theorem (which relates the measure F to Q) are valid, and we can formulate equivalent martingale measures much as in the standard case [14, 15, 16, 19]. We rewrite eq. (28) as

$$d\tilde{S} = \sigma\tilde{S}P^{\frac{1-q}{2}}dz \tag{29}$$

where the new driving noise term z is related to ω through

$$dz = \frac{(\tilde{\mu} - r)}{\sigma P_q^{\frac{1-q}{2}}}dt + d\omega \tag{30}$$

With respect to z, we thus obtain

$$d\tilde{S} = \sigma \tilde{S} d\Omega. \tag{31}$$

with

$$d\Omega = P_q^{\frac{1-q}{2}} dz \tag{32}$$

which is none other than a zero-mean Tsallis distributed generalized Wiener process, completely analogous to the one defined in eq. (6). Transforming back to S, we get

$$dS = rdt + \sigma S d\Omega. \tag{33}$$

Comparing this with eq (10), we see that the rate of return $\tilde{\mu}$ has been replaced with the risk-free rate r. This recovers the same result as in the standard risk-free asset pricing theory. Consequently, in the risk-free representation, eq. (12) becomes

$$\Omega(t) = \frac{1}{\sigma} \left(\ln \frac{S(t)}{S(0)} - rt + \frac{\sigma^2}{2} \int_0^t P_q^{1-q}(\Omega(s))ds \right) \tag{34}$$

where we have set $\tau = 0$ without loss of generality. This eliminates the implicit dependency on μ to which we alluded in the discussion of eq (17). However, through the variable transformation $\Omega(s) = \sqrt{\beta(t)/\beta(s)}\Omega(t)$, we can map the distributions $P_q(\Omega(s))$ onto the distribution $P_q(\Omega(t))$ and explicitly solve for $\Omega(t)$ as a function of $S(t)$ and r.

5   EUROPEAN OPTIONS

An option is termed *European* if it can be exercised only on the expiration date T. There are also *American* options, which can be exercised at any time during the life of the option. Other types of more complicated options include *Bermuda*, *Passport*, and *Parisian*. These are all just names and are traded on global exchanges, or over the counter at specific institutions. Probably the simplest options are European *put* options and European *call* options. A European put is the right to sell the underlying stock at a given strike price K on the expiration date T, and a European call is the right to buy the underlying stock at a given strike K on date T.

Typically, the possible payoff of the option is known at time T (for example, for a European call option the value at time T is either 0 or $S(T)-K$). Using this knowledge to infer appropriate boundary conditions at T, the generalized PDE (eq. (17)) can be solved numerically to obtain the fair price of the option. But it is sometimes more convenient to obtain the price by taking expectations in the risk-neutral representation. The two methods yield equivalent results. In the following we shall formulate an option pricing formula for a general European claim C, which could be a call or a put or some other structure. We shall then proceed to show that closed form solutions can be obtained for calls. (The closed form price of puts can also be derived, but we do not discuss that case here).

Suppose that the European claim C depends on $S(t)$, whose price f is given by its expectation value in a risk-free (martingale) world as

$$f(C) = E^Q[e^{-rT}C].$$
(35)

We assume the payoff on this option depends on the stock price at time T so that

$$C = h(S(T)).$$
(36)

After stochastic integration of eq. (29) to obtain $S(T)$, we get

$$f = e^{-rT} E^Q \left[h\left(S(0) \exp\left(\int_0^T \sigma P_q^{\frac{1-q}{2}} dz_s + \int_0^T (r - \frac{\sigma^2}{2} P_q^{1-q}) ds \right) \right) \right].$$
(37)

The key point in our approach now is that the random variable

$$\int_0^T P_q^{\frac{1-q}{2}} dz_s = \int_0^T d\Omega(s) = \Omega(T)$$
(38)

is distributed according to the Tsallis distribution eq. (8). This gives

$$f = \frac{e^{-rT}}{Z(T)}$$
$$\times \int_R h \left[S(0) \exp\left(\sigma\Omega(T) + rT - \frac{\sigma^2}{2}\alpha T^{\frac{2}{3-q}} + (1-q)\alpha T^{\frac{2}{3-q}} \frac{\beta(T)}{2}\sigma^2\Omega^2(T) \right) \right]$$
$$\times (1 - \beta(T)(1-q)\Omega(T)^2)^{\frac{1}{1-q}} d\Omega_T$$
(39)

with $\alpha = 1/2(3-q)((2-q)(3-q))c)^{(q-1)/(3-q)}$. To obtain this result we have utilized the fact that each of the distributions $P(\Omega(s))$ occurring in the latter term of eq. (37) can be mapped onto the distribution of $\Omega(T)$ at time T via the appropriate variable transformations $\Omega(s) = \sqrt{\beta(T)/\beta(s)}\,\Omega(T)$. A major difference to the standard case is the $\Omega^2(T)$-term which appears as a result of the noise-induced drift. In the special case of $q = 1$, the standard expression of the option price is recovered with this formula (see for example Oksendal [16]).

6   THE EUROPEAN CALL OPTION: AN EXACT SOLUTION

In the standard Black and Scholes theory, closed form solutions for European calls (and puts) can be obtained. We shall now show that the same is true in the generalized setting. As mentioned earlier, a European call is the right to buy the underlying stock at a given strike K on date T. The option is said to be *in the money* if there is a profit to be made at expiration. For a call option this would be if $S(T) > K$. The option is said to be *out of the money* if it instead expires worthless. For a call this is if $S(T) < K$. If it expires at the break even point, then it is said to be *at the money*, which for a put or a call is $S(T) = K$.

Equation (39) is valid for an arbitrary payoff h. For a European call, the payoff is $h = 0$ if it expires out of the money, and $h = S(T) - K$ otherwise. The value at time T must equal

$$C(T) = \max[S(T) - K, 0].\qquad(40)$$

The price c of such an option can be written as

$$c = E^Q[e^{-rT}C] = E^Q[e^{-rT}S(T)]_D - E^Q[e^{-rT}K]_D = J_1 - J_2 \qquad(41)$$

where the subscript D stands for the set $\{S(T) > K\}$. This condition is met if

$$-\frac{\sigma^2}{2}\alpha T^{\frac{2}{3-q}} + (1-q)\alpha T^{\frac{2}{3-q}}\frac{\beta(T)}{2}\sigma^2\Omega^2 + \sigma\Omega + rT > \ln\frac{K}{S(0)} \qquad(42)$$

which is satisfied for Ω between the two roots s_1 and s_2 of the corresponding quadratic equation. This is a very different situation from the standard case, where the inequality is linear and the condition $S(T) > K$ is satisfied for all values of the random variable greater than a threshold. In our case, due to the noise-induced drift, values of $S(T)$ in the risk-neutral world are not monotonically increasing as a function of the noise. As $q \to 1$, the larger root s_2 goes toward ∞, recovering the standard case. But as q gets smaller, the tails of the noise distribution get larger, as does the noise-induced drift, which tends to pull the system back. As a result we obtain

$$J_1 = S(0)\frac{1}{Z(T)}\int_{s_1}^{s_2}\exp\left(\sigma\Omega - \frac{\sigma^2}{2}\alpha T^{\frac{2}{3-q}} - (1-q)\alpha T^{\frac{2}{3-q}}\frac{\beta(T)}{2}\sigma^2\Omega^2\right)$$
$$\times(1 - (1-q)\beta(T)\Omega^2)^{\frac{1}{1-q}}d\Omega; \qquad(43)$$
$$J_2 = e^{-rT}K\frac{1}{Z(T)}\int_{s_1}^{s_2}(1 - (1-q)\beta(T)\Omega^2)^{\frac{1}{1-q}}d\Omega. \qquad(44)$$

Equation (41) with eqs. (43) and (44) constitute a closed form expression for the price of a European call option.

7   THE VOLATILITY SMILE: NUMERICAL AND EMPIRICAL RESULTS

Due to the fact that empirical distributions of real stock returns appear to be well modeled with a value of $q \approx 1.5$ [13, 17], we use this value of q to compare with results of the standard Black-Scholes model ($q = 1$). Figure 1 depicts the call option price as a function of the strike price for the standard Black-Scholes

Call Price

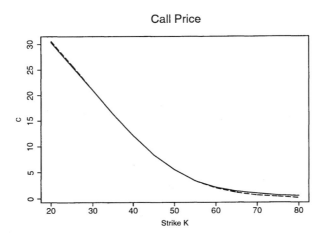

FIGURE 1   Call option price versus strike price, using $S(0) = 50$, $r = .06$, and $T = .6$, for $q = 1$ (dashed curve) and $q = 1.5$ (solid curve). For each q, σ was chosen so that the at-the-money options are priced equally ($\sigma = .3$ for $q = 1$ and $\sigma = .299$ for $q = 1.5$).

Call Price Difference

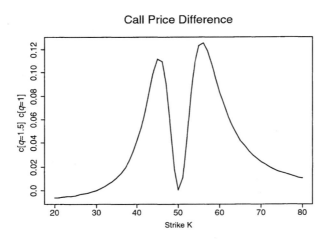

FIGURE 2   Calibrated so that at-the-money options are priced equally, the difference between the $q = 1.5$ model and the standard Black-Scholes model is shown, for $S(0) = 50$, $r = 0.06$, and $T = 0.05$, with $\sigma = .3$ for $q = 1$ and $\sigma = .41$ for $q = 1.5$. Times are expressed in years, r and σ are in annual units.

model and our model with $q = 1.5$, where σ is chosen such that the at-the-money prices are equal.

Figure 2 shows how the difference in call price between the two models varies as a function of the strike price. It is clear that both in-the-money and out-of-the money options are valued higher with $q = 1.5$, except for very deep in-the-money options which are valued lower. This behavior can be understood intuitively as follows. The distribution of Ω for $q = 1.5$ has fatter tails than the $q = 1$ model. Consequently, if the stock price gets deep out-of-the-money, then the noise may still produce shocks that can bring the stock back in the money again. This results in higher option prices for deep out-of-the-money strikes. Similarly, if the option is deep in the money, the noise can produce shocks to the underlying asset which can bring the price out of the money again. In addition, large shocks will increase the value of the noise-induced drift term which will decreases the probability of realizing higher stock prices. This results in lower option prices for deep in-the-money strikes. On the other hand, for intermediate values around the money, there will be a higher probability to land both in or out of the money, which leads to an increase in the option price, relative to the standard $q = 1$ model.

The volatilities that the standard Black-Scholes model must assume for each value of the strike in order to match the values obtained for the $q = 1.5$ model, are plotted in Figure 3, for $T = 0.1$ and $T = 0.4$. These implied volatilities form a smile shape, very similar to that which is implied by real market data. An indirect comparison between the theoretical $q = 1.5$ option prices and empirical ones is given by comparing empirical volatility smiles with the smile implied by matching the standard Black-Scholes model to the $q = 1.5$ one. Such results are shown in Figure 4, where the volatility smile for actual traded options on S&P 500 futures is shown together with the $q = 1.43$ smile (we use this value of q because it models well the empirical S&P 500 returns distribution [17]) A similar study for S&P 100 futures is shown in Figure 5. Based on the good agreement between the smiles, we infer that the $q \approx 1.5$ model can describe real option prices quite well, using just one value of σ across strikes. However, in reality there is a wide variety of smiles, and more empirical work is in progress in order to put these conclusions on a sounder statistical basis.

ACKNOWLEDGMENTS

Roberto Osorio and Jeremy Evnine are gratefully acknowledged.

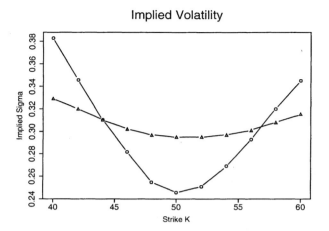

FIGURE 3   Using the $q = 1.5$ model (here with $\sigma = .3$, $S(0) = 50$, and $r = .06$) to generate call option prices, one can back out the volatilities implied by a standard $q = 1$ Black-Scholes model. Circles correspond to $T = 0.1$, while triangles represent $T = 0.4$. These implied volatilities capture features seen in real options data, with the smile more pronounced for small T.

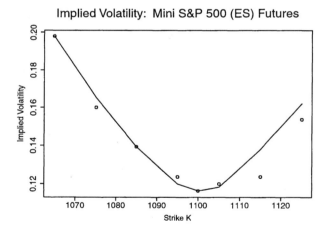

FIGURE 4   Implied volatilities for options on S&P 500 futures (ES May, last trading date May 17, 2002) [23] are plotted as a function of the strike (for $S(0) = 1100$, $r = .065$, $T = .0055$ [2 days]) (symbols). Also shown are the implied volatilities needed for a $q = 1$ Black-Scholes model to match option prices given by a $q = 1.43$ model using just one value of $\sigma = .2$ across all strikes (solid line). (Note that $q = 1.43$ also models well the empirical S&P 500 returns distribution [17].)

Implied Volatility: S&P 100 (OX) Futures

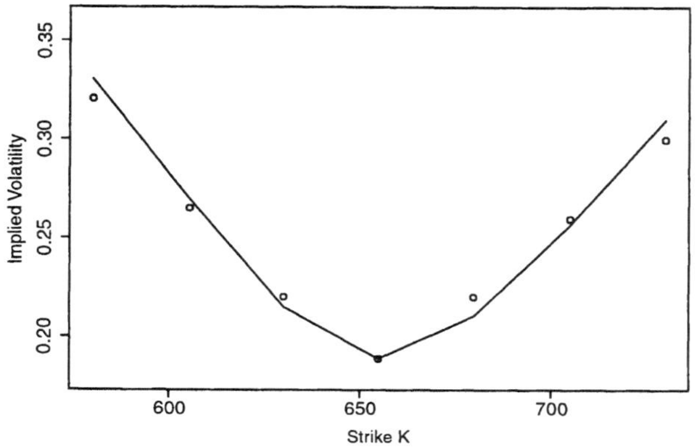

FIGURE 5   Implied volatilities for options on S & P 100 futures (OX June, last trading date June 15, 2001) [23] are plotted as a function of the strike (for $S(0) = 655$, $r = .045$, $T = .0274$ [10 days]) (symbols). Also shown are the implied volatilities needed for a $q = 1$ Black-Scholes model to match option prices given by a $q = 1.43$ model using just one value of $\sigma = .262$ across all strikes (solid line).

REFERENCES

[1] Black F., and M. Scholes. "The Pricing of Options and Corporate Liabilities." *J. Pol. Econ.* **81** (1973): 637–659.

[2] Borland, L. "The Microscopic Dynamics of the Nonlinear Fokker-Planck Equation: A Phenomenological Model." *Phys. Rev. E* **57** (1998): 6634.

[3] Borland, L. "Option Pricing Formulas based on a Non-Gaussian Stock Price Model." *Phys. Rev. Lett.* **89(9)** (2002): 098701. April 2002. arXiv.org e-Print Archive, Quantum Physics, Cornell University. September 2002. ⟨http://xxx.lanl.gov/abs/cond-mat/0204331⟩.

[4] Borland, L. "A Theory of Non-Gaussian Option Pricing." *Quant. Fin.* **2** (2002): 415–431. May 2002. arXiv.org e-Print Archive, Quantum Physics, Cornell University. September 2002. ⟨http://xxx.lanl.gov/abs/cond-mat/0205078⟩.

[5] Bouchaud, J. P., and M. Potters. *Theory of Financial Risks.* Cambridge: Cambridge University Press, 2000.

[6] Curado, E. M. F., and C. Tsallis. "Corrigenda." *J. Phys. A* **24** (1991): 3187.

[7] Curado, E. M. F., and C. Tsallis. "Corrigenda." *J. Phys. A* **25** (1992): 1019.

[8] Curado, E. M. F., and C. Tsallis. "Generalized Statistical Mechanics: Connection with Thermodynamics." *J. Phys. A* **24** (1991): L69.

[9] Eberlein, E., U. Keller, and K. Prause. "New Insights into Smile, Mispricing and Value at Risk: The Hyperbolic Model." *J. Bus.* **71(3)** (1998): 371–405.

[10] Gardiner, C. W. *Handbook of Stochastic Methods.* 2d ed. Berlin: Springer-Verlag, 1994.

[11] Hull, J. C. *Options, Futures, and Other Derivatives.* 3rd ed. New Jersey: Prentice-Hall, 1997.

[12] Merton R. C. "Theory of Rational Option Pricing." *Bell J. Econ. & Mgmt. Sci.* 4 (1973): 141–183.

[13] Micahel, F., and M. D. Johnson. "Financial Market Dynamics." August 2001. arXiv.org e-Print Archive, Quantum Physics, Cornell University. September 2002. ⟨http://xxx.lanl.gov/abs/cond-mat/0108017⟩.

[14] Musiela, M., and M. Rutkowski. *Martingale Methods in Financial Modelling.* Applications of Mathematics, vol. 36. Berlin: Springer, 1997.

[15] Neftci, S. N. *An Introduction to the Mathematics of Financial Derivatives.* San Diego, CA: Academic Press, 1996.

[16] Oksendal, B. *Stochastic Differential Equations.* 5th ed. Berlin: Springer-Verlag, 1998.

[17] Osorio, R., L. Borland, and C. Tsallis. "Distributions of High-Frequency Stock-Market Observables." This volume.

[18] Risken, H. *The Fokker-Planck Equation.* 2d ed. Berlin: Springer-Verlag, 1996.

[19] Shreve, S. *Lectures on Finance*, edited by P. Chalasani and S. Jha. ⟨http://www.cs.cmu.edu/~chal/shreve.html⟩.

[20] Tsallis, C. "Possible Generalization of Boltzmann-Gibbs Statistics." *J. Stat. Phys.* **52** (1988): 479.

[21] Tsallis, C., and D. J. Bukman. "Anomalous Diffusion in the Presence of External Forces: Exact Time-Dependent Solutions and Their Thermostatistical Basis." *Phys. Rev. E* **54** (1996): R2197.

[22] Tsallis, C., S. V. F. Levy, A. M. C. Souza, and R. Maynard. "Statistical-Mechanical Foundation of the Ubiquity of Levy Distributions in Nature." *Phys. Rev. Lett.* **75** (1995): 3589.

[23] ⟨http://www.pmpublishing.com⟩.

Distributions of High-Frequency Stock Market Observables

Roberto Osorio
Lisa Borland
Constantino Tsallis

Power laws and scaling are two features that have been known for some time in the distribution of *returns* (i.e., price fluctuations), and, more recently, in the distribution of *volumes* (i.e., numbers of shares traded) of financial assets. As in numerous examples in physics, these power laws can be understood as the asymptotic behavior of distributions that derive from nonextensive thermostatistics.

Recent applications of the q-Gaussian distribution to returns of exchange rates and stock indices are extended here for individual U.S. stocks over very small time intervals and explained in terms of a feedback mechanism in the dynamics of price formation. In addition, we discuss some new empirical findings for the probability density of low volumes and show how the overall volume distribution is described by a function derived from q-exponentials.

1   INTRODUCTION

In March 1900 at the Sorbonne, a 30-year-old student—who had studied under Poincaré—submitted a doctoral thesis [2] that demonstrated an intimate knowledge of trading operations in the Paris Bourse. He proposed a probabilistic method to value some options on *rentes*, which were then the standard French government bonds. His work was based on the idea that *rente* prices evolved according to a random-walk process that resulted in a Gaussian distribution of price differences with a dispersion proportional to the square root of time. Although the importance of Louis Bachelier's accomplishment was not recognized by his contemporaries [24], it preceded by five years Einstein's famous independent, but mathematically equivalent, description of diffusion under Brownian motion. The idea of a Gaussian random-walk process (later preferably applied to logarithmic prices) eventually became one of the basic tenets of most twentieth-century quantitative works in finance, including the Black-Scholes [3] complete solution to the option-valuation problem—of which a special case had been solved by Bachelier in his thesis.

In the times of the celebrated Black-Scholes solution, however, a change in perspective was already under way. Starting with the groundbreaking works of Mandelbrot [18] and Fama [11], it gradually became apparent that probability distribution functions of price changes of assets (including commodities, stocks, and bonds), indices, and exchange rates do not follow Bachelier's principle of Gaussian (or "normal") behavior. Instead, the consistent appearance of *fat tails* signals in different degrees an almost universal deviation from normality. In this sense the *normal* distribution becomes the *exception*, and is only a reasonable approximation for price changes over the long range (months or years). For shorter periods, extreme events occur with frequencies that can be many orders of magnitude larger than those expected from Gaussian probabilities.

Why are price changes over medium-range periods (say days) not normally distributed? Take the case of daily *returns*, which are defined as relative price differences. (More precisely, as explained in the next section, this discussion pertains to logarithmic returns.) Each daily return is the sum of a large number of, say, five-minute returns. If these five-minute returns were independent and identically distributed with a finite variance, the central limit theorem (CLT) would guarantee that daily returns are approximately Gaussian. (See, e.g., Mantegna and Stanley [19].) Since this is not so, one or more CLT assumptions are not satisfied. In fact, even though returns display exponentially decaying autocorrelations that, for all practical purposes, disappear in minutes, *squared* returns (and also *absolute values* of returns) show strong temporal autocorrelations that decay slowly as a power law [8, 10, 17]. The CLT assumption of independence is thus violated. There is no longer a basis for expecting normal distributions for daily returns.

We illustrate this point by showing the distribution of daily returns of the Standard & Poor's 500 (SP500) stock index, along almost eight years, or 2000

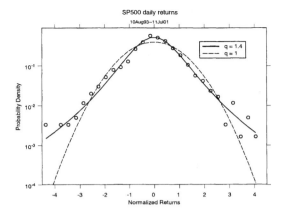

FIGURE 1 Empirical histogram (circles) for normalized de-meaned returns of the SP500 index, compared with a Gaussian distribution of zero mean and unit variance (dashed line) and a q-Gaussian with $q = 1.4$ (solid line).

trading days, in figure 1. The empirical histogram (in a log-linear plot) is compared to the normal distribution ($q = 1$), which obviously underestimates the frequency of extreme events. In figure 1, we anticipate the fact that the q-Gaussian function of generalized thermostatistics provides a better description of returns over their whole range.

Another important feature of the distributions of returns is *time scaling*: the same functional form describes returns measured over different time intervals, as already noticed by Mandelbrot [18]. Scaling and power laws are some of the results that strongly resonate with ideas of statistical mechanics. These results, and the recent availability for research of a large amount of high-frequency intraday trading data, brought the attention of an increasing number of physicists to this area. For the so-called "econophysicists" (see reviews in Bouchaud and Potters [7], Farmer [12], and Mantegna and Stanley [19]), financial markets are complex systems and price movements and trading volumes are the macroscopic collective effects of the microscopic interactions of a large number of trading agents.

Mandelbrot used the scaling property and the observation of power-law tails in a time series of cotton prices to propose a stable Lévy distribution (with infinite variance) for financial returns. Recent studies of data sets of high-frequency individual stocks returns [14, 22] suggest both that their variances are finite and that the exponent in the power-law tails lies outside the stable Lévy interval. This opens the door to the consideration of other models to describe these distributions.

The q-Gaussian distribution of nonextensive thermostatistics [25, 26, 27, 28, 29, 30, 31] was applied very recently with remarkable success to describe returns of exchange rates [23] and of the SP500 stock index [20] over a very wide range of values, including both the central regime, which is approximately Gaussian, and the power-law tails. In this work, we extend these applications to sets of individual stocks and relate the returns distribution to a feedback mechanism that determines the microscopic dynamics of price changes.

Volatility, which is a measure of the amplitude of price fluctuations, is another financial variable whose definition is based on price changes and has been recently explored in high-frequency data (see, e.g., Liu et al. [17]). There is another basic financial variable of interest, however, related to volatility but defined independently: the *trade volume*. This can be measured in terms of either the number of shares or the number of currency units traded in a given time interval. (The statistical properties are practically the same for both definitions.) Recent high-frequency studies of trade volume [15] showed a power-law decay for high volumes, as it is found for returns. Here we show empirical results for a higher range of normalized volumes, which include the novel finding of a possible different power law for low volumes.

The data set studied here comes from the "Trades and Quotes" database, which is distributed by the New York Stock Exchange (NYSE). We use the full year of 2001 for the study of returns and of volumes. A distribution of 1-min returns and volumes (both appropriately normalized, as described in the next sections) for the 20 top-volume stocks for each of the two major U.S. exchanges—the NYSE and the National Association of Security Dealers Automated Quotation (NASDAQ)—is shown in figure 2. This illustrates how returns and volumes are correlated. High volumes are usually accompanied by high absolute values of returns.

One of the ultimate goals of a complete theory of market activity would be to explain eventually this complete joint distribution of returns and volumes and its evolution at different time intervals. This still seems to elude the current approaches by researchers in the field. Meanwhile, the separate characteristics of the distributions of returns and of volumes remain of great interest. We discuss these properties in the next sections.

2 RETURNS

We approximate a time series of prices of a financial asset, recorded at each trade, by a continuous stochastic process $S(t)$. Given a time interval of length Δt, we use the logarithmic definition of returns

$$R(t, \Delta t) = \ln S(t) - \ln S(t - \Delta t). \tag{1}$$

This definition has the advantage of being additive over time. This means that $R(t, \Delta t)$ can be expressed as a sum of returns over smaller time intervals spanned

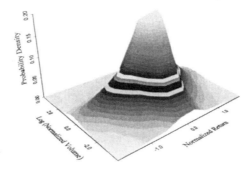

FIGURE 2   Empirical histogram for the joint distribution of normalized 1-min returns and volumes of 40 high-volume U.S. stocks in 2001.

by Δt. The first-order term in the Taylor expansion of the logarithmic definition is the fractional price change

$$R(t, \Delta t) \approx \frac{S(t) - S(t - \Delta t)}{S(t - \Delta t)}, \tag{2}$$

which is the traditional economic definition of the rate of return on an investment (or, for short, a return).

For a fixed Δt and a given stock, we then define *normalized returns* as

$$r(t) = \frac{[R(t) - \overline{R}]}{\sigma(R)}, \tag{3}$$

where $\overline{R}$ is the mean and $\sigma(R)$ is the standard deviation of $R(t)$ over the given time series. This normalization is useful in consolidating the data for different stocks. Figure 3 displays the 1-min normalized returns of the four NASDAQ stocks with the highest median number of trades per minute in the year 2001. The possibility that the four sets of data are sampled from the same scaled, underlying distribution is apparent.

We use the normalization of eq. (3) to examine separately the data for the ten top-volume stocks in each of the NASDAQ and the NYSE exchanges, during the year 2001, for $\Delta t = 1, 2, 3$ min. Each 1-min histogram results from $\sim 10^6$ data points. As seen in figures 4 and 5, the Gaussian distribution, represented by the dotted lines, is again obviously totally inadequate to describe the tail. Note that the curves for $\Delta t = 2$ and 3 min have been translated vertically for display purposes. For each curve, we present a fit for a q-Gaussian scaled distribution

$$P_q(r) = \frac{1}{Z}[1 + (q - 1)br^2]^{-1/(q-1)}. \tag{4}$$

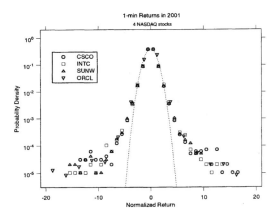

FIGURE 3   Empirical histogram for 1-min normalized returns of the four most highly traded (as measured by the median number of trades per min) stocks in the NASDAQ exchange. The legend identifies each stock by its "ticker symbol." The dotted line is the scaled Gaussian distribution.

Each fit minimizes the sum of the square errors of the logarithms of the probability density and uses q as the only adjustable parameter. This follows from the fact that the scaled distribution has unit variance, which implies that $b = 1/(5 - 3q)$ for $q < 5/3$, and that the normalization factor Z is also a function only of q. We should emphasize that it would be hard to justify a unique fitting scheme: the precise value of a fitted q depends on both the type (e.g., linear or logarithmic) and size of histogram bins and on the weights one decides to give to different regions in the histogram. The important result is that values close to $q \approx 1.4$ give a good description of the main characteristics of the distribution, including the power-law behavior at the tails—with exponent $2/(1-q)$—and the nearly Gaussian behavior at the center. (It should be noticed, however, that for very small absolute values of returns, there will be downward deviations from Gaussian behavior caused by the discreteness of price changes.)

A recent model [13] of trading agents produces universal "cubic laws" (i.e., power-law exponents -3) for *cumulative* distributions of returns. The corresponding exponent -4 for probability densities would follow, in our approach, from the asymptotic behavior of q-Gaussians with $q = 1.5$. Some variation of the power-law exponents in Plerou et al. [22] occurs from stock to stock. Temporal variations of q, which may result in different averaged values of q over the overall time periods of each data set (1993–2001 for the SP500, 2001 for the intraday data) may also be a concern. The other studies that apply generalized thermostatistics to financial distributions disagree on whether q should remain constant over different time intervals. Ramos and co-workers [23] use different q values to

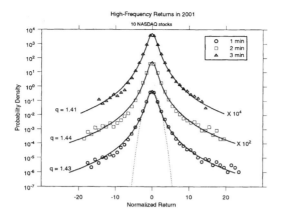

FIGURE 4 Empirical histograms for combined normalized returns of ten high-volume stocks in the NASDAQ exchange. The dotted line is the scaled Gaussian distribution.

fit foreign-exchange returns over different time periods, while Michael and Johnson [20] use a constant value of q to explain the evolution of the distribution of the SP500 index. In the latter approach, convergence to Gaussian behavior occurs only in the sense that the occurrence of tail events becomes rare as less data become available for larger (nonoverlapping) time intervals. The variation of q in different small time intervals in our study is not strong enough to lead us to reject the hypothesis that q is constant. More extensive empirical research may be needed to separate between the two possibilities.

Based on the empirical result that log-returns evolve according to a q-Gaussian, we propose the following model. Consider the evolution of prices $S(t)$ of a financial asset in the interval $[\tau, \tau + \Delta t]$ between two instants used to define an observed return.

The quantity

$$Y(t) = \ln \frac{S(\tau + t)}{S(\tau)} \tag{5}$$

represents the return evolving from its zero value at $t = 0$ to its observed value $R(\tau + \Delta t, \Delta t)$, in the notation of eq. (1), at $t = \Delta t$. We propose that the dynamics of $Y(t)$ inside this interval is described by the Langevin equation

$$dY = \mu dt + \sigma d\Omega, \tag{6}$$

with drift μ and noise coefficient σ. In a standard Wiener process σ is the volatility and $d\Omega$ will be $d\omega$, a Gaussian uncorrelated random variable sampled from a distribution with zero mean and variance dt. Instead, we assume here that $d\Omega$

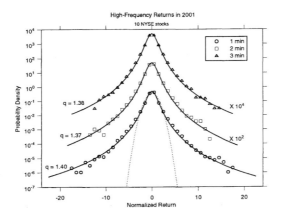

FIGURE 5   Empirical histograms for combined normalized returns of ten high-volume stocks in the NYSE exchange. The dotted line is the scaled Gaussian distribution.

follows the statistical feedback process [5]

$$d\Omega = P(\Omega)^{-\frac{q-1}{2}} d\omega , \qquad (7)$$

which, for $q > 1$, indicates that occurrences of high absolute values of the integrated noise Ω will cause higher variability, suggesting the observed phenomenon of volatility clustering. The probability distribution P then satisfies the nonlinear Fokker-Planck equation [21, 32]

$$\frac{\partial}{\partial t} P(\Omega, t) = \frac{1}{2} \frac{\partial}{\partial \Omega^2} P^{2-q}(\Omega, t) . \qquad (8)$$

Explicit time-dependent solutions for P are given by the q-Gaussians

$$P_q(\Omega, t) = \frac{1}{Z(t)} \{1 + (q-1)\beta(t)\Omega(t)^2\}^{-\frac{1}{q-1}} , \qquad (9)$$

where the coefficients $\beta(t)$ and $Z(t)$ are as in Borland [6]. Consequently, the observed returns $R = Y(\Delta t)$ follow the distribution

$$P_q(R) = \frac{1}{Z(\Delta t)} \{1 + (q-1)\tilde{\beta}(\Delta t)(R - \mu\Delta t)^2\}^{-\frac{1}{q-1}} , \qquad (10)$$

with $\tilde{\beta}(t) = \beta(t)/\sigma^2$. The variance of this distribution is inversely proportional to $\tilde{\beta}(t)$, which implies *scaling* if q is assumed constant: The normalized return r of eq. (3) will follow a single distribution, given by eq. (4), over different time scales Δt.

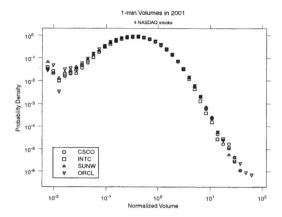

FIGURE 6 Empirical histogram for scaled 1-min volumes of the four most highly traded stocks in the NASDAQ exchange.

In the limit $q \to 1$ the standard theory is recovered, and eq. (10) becomes a standard Gaussian distribution. Besides providing a model that fits empirical stock returns well, the feedback process of Eqs. (6)–(7) has also been used to derive a non-Gaussian option pricing theory [4, 6] that reproduces some other empirical observations.

3 VOLUMES

A stock's *share volume* is defined as the number of shares traded over a given time interval Δt. This is obviously a positive quantity, in fact represented by very right-skewed distributions. It can be expressed as the product of the number of transactions in Δt with the average order size, but we treat here only the combined distribution. A recent analysis [15] of high-frequency data of individual stocks showed that volumes follow a power-law behavior at high values. Here we discuss these distributions over a range that includes their low-volume behavior.

Starting with a time series of volumes $V(t, \Delta t)$ for each stock, aggregated over time intervals of length Δt, we define normalized volumes $v = V/\overline{V}$, where the denominator $\overline{V}$ is the mean volume per time Δt in the time series of each stock. This definition differs from that of Gopikrishnan et al. [15] in that we use the mean instead of their median.

The resulting volume distribution for the four most highly traded stocks in the NASDAQ exchange in 2001 is shown in figure 6 in a log-log plot. The normalized volumes for these particular stocks appear to rise from the same

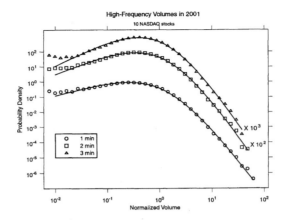

FIGURE 7 Empirical histogram for scaled volumes of ten high-volume stocks in the NASDAQ exchange.

underlying distribution. This is not true, however, for stocks with lower trading activity.

We show next the consolidated histograms for the ten top-volume stocks in each of the NASDAQ and the NYSE exchanges (figs. 7 and 8). The curves for $\Delta t = 2$ and 3 min have been translated vertically for display purposes. In addition to the previously recognized power-law regime at high values, these figures show a possibly novel regime where the probability density increases with the volume. The two regimes are separated by a parabolic maximum that signals log-normal behavior. This presence of a maximum is reminiscent of the behavior of a Poisson distribution, which would be an adequate description for volumes if trades occurred at random at a constant rate and all trade sizes were the same. For a large enough rate of arrival of events, a Poisson distribution also displays a maximum value.

The characteristic shape of the volume curve is similar to that of high-frequency volatilities, where a log-normal regime [1, 17] and a power-law high-value regime [17] have been discovered. The cross-sectional dispersion, or *variety*, of returns of stocks belonging to a given index seems to show similar behavior [16].

The solid lines in figures 7 and 8 show results of fitting a function

$$P_{\gamma,q}(v) = A \left(\frac{v}{v_0} \right)^\gamma \exp_q \left(\frac{-v}{v_0} \right) \tag{11}$$

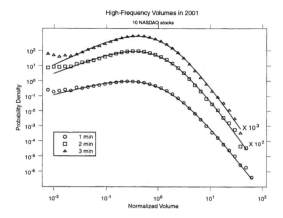

FIGURE 8 Empirical histogram for scaled volumes of ten high-volume stocks in the NYSE exchange.

TABLE 1 Best-fit parameters for the volume distribution $P_{\gamma,q}(v)$ of eq. (11).

Exchange	Δt (min)	γ	q	v_0
NASDAQ	1	0.93	1.19	0.23
	2	1.36	1.16	0.20
	3	1.69	1.15	0.17
NYSE	1	0.42	1.25	0.29
	2	1.05	1.20	0.19
	3	1.79	1.17	0.12

(where A is a normalization parameter defined by $\int_0^\infty P_{\gamma,q}(v)dv = 1$). This function is a modification of the q-exponential [9, 30, 29, 28, 25, 26, 27, 33]

$$\exp_q(x) = [1 + (1 - q)x]^{1/(1-q)} . \tag{12}$$

The values of the fitting parameters used in the curves in figures 7 and 8 are shown in table 1.

These curves describe well the behavior of the volume distributions, except for very low volumes. This case, which corresponds to few occurrences (note that this regime is exaggerated in log-log plots), is likely influenced by the discreteness of numbers of trades in a time interval and the discreteness of numbers of shares traded. (Stock shares are usually traded in multiples of 100.)

Although the mechanisms that lead to distributions of volume that follow eq. (11) are not yet well understood, we conjecture that, as in the case of the distributions of price variations, it reflects the occurrence of long-range memory in the activity of traders. High volumes tend to be followed by high volumes. A classical description, in this case based on a Poisson process for the arrival of orders, is not adequate to such systems. Further research is needed to model such distributions.

4   CONCLUDING REMARKS

We showed in this discussion how financial variables are amenable, similarly to many systems in the physical sciences, to descriptions in terms of distributions that follow from the concept of nonextensive entropy, and that, in some asymptotic limits, may provide universal power laws. Individual agents in financial markets interact "microscopically" in such a way as to provoke observable "macroscopic" effects. Temporal long-range autocorrelations in volatilities and volumes, for instance, result from mechanisms such as a herding attitude among traders. Such a behavior can be modeled by microscopic equations with feedback terms, which lead to the stochastic dynamics that result in non-Gaussian distributions of returns.

A similar mechanism may be in place in the construction of a theory for the distributions of volumes, which show some resonances with generalized thermostatistics. A more ambitious goal is a general theory of trading activity that may explain the common distribution of returns and volumes (such as in fig. 2), and how it evolves in different time scales.

ACKNOWLEDGMENTS

The authors are grateful to Jeremy Evnine for helpful discussions.

REFERENCES

[1] Andersen, T. G., T. Bollersev, F. X. Diebold, and H. Ebens. "The Distribution of Stock Return Volatility." *J. Fin. Econ.* **63** (2001): 43–76.
[2] Bachelier, L. "Théorie de la Spéculation." *Annales Scientifiques de l'Ecole Normale Supérieure* **III** (1900): 21–86; English translation in P. Cootner, ed., *The Random Character of Stock Market Prices* (Cambridge, MA: MIT Press, 1964).
[3] Black, F., and M. Scholes. "The Pricing of Options and Corporate Liabilities." *J. Pol. Econ.* **81** (1973): 637–654.
[4] Borland, L. "Closed-Form Option Pricing Formulas based on a Non-Gaussian Stock Price Model." *Phys. Rev. Lett.* **89** (2002): 098701.
[5] Borland, L. "Microscopic Dynamics of the Nonlinear Fokker-Planck Equation." *Phys. Rev E* **57** (1998): 6634–6642.

[6] Borland, L. "The Pricing of Stock Options." This volume.

[7] Bouchaud, J.-P., and M. Potters, *Theory of Financial Risks*. Cambridge, UK: Cambridge University Press, 2000.

[8] Crato, N., and P. J. de Lima. "Long-Range Dependence in the Conditional Variance of Stock Returns." *Econ. Lett.* **25** (1993): 281–285.

[9] Curado, E. M. F., and C. Tsallis, "Generalized Statistical Mechanics: Connection with Thermodynamics." *J. Phys. A* **24** (1991): L69–L72; Corrigenda: **24** (1991): 3187; and **25** (1992): 1019.

[10] Ding, Z., C. W. J. Granger, and R. F. Engle. "A Long Memory Property of Stock Market Returns and a New Model." *J. Emp. Fin.* **1** (1993): 83–106.

[11] Fama, E. F. "The Behavior of Stock Market Prices." *J. Bus.* **38** (1965): 34–105.

[12] Farmer, J. D. "Physicists Attempt to Scale the Ivory Towers of Finance." *Comp. Sci. & Eng. (IEEE)* **Nov/Dec** (1999): 26–39.

[13] Gabaix, X., P. Gopikrishnan, V. Plerou, and H. E. Stanley. "A Simple Theory of the 'Cubic' Laws of Stock Market Activity." *Quart. J . Econ.* (2001): submitted.

[14] Gopikrishnan, P., V. Plerou, Y. Liu, L. A. N. Amaral, X. Gabaix, and H. E. Stanley. "Scaling and Correlation in Financial Time Series." *Physica A* **287** (2000): 362–373.

[15] Gopikrishnan, P., V. Plerou, X. Gabaix, and H. E. Stanley. "Statistical Properties of Share Volume Traded in Financial Markets." *Phys. Rev. E* **62** (2000): R4493–R4496.

[16] Lillo, F., and R. Mantegna. "Variety and Volatility in Financial Markets." *Phys. Rev. E* **62** (2000): 6126–6134.

[17] Liu, Y., P. Gopikrishnan, P. Cizeau, M. Meyer, C. K. Peng, and H. E. Stanley. "The Statistical Properties of the Volatility of Price Fluctuations." *Phys. Rev. E* **60** (1999): 1390–1400.

[18] Mandelbrot, B. B. "The Variation of Certain Speculative Prices." *J. Bus.* **36** (1963): 394–419.

[19] Mantegna, R. N., and H. E. Stanley. *An Introduction to Econophysics*. Cambridge, UK: Cambridge University Press, 2000.

[20] Michael, F., and M. D. Johnson. "Financial Market Dynamics." *Physica A* to be published. Preprint cond-mat/0108017, 2001.

[21] Plastino, A. R., and A. Plastino. "Non-Extensive Statistical Mechanics and Generalized Fokker-Planck Equation." *Physica A* **222** (1995): 347–354.

[22] Plerou, V., P. Gopikrishnan, L. A. N. Amaral, M. Meyer, and H. E. Stanley. "Scaling of the Distribution of Price Fluctuations of Individual Companies." *Phys. Rev. E* **60** (1999): 6519–6529.

[23] Ramos, F. M., C. Rodrigues Neto, R. R. Rosa, L. D. Abreu Sá, and M. J. A. Bolzam. "Generalized Thermostatistical Description of Intermittency and Nonextensivity in Turbulence and Financial Markets." *Nonlinear Analysis* **23** (2001): 3521–3530.

[24] Taqqu, M. S. "Bachelier and His Times: A Conversation with Bernard Bru." *Fin. & Stochastics* **5** (2001): 3–32.

[25] Tsallis, C. "Entropic Nonextensivity: A Possible Measure of Complexity." *Chaos, Solitons and Fractals* **13** (2002) 371–391.

[26] Tsallis, C. "Nonextensive Statistical Mechanics: A Brief Review of its Present Status." *Annals Braz. Acad. Sci.* (2002): to appear

[27] Tsallis, C. "Nonextensive Statistical Mechanics: Construction and Physical Interpretation." This volumes.

[28] Tsallis, C. "Nonextensive Statistical Mechanics and Thermodynamics: Historical Background and Present Status." In *Nonextensive Statistical Mechanics and Its Applications*, edited by S. Abe and Y. Okamoto. Lecture Notes in Physics Series. Heidelberg: Springer-Verlag, 2001.

[29] Tsallis, C. "Nonextensive Statistics: Theoretical, Experimental and Computational Evidences and Connections." *Braz. J. Phys.* **29** (1999): 1–35.

[30] Tsallis, C. "Non-Extensive Thermostatistics: Brief Review and Comments." *Physica A* **221** (1995): 277–290.

[31] Tsallis, C. "Possible Generalization of Boltzmann-Gibbs Statistics." *J. Stat. Phys.* **52** (1988): 479–487.

[32] Tsallis, C., and D. J. Bukman. "Anomalous Diffusion in the Presence of External Forces: Exact Time-Dependent Solutions and their Thermostatistical Basis." *Phys. Rev. E* **54** (1996): R2197–R2200.

[33] Tsallis, C., R. S. Mendes, and A. R. Plastino. "The Role of Constraints within Generalized Nonextensive Statistics." *Physica A* **261** (1998): 534–554.

Entropic Subextensivity in Language and Learning

Lukasz Debowski

1 INTRODUCTION

In this chapter, we identify possible links between theoretical computer science, coding theory, and statistics reinforced by subextensivity of Shannon entropy. Our specific intention is to address these links in a way that may arise from a rudimentary theory of human learning from language communication.

The semi-infinite stream of language production that a human being experiences during his or her life will be called simply *parole* (= "speech," [7]). Although modern computational linguistics tries to explain human language competence in terms of explicit mathematical models in order to enable its machine simulation [17, 20], modeling *parole* itself (widely known as "language modeling") is not trivial in a very obscure way. When a behavior of *parole* that improves its prediction is newly observed in a finite portion of the empirical data, it often suggests only minor improvements to the current model. When we use larger portions of parole to test the freshly improved model, this model always fails seriously, but in a different way. How we can provide necessary updates to, with-

Nonextensive Entropy—Interdisciplinary Applications
edited by Murray Gell-Mann and Constantino Tsallis, Oxford University Press

out harming the integrity of, the model is an important problem that experts must continually solve. Is there any sufficiently good definition of *parole* that is ready-made for industrial applications?

Although not all readers of human texts learn continuously, *parole* is a product of those who can and often do learn throughout their lives. Thus, we assume that the amount of knowledge generalizable from a finite window of *parole* should diverge to infinity when the length of the window also tends to infinity. Many linguists assume that a very distinct part of the generalizable knowledge is "linguistic knowledge," which can be finite in principle. Nevertheless, for the sake of good modeling of *parole* in practical applications, it is useless to restrict ourselves solely to "finite linguistic knowledge" [6, 22].

Inspired by Crutchfield and Feldman [5], we will call any processes (distributions of infinite linear data) "finitary" when the amount of knowledge generalizable from them is finite, and "infinitary" when it is infinite. The crucial point is to accurately define the notion of knowledge generalized from a data sample. According to the principle of minimum description length (MDL), generalizable knowledge is the definition of such representation for the data which yields the shortest total description. In this case, we will define infinitarity as computational infinitarity (CIF).

It is not so straightforward to point out an *easily* computable function of data sample that is co-divergent with the amount of generalizable knowledge. Perhaps some function of a sequence of entropies might be this needed observable, but first let us point out some functions that are not co-diverging.

Conserving infinitarity, *parole* can be written down as a string of discrete symbols from a finite alphabet. It is known that the relative frequencies of single symbols for any text in a fixed ethnic language approach constants like those in a Bernoulli process, i.e., a string of independent, identically distributed (IID) variables [15]. Nevertheless, Bernoulli processes are finitary. Also, Zipf's law for words in a simplified form (counts of words are proportional to $-(1 + \epsilon)$th power of their count ranks) can be explained as an effect of the Bernoulli process for characters [1, 12].

Parole does not resemble stationary Gaussian processes with long memory in the sample mean, either. For those processes, the autocorrelation function $\rho(n)$ between two positions in the series, separated by $(n-1)$ ones, decays like $1/n$ or slower [2]. For Gaussian processes, mutual information between the positions is $I(S_1; S_{n+1}) = -\frac{1}{2} \log_2[1 - \rho(n)^2]$. For *parole*, $I(S_1; S_{n+1})$ decays probably only like $1/n^3$ [19].

2   THE PRINCIPLE OF MINIMUM DESCRIPTION LENGTH

Let us assume that we have a single semi-infinite string S_1^∞ of symbolic data rather than a probability distribution for all semi-infinite strings. Let S_1^N be the prefix of S_1^∞ of length N. (We write $S_M^N = S_M S_{M+1}...S_N$ where S_i are the individual symbols.) Following Debowski [8, 9], we will show a *suggestive* differential equation linking the amount of knowledge generalizable from a finite prefix S_1^N and the length of the whole optimal representation for S_1^N. We have adopted the MDL meaning of generalizable knowledge. Let $D(N, C)$ be the total description length using representation definition C,

$$D(N, C) = \Delta(N, C) + \Theta(C), \tag{1}$$

where $\Delta(N, C)$ is the length of the coded data and $\Theta(C)$ is the length of representation definition C.

Let $C(N)$ stand for that C which is applied to yield the shortest total description. Adding any constraints for the admissible descriptions does not influence the further reasoning. Let us assume that N and C can be approximated as continuous, and $\Delta(N, C)$ and $\Theta(C)$ are differentiable functions of their parameters. Minimality of $D(N, C)$ for $C = C(N)$ is expressed as

$$\left. \frac{\partial D(N, C)}{\partial C} \right|_{N=\text{const}, C=C(N)} = 0. \tag{2}$$

Let us assume that for any constant representation definition C, coded data grow proportionally to the original data size,

$$\left. \frac{\partial D(N, C)}{\partial N} \right|_{C=\text{const}} = \frac{\Delta(N, C)}{N}. \tag{3}$$

Equation (3) is crucial for our further reasoning; nevertheless, one should expect equality (3) for sufficiently large N if data representations are asymptotically local.

The equation $D(N) := D(N, C(N))$ is the so-called minimum description length (MDL), $\Theta(N) := \Theta(C(N))$ is the amount of generalizable knowledge, and $\Delta(N) := \Delta(N, C(N))$ is the length of the optimal data encoding. Combining Eqs. (1), (2), and (3) yields

$$\Theta(N) = D(N) - ND'(N). \tag{4}$$

If $\Delta(N) = \epsilon_\mu N^\mu$, then

$$\Theta(N_2) - \Theta(N_1) = \frac{1 - \mu}{\mu} \epsilon_\mu \left[N_2^\mu - N_1^\mu \right]. \tag{5}$$

If $\Delta(N)$ contains several power-law components, one just needs to add the respective solutions from eq. (5). Then $\mu \leq 1$ holds themselves since the shortest representation for any data cannot grow faster than the data themselves.

For $0 < \mu < 1$, the amount of generalizable knowledge diverges to infinity like the μth power of the amount of unencoded data. If $\mu = 0$ or $\mu = 1$, the growth of the generalizable knowledge can be only logarithmic with the unencoded data size. Finally, let us note that the reasoning presented in this section does not depend on the assumed computational power and reversibility of the coding procedures.

3 COMPRESSION BY DEFINING AND POINTING

The formal notion of a minimum description length $D(N)$ for a finite symbolic string S_1^N can be approached in many ways. Let us present two of the most popular ones [4]:

1. With some Turing machine T we have to generate string S_1^N. The Kolmogorov complexity $D_K(N, T, t)$ is the length of the shortest program for the machine to output S_1^N and stop before t operations. Semi-infinite strings S_∞, such that $\lim_{N,t \to \infty} D_K(N, T, t) = \infty$, do exist and are called incompressible.
2. S_1^N is split into partition Π, i.e., a string of substrings π_i whose concatenation is S_1^N. We look for the set of uniquely decodable binary strings b_i for substrings π_i such that the sum of lengths l_i of all identifiers b_i in S_1^N is minimal. The Huffman complexity is $D_H(N, \Pi) = \sum_i c_i l_i$, where c_i is the count of π_i in Π and l_i is the length of optimal b_i (the solution for b_i is known as Huffman code). For the optimal b_i, the condition $\sum_i c_i l_i = \min$ and Kraft's inequality $\sum_i 2^{-l_i} \leq 1$ (the condition for unique decodability) yield the constraint

$$\frac{-\log_2 c_i}{M} \leq l_i < \frac{-\log_2 c_i}{M} + 1, \quad M = \sum_i c_i. \tag{6}$$

We may need up to $\propto A^N t$ operations to compute $D_K(N, T, t)$ exactly. Instead, we will discuss the practical perspectives of using Huffman complexity, the latter being much faster to compute.

A very special case for Huffman coding occurs when the types of predefined substrings π_i are fixed and their relative counts c_i/M converge to constants p_i for $M \to \infty$ ($\Leftrightarrow$ the law of large numbers is met for c_i/M). This is the case of *parole* if π_i are single symbols (letters or phonemes). An even more special case is when the occurrences of substrings π_i resulting from some partition Π of S_1^N are IID variables. Then, the minimum of average Huffman complexity per substring in Π applied to any partition Π' that is not more dense than Π converges to Shannon entropy for atomic substrings in Π,

$$\lim_{M \to \infty} \min_{\Pi' \succ \Pi} \frac{\langle D_H(N, \Pi') \rangle}{M} = H(\{p_i\}) := -\sum_i p_i \log_2 p_i. \tag{7}$$

(Π' is not more dense than Π, $\Pi' \succ \Pi$, when any occurrence of substring π'_j from Π' is fully contained in some occurrence of π_i from Π.) Tuning the Huffman coding to the Shannon limit requires absurd amounts of symbolic processing compared to the profits. One needs to test all possible partitions Π' to find the best one, while earning less than 1 bit per substring in partition Π. The result should be read differently: In constrained optimal coding, perceiving IID randomness of entities is roughly equal to using information about the counts of entities and to being incapable of using information about their linear order.

Using more information about the particular linear order of symbols in S_1^N can only decrease its Huffman complexity.

Let us suppose that codebook C consists of a list of correspondences: "binary codeword for pointing to entity" $\leftrightarrow$ "entity worth pointing out separately." When the codebook length $\Theta(C)$ grows infinitely with N for the optimal $C = C(N)$, we should note three things:

1. Equation (5) states that for $\epsilon_1 = 0$ the size of the codebook is always comparable to the size of the encoded data.
2. Since the codebook can be large, entities in the codebook should be defined in terms of binary identifiers for subentities rather than in terms of subentities themselves. To find the optimal binary codewords b_i, occurrences of π_i should be counted not only in the data but in the codebook as well.
3. A global search for the codebook yielding minimal Huffman complexity for S_1^N consumes time, exponential in N. In a cheaper local search approach, each fixed state of the codebook and the encoded string could be represented as a node σ in the graph where each node σ is annotated with its Huffman complexity $l(\sigma)$. A link from node σ to node σ' exists if and only if description σ' is yielded by applying to the description σ some symbolic transformation from the predefined finite set of transformations X. If reversibility of the compression is assumed, then for each link from σ to σ' there must be a path from σ' to σ. The search starts in the node with an empty codebook and encoded string equal to unencoded string S_1^N. Then the search moves from one node σ to the next node σ' with minimal $l(\sigma')$, always along existing links, until $l(\sigma) > l(\sigma')$.

One implementation of the presented principles is known as a de Marcken algorithm [6]. In this algorithm, set X consists of two kinds of symbolic transformations: (a) defining a concatenation of two codewords with undefining 0, 1, or 2 elements of the pair, and (b) undefining a codeword. Defining an object means introducing a new codeword, replacing all occurrences of the object with the codeword and enrolling the definition of the codeword, in lieu of the object, into the codebook. Undefining the codeword is the reverse. For English texts, the resulting codebook consists of recursive and mostly meaningful definitions of frequently used syllables, morphemes, words, and fixed phrases. Since many patterns in *parole* cannot be expressed as pure concatenations, the growth of the codebook slows when the compression rate stabilizes at about 2 bits/character. A short example from de Marcken [6] is

the un it ed   st at e s   of a me r ic a   .

The bars overbrace the concatenations that were defined as codewords. The text was deprived of spaces, capitalization, and punctuation.

Une langue est un système où tout se tient [7]. Can it be that a linguistic entity arises only if its frequency roughly exceeds the product of frequencies of its parts? Except for meaningless "atomic symbols" of perception and basic "symbolic operations," there may be hardly any entities of language processing in the human brain before their frequencies are really observed. Everything else would result from the "trade" between feasible definitions (commodities) and available pointers to them (prices). The continuous space of all frequency distributions of strings is cut into discrete basins of optimal formal grammars for them. The same system of discrete entities can be inferred even for highly contextually dependent frequency distributions, such as frequencies of content words. That property would enable quite free communication among adults while a child carefully listening to them could still infer a very similar system for language processing.

Further progress in machine language learning concerns probably the question of more complex reversible symbolic transformations of the descriptions. The de Marcken algorithm rewrites its input as a nearly shortest context-free grammar (CFG). This CFG is a restricted L-system; i.e., codeword definitions are functional rewriting rules $b_i \mapsto \phi(b_i)$, but the rewriting rules cannot contain any cycles. (The letter is necessary to provide only one derivation.) The rewriting of the initial symbol (axiom) is the encoded data. Theoretically, the codebook might be a context-sensitive grammar (CSG) with rules rewriting any binary string as a longer binary string. In practice, such a format would be both too unrestricted and too restricted. (The format cannot contribute to acquiring, for example, inflectional paradigms in a non-redundant form [14].)

4 COMPUTATIONAL AND STATISTICAL INFINITARITY

We say that sequence S_1^∞ is computationally infinitary when $\mu_C > 0$, where

$$\mu_C = \lim_{N \to \infty} \frac{\log [D(N) - N\epsilon_1]}{\log N} < 1, \tag{8}$$

$$\epsilon_1 = \lim_{N \to \infty} \frac{D(N)}{N} = \lim_{N \to \infty} [D(N) - D(N-1)] . \tag{9}$$

$D(N)$, as previously introduced, is interpreted as the average minimum description length for N-tuple of symbols defined alone, without using the properties of any larger ensemble of data. On the other hand, Shannon entropy $H(N)$ is almost the average length of the Huffman code for N-tuple of symbols immersed in the infinite ensemble that meets the law of large numbers for frequencies of N-tuples. It has been shown that for strings of IID variables (and more general cases as well), $\lim_{N \to \infty} \langle D_K(N, T, \infty) \rangle / H(N) = 1$, where $\langle D_K(N, T, \infty) \rangle$ is the mean of Kolmogorov complexity $D_K(N, T, \infty)$ for all N-tuples in the process [4].

The computational mechanics group from Santa Fe Institute has developed a parallel definition of (statistical) infinitarity that resembles our computational infinitarity. In Crutchfield and Feldman [5], (statistically) infinitary processes are defined such that $E = \infty$, where excess entropy E for conditionally stationary processes $S_{-\infty}^{\infty}$ is defined as

$$E = \lim_{N \to \infty} I(S_1^N; S_{N+1}^{2N}) = \lim_{N \to \infty} [H(N) - Nh_1] \, , \qquad (10)$$

$$h_1 = \lim_{N \to \infty} \frac{H(N)}{N} = \lim_{N \to \infty} [H(N) - H(N-1)] \, . \qquad (11)$$

(For $D(N) = D_K(N, T, \infty)$ and the Bernoulli process, $\langle \epsilon_1 \rangle = h_1$.) Still we do not know if, in the general case, $E = \infty$ and $\langle \mu_C \rangle > 0$ imply each other. We ignore if there is a simple relation between $\langle \mu_C \rangle$ and μ_S, where $\mu_S = \lim_{N \to \infty} \log [H(N) - Nh_1] / \log N < 1$. Constant inequilibrium between the resources of definitions and pointers in infinitary processes may imply $\langle \mu_C \rangle > \mu_S$.

A positive result of computational mechanics is the construction of ϵ-machine [23]. The ϵ-machine is the deterministic infinite state automaton whose states are minimal sufficient statistics of $S_{-\infty}^t$ for predicting S_{t+1}^{∞}. Transition from the state at t to the state at $t+1$ is deterministic, given S_{t+1}. The entropy of ϵ-machine's state, called statistical complexity C_μ, meets $C_\mu > E$. If $E = \infty$, also $C_\mu = \infty$.

5   SHANNON ENTROPIES OF PAROLE

Statistical infinitarity of *parole* ($E = \infty$) can be deduced from experiments more easily than computational infinitarity ($\Theta = \infty$) can be. Ebeling and Nicolis [10] and Ebeling and Pöschel [11] measured Shannon entropy $H(N)$ for N-tuple of symbols and reported that it can be fitted by formula

$$H(N) = h_0 + h_\mu N^\mu + h_1 N \, , \qquad (12)$$

with $\mu = 1/2$ for natural language texts and $\mu = 1/4$ for music transcripts. For English and German texts $H(N)$ could be safely estimated up to $N \approx 30$ characters with $h_0 \approx 0$, $h_{1/2} \approx 3.1$ bits, and $h_1 \approx 0.4$ bits. Experiments with prediction by native speakers suggest that eq. (12) can be extrapolated at least up to $N \approx 100$ [16].

Coexistence of inequalities $h_{1/2} > 0$ and $h_1 > 0$ for *parole* deserves a qualitative comment. The inequality $h_{1/2} > 0$ may be a consequence of the fact that infinite learning of new associations from *parole* is possible. The inequality $h_1 > 0$ might be a consequence of the fact that some part of human communication does not require learning. For this part of communication, the listener can be reduced to an ϵ-machine, a deterministic, infinite-state automaton (possibly a transducer) interpreting *parole* one character/phoneme after another. Then the speaker can communicate completely random commands without risking being misunderstood.

Does $\mu = 1/2$ hold for $N \gg 100$ characters still? Does human learning last a lifetime at a regular rate? For small N $\mu = 1/2$ may be only a trace of a brain device to speed up the learning of a general lexicon, grammar, and other knowledge by children listening to casual adult speech. (Same device might be used for restimulating the adults.) For large N, μ may decrease to much smaller values depending on the text and the amount of generalizable knowledge still inferable by an adult.

6   SUBEXTENSIVE ENTROPIC EXPONENT

In Debowski [8] we tried to explain why exactly $\mu = 1/2$ holds for *parole* if, for all $0 < \mu < 1$, the optimal codebook grows infinitely. The first hypothesis was that the language learner memorizes only the optimal codebook. The optimality criterion would be $\Theta(N) = \max$, which yields the optimal μ dependent on N with $\mu \to 1$ for $N \to \infty$. The largest codebook could be extracted from sufficiently large amounts of almost complete noise! Nevertheless, in order to enlarge the codebook, one needs to store more and more previously encoded data because the newly extracted portion of the codebook may associate occurrences of some objects in the previously encoded data and in the fresh ones.

The second hypothesis in Debowski [8] was that, for $\epsilon_1 = 0$, the learner would memorize both the codebook and the encoded data. For $D(N) \propto N^\mu$, $\mu = 1/2$ could correspond to the solution of constraint

$$\frac{N}{D(N)} \Theta'(N) \approx \mu(1 - \mu) = \max . \tag{13}$$

Such constraint assumes the tendency of the learner to maximize his or her overall compression factor $N/D(N)$ and the effective tendency of *parole* to maximize the learner's rate of effective codebook acquisition $\Theta'(N)$.

Yet an alternative explanation of $\mu \approx 0.5$ uses the low precision of the estimates of μ. Conditional mutual information is defined as

$$
\begin{aligned}
I(N_1; N_3 | N_2) &:= I(S_1^{N_1}; S_{N_1+N_2+1}^{N_1+N_2+N_3} | S_{N_1+1}^{N_1+N_2}) \\
&= H(N_1 + N_2) + H(N_2 + N_3) - H(N_1 + N_2 + N_3) - H(N_2) .
\end{aligned}
$$

For first-order Markov chains, $I(1; 1|N) = 0$ if $N \geq 1$. $I(N; N|N)/H(N) = \max$ for $H(N) \propto N^\mu$ means $2 \cdot 2^\mu - 3^\mu - 1 = \max$ and yields $\mu \approx 0.574$. *Parole* may be well optimized for uniform departure from Markov chains in all time scales. When speaking, we try to delay the independent partial explanations of our current speech acts. On the other hand, the delays can be tolerable since in each time scale *parole* would resemble a second-order Markov chain. (Compare these second-order models to those in Manning and Schutze [20].)

7   FORMAL AND PROBABILISTIC FORMAL GRAMMARS

Possible interactions between the values of μ_S and the formal constraints in the human language might be explained by the theory of formal languages. Formal language L is any subset of the set of all strings over a finite alphabet. One of the most powerful methods of the theory is to map formal properties of the language L into analytic properties of its generating function $G(z) = \sum_{n \geq 0} g(n)z^n$, where $g(n)$ is the number of strings of length n belonging to L. For example, the classic result [3] says that $G(z)$ is an algebraic function of z if language L is generated by a context-free grammar that is unambiguous (each string generated by the grammar has only one derivation tree). If language L consists of all N-tuples appearing in some string S_1^∞ and all N-tuples appear equally often for given N, then $\log_2 g(N) = H(N)$.

Intriguingly, Flajolet [13] asks whether languages with $\log_2 g(N) \propto \sqrt{N}$ can be generated by context-free grammars at all. Still, there is a need for a rigorous proof of whether statistically infinitary processes can be generated by probabilistic context-free grammars (PCFGs). Crutchfield and Feldman [5] seem to suspect that the answer is positive, while our first impression in Debowski [9] was the opposite. Kuich [18] and Miller and O'Sulllivan [21] consider the entropy rate h_1 of PCFGs, but we have not come across any similar publication on excess entropy E.

8   CONCLUSIONS

Infinitarity of human language production may be its vital property which cannot be ignored in almost any application of linguistics. Infinitarity may be linked with power-law sublinear growth of different measures of description length explored in the various domains of science. Theoretical linguistics could profit a lot from a rigorous comparison of all these approaches. On the other hand, it may also contribute to some new insights.

ACKNOWLEDGMENTS

We would like to thank Cosma Shalizi for his review and remarks. The longer version of this article is available from ⟨http://www.ipipan.waw.pl/~ldebowsk⟩.

REFERENCES

[1] Belevitch, V. "Théorie de l'information et statistique linguistique." *Académie royale de Belgique. Bulletin de la classe des sciences* (1956): 419.

[2] Beran, J. *Statistics for Long-Memory Processes.* Chapmann & Hill, 1994.

[3] Chomsky, N., and M. P. Schützenberger. "The Algebraic Theory of Context-Free Languages." In *Computer Programming and Formal Systems*, edited by P. Bradford and D. Hirschberg. Amsterdam: North-Holland, 1963.

[4] Cover, T. M., and J. A. Thomas. *Elements of Information Theory.* New York: John Wiley & Sons, Inc., 1991.

[5] Crutchfield, J. P., and D. P. Feldman. "Regularities Unseen, Randomness Observed: Levels of Entropy Convergence." Working Paper 01-02-012, Santa Fe Institute, Santa Fe, NM, 2001.

[6] de Marcken, C. G. "Unsupervised Language Acquisition." Ph.D. thesis, Massachussetts Institute of Technology, 1996, unpublished.

[7] de Saussure, F. *Cours de linguistique générale.* Paris: Payot, 1916.

[8] Debowski, L. "Quantitative Considerations on Finding the Shortest Descriptions for Meaningful Symbolic Sequences." ICS PAS Reports, Nr 924, Instytut Podstaw Informatyki PAN, 2001.

[9] Debowski, L. "A Revision of Coding Theory for Learning from Language." Paper presented at Formal Grammar/Mathematics of Language—FGMOL 2001, held August 10–12, 2001 in Helsinki. Electronic proceedings under construction.

[10] Ebeling, W., and G. Nicolis. "Word Frequency and Entropy of Symbolic Sequencies: A Dynamical Perspective." *Chaos, Solitons and Fractals* **2** (1992): 635.

[11] Ebeling, W., and T. Pöschel. "Entropy and Long-Range Correlations in Literary English." *Europhys. Lett.* **26** (1994): 241.

[12] Ferrer, R., and R. V. Solé. "The Small-World of Human Language." Working Paper 01-03-016, Santa Fe Institute, Santa Fe, NM, 2001.

[13] Flajolet, P. "Analytic Models and Ambiguity of Context-Free Languages." *Theor. Comp. Sci.* **49** (1987): 283.

[14] Goldsmith, J. "Unsupervised Learning of the Morphology of a Natural Language." *Comp. Ling.* **27** (2001): 153.

[15] Herdan, G. *Quantitative Linguistics.* London: Butterworths & Co., 1964.

[16] Hilberg, W. "Der bekannte Grenzwert der redundanzfreien Information in Texten—eine Fehlinterpretation der Shannonschen Experimente?" *Frequenz* **44** (1990): 243.

[17] Jurafsky, D., and J. H. Martin. *Speech and Language Processing: An Introduction to Natural Language Processing, Computational Linguistics, and Speech Recognition.* Englewood Cliffs: Prentice Hall, 2000.

[18] Kuich, W. "On the Entropy of Context-Free Languages." *Infor. & Control* **16** (1970): 173.

[19] Li, W. "Mutual Information Functions versus Correlation Functions." *J. Stat. Phys.* **60** (1990): 823.

[20] Manning, C. D., and H. Schütze. *Foundations of Statistical Natural Language Processing.* Cambridge, MA: MIT Press, 1999.

[21] Miller, M. I., and J. A. O'Sullivan. "Entropies and Combinatorics of Random Branching Processes and Context-Free Languages." *IEEE Trans. Infor. Theor.* **38** (1992): 1292.

[22] Möbius, B. "Rare Events and Closed Domains: Two Questionable Concepts in Speech Synthesis." 2001. CiteSeer, NEC Research Institute. Sept. 2002. ⟨http://citeseer.nj.nec.com/421578.html⟩.

[23] Shalizi, C. R. "Causal Architecture, Complexity and Self-Organization for Time Series and Cellular Automata." Ph.D. thesis, University of Wisconsin-Madison, 2001, unpublished.

A Generalization of the Zipf-Mandelbrot Law in Linguistics

Marcelo A. Montemurro

1 INTRODUCTION

Human language evolved by natural mechanisms into an efficient system capable of coding and transmitting highly structured information [12, 13, 14]. As a remarkable complex system it allows many levels of description across its organizational hierarchy [1, 11, 18]. In this context statistical analysis stands as a valuable tool in order to reveal robust structural patterns that may have resulted from its long evolutionary history.

In this chapter we shall address the statistical regularities of human language at its most basic level of description, namely the rank-frequency distribution of words.

Around 1932 the philologist George Zipf [6, 19, 20] noted the manifestation of several robust power-law distributions arising in different realms of human activity. Among them, the most striking was undoubtedly the one referring to the distribution of words frequencies in human languages. The best way to introduce Zipf's law for words is by means of a concrete example. Let us take a literary

Interdisciplinary Applications of Ideas from Nonextensive. . .
edited by Murray Gell-Mann and Constantino Tsallis, Oxford University Press.

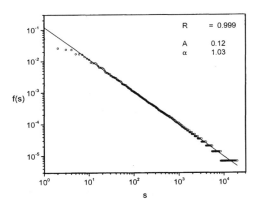

FIGURE 1 Frequency rank distribution of words in James Joyce's Ulysses. For this text the total number of words is $N = 268,112$, and $V = 28,838$ different word forms.

work, say, James Joyce's *Ulysses*, and perform some basic statistics on it, which simply consists in counting all the words present in the text and noting how many occurrences each distinct word form has. For this particular text we should arrive at the following numbers: the total number of words $N = 268,112$, and the number of different word forms $V = 28,838$. We can now order the list of different words according to decreasing number of occurrences, and we can assign to each word a rank index s equal to its position in the list starting from the most frequent word. Some general features of the rank-ordered list of words can be mentioned at this point. First, the top-rank words are functional components of language devoid of direct meaning, such as the article *the* and prepositions, for instance. A few ranks down the list, words more related to the contents of the text start to appear. Zipf's law for words is a simple mathematical statement that relates the number of occurrences of a word and its corresponding rank index s. In order to make apparent that relation we can plot in a double logarithmic graph the normalized frequency $f(s) = n(s)/N$ versus the rank s of each word and the result can be seen in figure 1. From the figure it is clear that the words tend to fall on a straight line in the logarithmic scale indicating the following power-law relation between rank and frequency, which constitutes Zipf's law for words: $f(s) = A/s^\alpha$.

In the original formulation of this empirical law, the exponent α was taken to be exactly 1. If instead the exponent is assumed as a parameter and fitted to the data, it usually takes on values slightly greater than unity, as is the case for Joyce's text where it takes $\alpha \approx 1.03$. However, the most striking feature of Zipf's law in not merely its simplicity but its ubiquitous validity corroborated over a large number of human languages. In table 1 we show the values of the

TABLE 1 Values of α for different text sources.

Source	α
Ulysses	1.03
Don Quijote	1.09
Iliad	1.06
La Divina Commedia	1.07
David Copperfield	1.12
Hamlet	1.08

exponents in the power law that best fits the data in the central region of the rank axis for a small set of literary works in different languages.

A close look at the rank-frequency distribution depicted in figure 1, reveals some few common patterns. On the one hand, the most frequent words deviate from the power-law behavior showing frequencies slightly smaller than those corresponding to the extrapolation of the power-law region. On the other, the lack of sufficient statistics for high-rank words makes it difficult to conclude about the form of the distribution for the infrequent words.

Based on some simplified arguments on the structure of language, Benoit Mandelbrot proposed the following generalization to the original Zipf's law [8]: $f(s) = A/(1 + Cs)^{\alpha}$, where C is a second parameter that needs to be adjusted to fit the data. Ever since the discovery of the law there have been opposite views regarding its origin and significance. It has been shown that this form of the law is also obeyed by random processes that can be mapped onto texts [7]; hence, it rules out any sufficient character for linguistic depth inherent to the Zipf-Mandelbrot law. Nevertheless, it has been argued that it is possible to discriminate between human writings and stochastic versions of texts precisely by looking at statistical properties of words that fall beyond the scope where Mandelbrot's generalization holds [2].

2 WORD FREQUENCIES IN VERY LARGE TEXT SAMPLES

The generalized expression proposed by Mandelbrot describes very well the statistical behavior of words in a range from the highest frequencies (low ranks) to the Zipf's regime, characterized by the power-law region of the distribution. However, the behavior of very low frequency words cannot be correctly assessed from individual text samples, since much larger texts should be required to resolve the statistical behavior of uncommon words.

In figure 2 we show the Zipf's graphs obtained for four large corpora, each gathering several works from four different authors respectively. The points in the graphs represent local averages on windows of constant width in the logarithmic

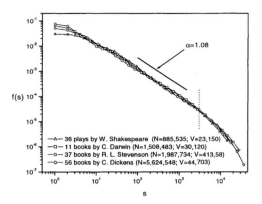

FIGURE 2 Frequency-rank distribution of words for four large text samples. The vertical dash line is placed approximately where Zip's law ceases to hold.

scale. This was done solely to smooth fluctuations in the data. It is apparent that there are three clearly identifiable regimes in the rank-frequency distribution. The first one corresponds to very frequent words and show variations for each of the corpora considered. The second one is the Zipfian regime where all the curves show the same power-law behavior from $s \approx 10$ to $s \approx 2000$ to 3000. At that point the third regime appears that is characterized by a faster decay. It is remarkable that all the curves start to deviate from the power-law behavior at approximately the same value of the rank. This suggests that regardless of the different sizes of the texts considered, the vocabularies can be divided into two parts of distinct nature [4, 5]: one of basic usage whose overall linguistic structure leads to the Zipf-Mandelbrot law, and a second part containing more specific words with a less flexible syntactic function.

In figure 3 we show the frequency-rank distribution of words in a very large corpus made up of 2,606 books written in English comprising nearly 1.2GB of ASCII data. The total number of tokens in this case rose to 183,403,300 with a vocabulary size of 448,359 different words. It is remarkable that the point at which the departure from Zipf's law takes place has just moved to $s \approx 6000$ despite the increase in sample size of nearly two orders of magnitude. However, the striking new feature is that the form of the distribution for high ranks reveals as a second power-law regime [5].

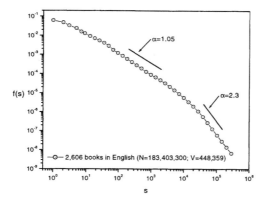

FIGURE 3   Zipf's plot for a large corpus comprising 2,606 books in English, mostly literary works and some essays. The straight lines in the logarithmic graph show pure power laws as a visual aid.

3   QUANTITATIVE DESCRIPTION OF THE RANK-FREQUENCY DISTRIBUTION OVER THE WHOLE RANGE OF RANKS

Nonextensive statistical mechanics applies successfully to systems that exhibit some kind of (multi)fractality and long-range correlations. In this sense, it is worth mentioning here that the Zipf-Mandelbrot law for words has been related to Tsallis' generalized thermodynamics by means of arguments based on the fractal structure of symbolic sequences with long-range correlations [3]. Recently, a direct measurement of long-range fractal correlations in written language has been accomplished by mapping texts onto time series and preserving words as fundamental linguistic units [10]. These results strongly motivate the application of ideas from nonextensive statistical mechanics to human language.

As a first step in that direction we shall discuss a phenomenological model for a generalization of Zipf-Mandelbrot law inspired on the nonextensive formalism proposed by C. Tsallis [17]. We start from the observation that the Zipf-Mandelbrot law satisfies the following first-order differential equation,

$$\frac{df}{ds} = -\lambda f^q . \tag{1}$$

The solutions to eq. (1) asymptotically take the form of pure power laws with decay exponent $1/(q-1)$. It is possible to generalize this expression to include a crossover to a second regime, as follows:

$$\frac{df}{ds} = -\mu f^r - (\lambda - \mu)f^q , \tag{2}$$

where we have added a new parameter and a new exponent. In the case $1 \leq r < q$ and $\mu \neq 0$ the new additions allow the presence of two global regimes characterized by the dominance of either exponent depending on the particular value of f. The use of this equation in the realm of linguistics was originally suggested by C. Tsallis [15], and it had previously been used to describe experimental data on the re-association in folded proteins [16] within the framework of nonextensive statistical mechanics [17].

We can distinguish three qualitatively different cases in the solutions of eq. (2) according to the values assumed by the parameters. The first case corresponds to the recovery of Zipf-Mandelbrot law by taking $r = q > 1$, or which has the same effect $\mu = 0$ and $q > 1$. In this case the solutions of eq. 2 is the q-exponential distribution:

$$f(s) = \frac{1}{\left[1 + (q-1)s\lambda\right]^{\frac{1}{q-1}}}, \tag{3}$$

where we chose $f(0) = 1$.

As it was mentioned above, Zipf-Mandelbrot law fits correctly the distribution of words for single texts, but fails when large volumes of data are analyzed. However, we shall show that the solutions of eq. (2) describe accurately the rank-frequency distribution of words in all possible situations encountered. If we now take $r = 1$ and $q > 1$, we obtain

$$f(s) = \frac{1}{\left[1 - \frac{\lambda}{\mu} + \frac{\lambda}{\mu}e^{(q-1)\mu s}\right]^{\frac{1}{q-1}}}. \tag{4}$$

This expression shows a very interesting behavior for $\mu \ll \lambda$, since for small values of s it reduces to eq. (3) and then for larger values of s it undergoes a crossover to an exponential decay. Equation (4) describes with great accuracy the rank-frequency distribution obtained for large copora for individual authors. As an example, in figure 4 we can see the excellent fit obtained with eq. (4) for one of the large text samples already used in figure 2. Whereas Zipf-Mandelbrot law would have only fitted a small fraction of the total vocabulary present in these corpora, eq. (4) captures the behavior of the rank-frequency distribution along the whole range of the rank variable.

Finally, for the more general situation $1 < r < q$, the solution of eq. (2) yields an following involved expression [9, 16]:

$$s = \frac{1}{\mu}\left\{\frac{f^{-(r-1)} - 1}{r-1} - \frac{(\lambda/\mu) - 1}{1 + q - 2r}\left[H(1; q - 2r, q - r, (\lambda/\mu) - 1)\right.\right.$$
$$\left.\left. -H(f; q - 2r, q - r, (\lambda/\mu) - 1)\right]\right\}, \tag{5}$$

with the definition

$$H(f; a, b, c) = f^{1+a}\, {}_2F_1\left(1; \frac{1+a}{b}; \frac{1+a+b}{c}; -f^b c\right), \tag{6}$$

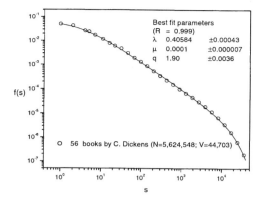

FIGURE 4   Frequency-rank distribution for a corpus made up of 56 books by Charles Dickens (circles) together with a fit (full line) by eq. (4).

where $_2F_1$ is the hypergeometric function. However, it is possible to derive a much simpler relation for the probability density function $p_f(f)$. The value of the rank for a word with a normalized frequency of occurrence f, can be written in the following way:

$$s(f) = \int_f^\infty Np_f'(f')\,df'\,. \tag{7}$$

This can be seen by noting that $Np_f'(f')$ gives the number of words that appear with normalized frequency f', thus the corresponding position in the rank list of a word with frequency f equals the total number of words that have frequency greater or equal than f. From this we can write

$$p_f(f) \propto -\frac{ds}{df}\,. \tag{8}$$

Consequently, in the probability density representation, the solution is

$$p_f(f) \propto \frac{1}{\mu f^r + (\lambda - \mu)f^q}\,. \tag{9}$$

This result is particularly interesting in view of the mathematical simplicity of eq. (9). In figure 5 we show the rank-frequency distribution for the very large corpus of 2606 books in English compared with a plot of eq. (5) using the best fit parameters obtained by fitting the data with eq. (9).

Finally, in table 2, a summary collecting the value of the two main exponents in the distribution is presented for the text corpora analyzed in this work with the addition of a set of seventy books written in classical Latin. In the table we can see clearly that single-author corpora yield $r = 1$, which means exponentially

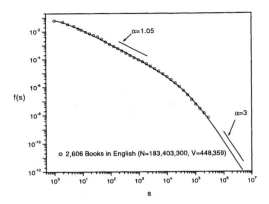

FIGURE 5   Actual data from the corpus of $2,606$ books in English together with a plot of eq. (5) with the following parameters obtained after fitting the data with eq. (9): $q \approx 1.95$, and $r \approx 1.32$

TABLE 2   Values of the exponents q and r for different large corpora.

Source	q	r
36 plays by William Shakespeare	1.89	1
11 books by Charles Darwin	1.90	1
56 books by Charles Dickens	1.89	1
37 books by Robert L. Stevenson	1.91	1
2606 books in English	1.95	1.32
70 books in Classical Latin	2.05	1.23

decaying tails in the rank-frequency plot. However, when different authors are combined, the final regimes becomes a power law, and consequently $r > 1$. The understanding of the full implications of these observations requires further interdisciplinary research.

4   CONCLUSIONS

In this chapter we have presented a generalization of Zipf-Mandelbrot law for words with ideas drawn from nonextensive statistical mechanics. After analyzing the rank-frequency distributions of words in large text samples we found strong systematic deviations to Zipf-Mandelbrot law that emerged as robust statistical features. These statistical regularities make a complex scenario of different regimes in the distribution, and they could be all encompassed quantitatively by considering the set of solutions of a first-order differential equation (eq. (2)).

More investigation is required in order to propose plausible microscopic mechanisms for the emergence of the complex phenomenology described in this work.

ACKNOWLEDGMENTS

The author is very grateful to both Constantino Tsallis and Damián Zanette for their valuable and insightful observations. The text files used in this work are from the *Project Gutenberg* etex archive [21].

REFERENCES

[1] Aknajian, A., R. A. Demers, A. K. Farmer, and R. M. Harnish. *Linguistics: An Introduction to Language and Communication.* Cambridge, MA: MIT Press, 1992.

[2] Cohen, A., R. N. Mantegna, and S. Havlin. "Can Zipf Analyses and Entropy Distinguish between Artificial and Natural Language Texts?" *Fractals* **5** (1997): 95.

[3] Denisov, S. "Tsallis Thermodynamics and Zipf's Law." *Phys. Lett. A* **235** (1997): 447.

[4] Ferrer, R., and R. V. Solé. "The Small World of Human Language." *Proc. Roy. Soc. Lond. B* **268** (2001): 2261–2266.

[5] Ferrer, R., and R. V. Solé. "Two Regimes in the Frequency of Words and the Origins of Complex Lexicons: Zipf's Law Revisited." *J. Quant. Ling.* (2002): in press.

[6] Gell-Mann, M. *The Quark and the Jaguar: Adventures in the Simple and the Complex.* New York: Freeman, 1995.

[7] Li, W. "Random Texts Exhibit Zipf's-Law-like Word Frequency Distribution." *IEEE Trans. Inform. Theory* **38(6)** (1992): 6.

[8] Mandelbrot, B. *The Fractal Structure of Nature.* Freeman: New York, 1983.

[9] Montemurro, M. A. "Beyond the Zipf-Mandelbrot Law in Quantitative Linguistics." *Physica A* **300** (2001): 567–578.

[10] Montemurro, M. A., and P. A. Pury. "Long-Range Fractal Correlations in Literary Corpora." *Fractals* (2002): in press.

[11] Montemurro, M. A., and D. H. Zanette. "Entropic Analysis of the Role of Words in Literary Texts." *Adv. Compl. Sys.* **5(1)** (2002): 7–17.

[12] Nowak, M. A., J. B. Plotkin, and V. A. A. Jansen. "The Evolution of Syntactic Communication." *Nature* **404** (2000): 495.

[13] Pinker, S. *The Language Instinct.* New York: Harper Collins, 2000.

[14] Pinker, S. *Words and Rules.* New York: Harper Collins, 2000.

[15] Tsallis, C. Private communication.

[16] Tsallis, C., G. Bemski, and R. S. Mendes. "Is Re-association in Folded Proteins a Case of Nonextensivity." *Phys. Lett. A* **257** (1999): 93.

[17] Tsallis, C. "Possible Generalization of Boltzmann-Gibbs Statistics." *J. Stat. Phys.* **52** (1988): 479.

[18] Van Dijk, T. A. "Semantic Macro-Structures and Knowledge Frames in Discourse Comprehension." In *Cognitive Processes in Comprehension*, edited by Marcel Adam Just and Patricia A. Carpenter, 3–32. Hillsdale, NJ: Lawrence Erlbaum, 1977.

[19] Zipf, G. K. *Human Behavior and the Principle of Least Effort.* Reading, MA: Addison-Wesley, 1949.

[20] Zipf, G. K. *The Psycho-Biology of Language, an Introduction to Dynamic Philology.* Cambridge MA: MIT Press, 1965.

[21] Official Home Page in the Internet. ⟨http://promo.net/pg/⟩.

Coarse-Graining, Scaling, and Hierarchies

Juan Pérez–Mercader

1 INTRODUCTION

We present a scenario that is useful for describing hierarchies within classes of many-component systems. Although this scenario may be quite general, it will be illustrated in the case of many-body systems whose space-time evolution can be described by a class of stochastic parabolic nonlinear partial differential equations. The stochastic component we will consider is in the form of additive noise, but other forms of noise such as multiplicative noise may also be incorporated. It will turn out that hierarchical behavior is only one of a class of asymptotic behaviors that can emerge when an out-of-equilibrium system is coarse grained. This phenomenology can be analyzed and described using the renormalization group (RG) [6, 15]. It corresponds to the existence of complex fixed points for the parameters characterizing the system.

As is well known (see, for example, Hochberg and Perez-Mercader [8] and Onuki [12] and the references cited there), parameters such as viscosities, noise couplings, and masses evolve with scale. In other words, their values depend on

Nonextensive Entropy—Interdisciplinary Applications
edited by Murray Gell-Mann and Constantino Tsallis, Oxford University Press   357

the scale of resolution at which the system is observed (examined). These scale-dependent parameters are called *effective parameters*. The evolutionary changes due to coarse graining or, equivalently, changes in system size, are analyzed using the RG and translate into differential equations for the probability distribution function [8] of the many-body system, or the n-point correlation functions and the effective parameters. Under certain conditions and for systems away from equilibrium, some of the fixed points of the equations describing the scale dependence of the effective parameters can be complex; this translates into complex anomalous dimensions for the stochastic fields and, therefore, the correlation functions of the field develop a complex piece. We will see that basic requirements such as reality of probabilities and maximal correlation lead, in the case of complex fixed points, to hierarchical behavior.

This is a first step for the generalization of extensive behavior as described by real power laws to the case of complex exponents and the study of hierarchical behavior.

A system may exhibit many possible asymptotic behaviors, including hierarchical and standard power-law behavior. Which behavior the system attains on coarse graining depends on where the initial values of the physical parameters are located in the basins of attraction of the fixed points. Thus, if their initial values are within the appropriate region of parameter space, one could start with (real) power-law behavior and, via coarse graining, enter into the domain where the complex fixed point dominates and the system develops a phase characterized by hierarchical behavior. Thus, the various behaviors of a system can *emerge* as scale-dependent collective phenomena.[1] In principle, for a given system, all possible combinations can occur: not only from real to complex, but from real to real, from complex to real or from complex to complex. What and how it happens depends on the microscopic dynamics that underlie the system, the nature of the stochastic component and, as already mentioned, on the initial values for the physical parameters at a particular scale.

We will conclude that hierarchical behavior is a class of *emergent* critical behavior in many-body systems where there is, *in addition to complex fixed-points, maximal correlation* between the components in the system.

After some required definitions and examples, we will dwell on the phenomenology of hierarchical behavior, develop the theoretical basis to describe it and its emergence, and illustrate this through some examples existing in nature. All our work is analytic, there are no computer simulations, and the results can be derived from first principles.

---

[1]The scale can be a scale of length, differences in temperatures, or any other suitable dimensional parameter, as in the applications of the renormalization group to critical phenomena.

2 HIERARCHIES: DEFINITIONS AND EXAMPLES

2.1 DEFINITIONS

A hierarchy is a manifestation of ordered behavior in a given structure or physical system. We find such behavior in many examples and, broadly speaking,
it is characterized by the property that these systems organize into sequential
subsystems. These subsystems are not isolated, but are contingent upon both
a larger subsystem and a smaller subsystem. In a hierarchic system the whole
system is "composed of interrelated subsystems, each of the latter being in turn
hierarchic in structure until we reach some lowest level of elementary subsystem." This definition, from Simon, is the one we will use for a hierarchy and a
hierarchical system.[2]

Hierarchies are present in social organizations such as an army, a corpora-
tion, or a university; in biological systems such as cells or ecologies; and in the
realm of astrophysical objects or in the forces known to control the universe as
we know it today. They are commonplace in *complex systems*. From the presence
of hierarchical organization in nature ranging from nuclei to atoms, molecules,
macromolecules, cells, organisms, ecologies, planets, planetary systems, interstellar clouds, groups of stars, clusters of stars, galaxies, clusters of galaxies, and the
whole universe, we infer that hierarchical behavior must be related to *differentiation, evolution, and adaptation* of the system to a given environment, which
can be random in its nature and which could, itself, be hierarchical!

2.2 EXAMPLES

Perhaps the simplest and most illustrative example of a hierarchy is provided by
the traditional Russian dolls known as "matryushky" (fig. 1). This traditional toy
consists of a succession of hollow lacquered wooden dolls which, when opened,
contain inside a similar doll, which once opened contains yet another doll and
so on. There is a largest doll (limited in size only by how big the craftsman
can afford to make it) and a smallest doll (limited in size by the skill of the
craftsman). Seen from far away each of the dolls "looks" like the others except
that it is built at a different scale; however, on closer inspection, they can be
different not only in size but in the *details* of their painted dress and so on.
The "matryushky" clearly fit the definition given above for a hierarchical system
and visually illustrate that the members of a hierarchy can be different in their
particular details.

Another example of a hierarchy is provided by the known forces and their
strengths: the strong, electromagnetic, weak, and gravitational forces have very
different strengths and the realms where they apply do not coincide.

---

[2]For a classical description of what is understood by a hierarchy see Simon [13].

FIGURE 1   The traditional Russian toy "matryushky" illustrates the notion of hierarchy.

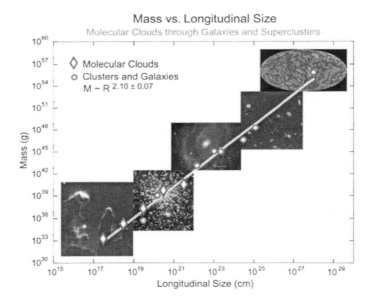

FIGURE 2   Plot of the data corresponding to the known classes of structures in the universe (larger than planetary systems). We show the average mass vs. average longitudinal size of structures from molecular clouds through galaxies to superclusters of galaxies and the cosmic microwave background radiation as well as their fit to a power law (exponent = 2.10 ± 0.07).

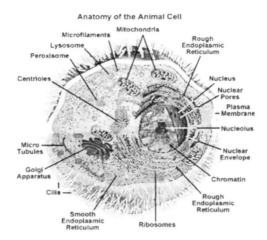

Anatomy of the Animal Cell

FIGURE 3   The eukaryotic cell. Today it is viewed as a complex network with multiple connections and functions, with hierarchical organization playing a basic role that could be fundamental to understanding its operation.

Yet they all are forces. In fact, their current observed properties suggest that they may all have derived from a unique unified force existing early in the history of the universe, and that the forces differentiated during the evolution of the universe. Their differentiation gave rise to the hierarchy observed today. The coarse graining in this example may have been provided (in partially understood ways) by the expansion of the universe together with the concomitant decoupling [9] of degrees of freedom.

But perhaps one of the clearest examples of hierarchical organization is provided by the universe at scales where gravity is important. We know that clusters of galaxies contain galaxies (of various ages and morphologies), which in turn contain huge molecular clouds, and globular and open clusters of stars. In some of these stars (although it is suspected for many of them) there are planetary systems which, as we will briefly discuss, are also hierarchically organized (fig. 2).

Many examples of hierarchical organization are known in biological systems: the eukaryotic cell itself (fig. 3) with its nucleus, ribosomes, and other corpuscles inside its cytoplasm, all contained within the cell membrane and wall, as well as the set of metabolic cycles and chemical reactions which take place to make it and keep it alive, are all magnificent examples of hierarchies of various types. And, of course, they provide extraordinary challenges when we attempt their mathematical/physical descriptions from first principles!

3 PHENOMENOLOGY OF HIERARCHIES IN CONDENSED SYSTEMS

Next we give a succinct description of the basic aspects of the phenomenology associated with hierarchies. These will help us identify the essential features present in hierarchical behavior. We will use them as guidelines to identify a theoretical framework appropriate for the description of systems and the conditions in which hierarchical behavior appears.

From the definition of hierarchy given above, it follows that they occur in many-body systems, in other words, in systems where there are many components interacting according to some physical rules. The rules summarize the effects of some force or they could describe the nature of some situation constrained by geometry or dynamical conditions, such as, in the case of nearest-neighbor interactions or near-to-nearest neighbor, or any other form of connection between system components that makes the set of many components into a many-body system.

Another generic feature of hierarchical behavior is that it manifests in systems where "the whole is more than the sum of its parts," that is, in systems experiencing *emergence*, a behavior that cannot be expected on the basis of the superposition (simple sum) of the individual properties attributable to each of the parts. For example, the emergence of the aroma in coffee comes from the interaction of the molecules of the alcaloids it contains.

These two basic tenets are, of course, not unique to hierarchical systems. There are other *systematic* properties that one associates with hierarchical systems. More specifically, hierarchical organization appears in *evolving* dynamical systems where the system undergoes some kind of transition from a nonhierarchical to a hierarchical organization. This organization is the result of the reshuffling of system components during the evolutionary process, and results as free energy is exchanged between system parts [2]. When we put this together with the association with emergence, we infer that hierarchies are related to (i) the spatiotemporal evolution of the complete system both of its parts and of the system as a collective, when all the components are involved, even if in different degrees. In hierarchical behavior, slow and fast, small and large size scales are simultaneously involved, although they may be *decoupled*, in the sense that fast degrees of freedom at a lower level in the hierarchy are averaged out and the slow motions one level up are "constant." In other words, (ii) scale diversity and scale changes are important when hierarchical behavior is at work. (Note that over the above qualitative description hovers the notion of "nonequilibrium.")

The association of spatiotemporal evolution, the decoupling of degrees of freedom, and scale changes implies that hierarchies should arise as a system is "coarse grained" [15], that is, as we change the resolution scale at which we study the evolving system. When this happens, and as is well known from the application of the RG to critical phenomena, clusters form within the system. This implies the existence of interfaces which can, in some instances, be energy rich and lead to emergent behavior at the interfaces. For dynamical critical phenomena, the reordering associated with coarse graining can *in some instances* also be related to a form of hierarchical behavior. Next we will study how this comes about in situations where complex fixed points of the renormalization group are present.[3]

In order to be self-contained, we will briefly discuss scaling in a physical system (section 4); then we will focus on some consequences of renormalization group scaling (section 5), and complex renormalization group exponents and hierarchies (section 6), then we will consider some applications (section 7); and finally we offer some conclusions.

4   SCALING IN A PHYSICAL SYSTEM

We say that a system displays scaling behavior when basic physical variables of the system, such as the free energy, satisfy a relation of the form

$$f(x) = \frac{1}{a} f(\lambda x) \,, \tag{1}$$

where f represents the physical variable, and x collectively represents the variables and parameters characterizing the physical system such as temperature or size; λ and a are numbers (for simplicity we will assume that $a \neq 1$). The number λ defines a change of scale from x to λx. The above functional equation has the solution

$$f(x) = C \, x^\sigma \,. \tag{2}$$

That is, $f(x)$ has power-law behavior in x with exponent σ. Substituting eq. (2) into eq. (1) we see at once that

$$\lambda^\sigma = a = e^{2\pi i n} \cdot a, \quad \text{with} \ \ n = 0, \pm 1, \pm 2, \ldots \tag{3}$$

where we have simply rewritten the unit as $e^{2\pi i n}$. Hence,

$$\sigma = \frac{\ln a}{\ln \lambda} + i \frac{2\pi}{\ln \lambda} \cdot n \equiv \sigma_R + i\sigma_I \,. \tag{4}$$

In other words, the exponent σ can be complex, with a *discrete* ("quantized") imaginary part proportional to $n = 0, \pm 1, \pm 2, \ldots$.

This can happen in many-body systems through the combined effects that the "regular" (interactions and causal evolution) and the "random" (fluctuations and uncertainties of various kinds) have on the probability distribution function

---

[3]This happens for nonunitary systems, where the Wallace-Zia theorem does not apply. Dynamical critical systems provide an explicit example of this. The author thanks R. Stinchcombe for a discussion on this point.

(pdf) for the system under consideration. In a stochastic field theory, for example, interactions and fluctuations (a) modify the n-point correlation functions of the fields at different space-time points, and (b) lead to divergences in the original set of parameters such as diffusion constants, masses, coupling constants, or noise parameters which serve phenomenologically to characterize the system. It follows from (a) and (b) that the originally Gaussian pdf is modified in a way which is scale dependent [8].[4] It can be shown that the system enters asymptotic scaling regimes with exponents which are calculable from the original equations and boundary or initial conditions describing the regular, the random, and their interrelation with the environment where evolution takes place.

The pdf is related to $Z[J]$, the characteristic functional (also known as the "generating functional") for the n-point correlation functions. In field theory, this is obtained from the action $S[\phi]$ by performing a path integral over all field configurations:

$$Z[J(\vec{x}, t)] = \int [d\phi] e^{iS[\phi] + i \int dx^d \, dt \phi(\vec{x}, t) J(\vec{x}, t)} . \tag{5}$$

Here $\phi(\vec{x}, t)$ represents the field value at a space-time point with coordinates $(\vec{x}, t)$ in a $(d + 1)$-dimensional space time and $J(\vec{x}, t)$ is the classical source for the field. When dealing with systems away from equilibrium, there are many technical complications for the application of these techniques, but the essence of this part of the procedure is completely captured by eq. (5). The correlation functions are essentially obtained by taking functional derivatives of $Z[J]$ [16].

4.1   SCALING AND THE RENORMALIZATION GROUP

Requiring independence of the physics from the choice of arbitrary length or momentum[5] scale [6], immediately leads to a partial differential equation satisfied by the pdf, and, therefore, by the Green's functions (see eq. (6) below) that describe the statistical properties of the many-body system. The coefficient functions of this partial differential equation are related to the ordinary differential equations satisfied by the effective parameters. The scale dependence induced on the parameters is precisely the one necessary to cancel the overall scale dependence of the Green's functions. The RG equation satisfied by the n-point correlation function $G^{(n)}(r_1, ..., r_n; \alpha_j(\lambda); \lambda)$ of the stochastic field ϕ is

$$\left[-\lambda \frac{\partial}{\partial \lambda} + \sum_i \beta_{\alpha_i} \frac{\partial}{\partial \alpha_i} + \frac{n}{2} \gamma(\alpha_j) \right] G^{(n)}(r_1, ..., r_n; \alpha_j(\lambda); \lambda) = 0 \tag{6}$$

---

[4]Divergences need to be subtracted. This introduces an arbitrary reference scale into the system. Requiring that the physics remains independent of this arbitrary scale is the statement mathematically represented by the Renormalization (semi) Group [6].

[5]One can also perform these operations in momentum space and instead use a momentum scale $\mu \sim 1/\lambda$. This is often more convenient. We are using λ in order to keep closer to the more intuitive notion of "scale" as identified with "size."

and the effective couplings α_i satisfy

$$\lambda \frac{d\alpha_i}{d\lambda} = -\beta_{\alpha_i}(\alpha_j) \, . \tag{7}$$

These β-functions (one per independent coupling α_i) are calculable in perturbation theory, and they summarize the effects of interactions and fluctuations on the couplings. The solutions to eq. (7) are the effective parameters. Just as the couplings become scale dependent, so do the fields themselves, which have a scale-dependent dimension and, therefore, deviate from the canonical value for free (noninteracting) fields. The quantity $\gamma(\alpha_j)$ in eq. (6) denotes that deviation, and for that reason receives the name of the *anomalous* dimension of the field ϕ. Put differently, γ is to the stochastic field ϕ what the $\beta_{\alpha_i}(\alpha_j)$ are to the couplings α_i.

The RG equation (6) together with eq. (7) can be immediately integrated using the method of characteristics. One finds that

$$G^{(n)}(r_1, r_2, ..., r_n; \{\alpha_j(\lambda)\}; \lambda) = \exp\left(-\frac{n}{2} \int_1^s \frac{d\lambda'}{\lambda'} \, \gamma_\phi(\lambda')\right)$$
$$\times G^{(n)}(r_1, r_2, ..., r_n; \{\alpha_j(\lambda_0)\}; \lambda_0) \, . \tag{8}$$

This form of the solution will be important later, when we discuss the connection between evolution, emergence, and scaling behavior leading, in some cases, to hierarchical behavior.

Many out-of-equilibrium stochastic field theories contain various typical terms: a combination of reaction-diffusion terms and stochastic forcing. In other words, they contain a form of the standard parabolic diffusion partial differential equation modified by several local monomials of the field. The monomials can have stochastic couplings and the system is said to be subject to multiplicative noise. When the stochastic term is additive, the contribution from the noise can be interpreted as a stochastic forcing term. In a many-body system, where we do not have access to *all* the variables, one adopts the strategy of following the deterministic evolution of a small subset of all possible variables, but the evolution of these variables may depend *also* on the evolution of the variables over which we have averaged and given up following their deterministic evolution. Thus, the initial values of time $t = 0$ of the variables that we follow do not fully determine their evolution. The stochastic terms are a means of incorporating into the dynamics the effects from degrees of freedom that may not have fully decoupled from the dynamics. We do that by representing the noise through the statistical properties of *its* probability distribution function.[6] These equations, as is well known, describe an amazing range and variety of systems.

It is known that complex fixed points do appear in these systems. We will illustrate this using an important example.

---

[6]We note here that the effects of initial and boundary conditions can also be incorporated into the noise.

4.2   EXAMPLE: A FORCED FLUID AND DYNAMICAL CRITICAL PHENOMENA

Many of the features just mentioned are represented in some regimes of the Navier-Stokes equations for the hydrodynamics of a stirred fluid subject to damping proportional to the velocity. A particularly important example is provided in the case of stirred and damped potential flow. Here, the Navier-Stokes equations take the form

$$\frac{\partial \vec{v}}{\partial t} + \lambda \left(\vec{v} \cdot \vec{\nabla} \right) \vec{v} = \nu \nabla^2 \vec{v} + \left(\frac{\zeta}{\rho} + \frac{1}{3}\nu \right) \vec{\nabla}(\vec{\nabla} \cdot \vec{v}) - \frac{1}{\rho}\vec{\nabla}p + \vec{f}(\vec{x}, t) \,, \quad (9)$$

with the standard terms describing the evolution of the velocity field, but with an extra term $\vec{f}(\vec{x}, t)$ containing both linear damping and a stirring force according to

$$\vec{f} = -m^2 \vec{v}(\vec{x}, t) - \vec{\nabla}\xi(\vec{x}, t) \,. \quad (10)$$

(The convective coupling λ in eq. (10) gets modified under renormalization, and should not be confused with the length scale λ introduced above.) Here ρ is the density, m is a masslike damping coefficient (which introduces an explicit length scale in the problem), and $\xi(\vec{x}, t)$ is the stirring potential, which is a Gaussian stochastic field defined by its correlation function which, in this example, we take as (using its Fourier decomposition)

$$\langle \xi(\vec{k}, \omega) \rangle = 0 \,, \quad (11)$$

and

$$\langle \xi(\vec{k}, \omega)\xi(\vec{k}', \omega') \rangle = 2\tilde{D}(\vec{k}, \omega)(2\pi)^{d+1}\delta^{(d)}(\vec{k} + \vec{k}')\delta(\omega + \omega') \quad (12)$$

together with

$$\tilde{D}(\vec{k}, \omega) = D_0 + D_\theta \left(\frac{k}{\Lambda} \right)^{-2\rho} \left(\frac{\omega}{\nu\Lambda^2} \right)^{-2\theta} \,. \quad (13)$$

The two couplings D_0 and D_θ control the amplitude of the noise, which we have taken as colored in space and in time with exponents ρ and θ. The quantity Λ is a momentum cutoff.

These equations play an important role in many problems, ranging from astrophysics, the chemistry of the origin of life, to directed percolation to polymers. They can be coarse grained [3, 4] using the standard techniques of the dynamical renormalization group (DynRG) [12]. This allows one to follow the changes in scale of the various parameters appearing in the above equations.

Even for the simplest case of generalized potential flow ($\vec{v} \propto \nabla\psi$) and a simple equation of state relating p and ρ, and relatively simple densities, these equations find application in many problems. For the purpose of applying the RG to the resulting equations it is convenient to select as variables the following dimensionless combinations of the couplings ($K_d = S_d/(2\pi)^d$ and S_d is the volume of the d-dimensional sphere)

$$V = 1 + \frac{m^2}{\nu \Lambda^2} \, , \tag{14}$$

$$U_0 = \lambda^2 D_0 K_d \frac{\Lambda^{d-2}}{\nu^3} \, , \tag{15}$$

$$U_\theta = \lambda^2 D_\theta K_d \frac{\Lambda^{d-2-2\rho-4\theta}}{\nu^{3+2\theta}} \, , \tag{16}$$

which become the effective parameters satisfying the RG equations from which one can examine the behavior of the system on coarse graining. Then one can determine the fixed points toward which the system tends asymptotically.

The equations for the evolution of the couplings as functions of the scaling variable s defined through $\vec{x} \to s\vec{x}$, are given in this example by

$$\frac{d\nu}{d\log s} = \nu \left[-\frac{d-2}{4d} U_0 - \frac{d-2-2\rho}{4d} U_\theta (1 + 2\theta) \sec(\theta \pi) \right] \, , \tag{17}$$

$$\frac{d\lambda}{d\log s} = \lambda \left[-\frac{U_\theta}{d} \theta (1 + 2\theta) \sec(\pi \theta) \right] \, , \tag{18}$$

$$\frac{dU_0}{d\log s} = (2 - d)U_0 + \frac{U_0^2}{2d}(2d - 3) + \frac{U_\theta^2}{4}(1 + 4\theta)\sec(2\pi\theta) \tag{19}$$

$$+ \frac{U_0 U_\theta}{4d}(5d - 8\theta - 6\rho - 6)(1 + 2\theta)\sec(\pi\theta) \, ,$$

$$\frac{dU_\theta}{d\log s} = (2 - d + 2\rho + 4\theta)U_\theta + \frac{U_0 U_\theta}{4d}(d - 2)(3 + 2\theta) \tag{20}$$

$$+ \frac{U_\theta^2}{4d}[-8\theta + (d - 2 - 2\rho)(3 + 2\theta)](1 + 2\theta)\sec(\pi\theta) \, .$$

For some ranges of values of the noise parameters this system of equations has both real and complex solutions [3, 4].

It follows from eq. (8) that when approaching a fixed point the exponential prefactor in the n-point function becomes

$$\exp\left(-\frac{n}{2} \int_1^s \frac{d\lambda'}{\lambda'} \gamma_\phi(\lambda') \right) \longrightarrow \exp\left(-\frac{n}{2} \gamma_\phi^* \int_1^s \frac{d\lambda'}{\lambda'} \right)$$

$$= \exp\left(-\frac{n}{2} \gamma_\phi^* \ln s \right) = \left(\frac{\lambda}{\lambda_0} \right)^{-\frac{n}{2}\gamma_\phi^*} \tag{21}$$

where γ_ϕ^* is the value of the anomalous dimension of the stochastic field at the fixed point. In the case of a complex fixed point, γ_ϕ^* is a complex number.

As will be seen below, this limit is important in understanding how hierarchies are generated through coarse graining.

5 SOME CONSEQUENCES OF RENORMALIZATION GROUP SCALING

The fixed points follow from eqs. (17)–(20) by simply setting them equal to zero and finding the roots of the resulting system of nonlinear algebraic equations. The physical parameters characterizing the physical system will be attracted to fixed points in the UV-regime (short distance or $s \to 0$) or in the IR-regime (long distance or $s \to \infty$) depending on whether the fixed point is UV attractive or IR attractive. As with any autonomous set of differential equations, other possibilities do exist. From the fixed points and their asymptotic behaviors follows the behavior of the correlation function as it is implicit in eq. (8).

Which fixed point the system is attracted to or repelled from depends on the properties of the basin of attraction where the initial conditions for the DynRG equations are located in parameter space.

For example, one may start at a given scale with a particular behavior at this system size; as the system is coarse grained, the parameters can evolve into a completely different set of values which correspond not only to a completely different quantitative behavior, but also to a completely different qualitative behavior. In this case we can talk of the *emergence* of "novel" behavior.

How can this *emergent* behavior be phenomenologically detected? Since the values of the couplings at the fixed points determine the anomalous dimensions, a strategy is to measure the correlation functions in the scaling regime associated with the approach to the fixed point. As is seen from eq. (8) and the discussion following (2), if the anomalous dimensions are complex, then there is some form of "wiggly" behavior (log-periodic) in the scale dependence of the correlation function. Thus the various *local maxima* of the correlation function would correspond to the regions where the physical system has maximal correlation. These maxima would happen at *discrete scales* which could be classified as members of a discrete *sequence*. We also note that the "wiggly" behavior is modulated by a power law. We now explore this in some detail.

6 COMPLEX RENORMALIZATION GROUP FIXED POINTS, THE TWO-POINT CORRELATION FUNCTION, AND HIERARCHIES

Let us now focus our attention on the two-point correlation function. This correlation function is the easiest to measure in a many-body system. Its phenomenological interpretation is that it represents *the joint probability of finding two objects located in two independent volume elements*. It is important to note that since the two-point correlation function is interpreted as a probability, it *must* then be *real*.

6.1 ASYMPTOTIC BEHAVIOR OF THE TWO-POINT CORRELATION FUNCTION

·We ask the question, "how does the two-point correlation function behave when, under coarse graining, the system enters the basin of attraction of a complex fixed point?" From eqs. (8) and (21), in the neighborhood of a fixed point the two-point correlation function has the limit

$$G^{(2)}(r_1, t_1; r_2, t_2) \propto |r_1 - r_2|^{2\chi} F\left(\frac{|t_1 - t_2|}{|r_1 - r_2|^z}\right), \qquad (22)$$

where the so-called "roughening" and "dynamic" exponents χ and z are simple arithmetic combinations of anomalous, engineering and the number of spatial dimensions [1]. The function $F(u)$ is a universal function with the following behavior as its argument approaches the indicated limits

$$\lim_{u \to 0} F(u) \propto \text{constant}, \qquad (23)$$

$$\lim_{u \to \infty} F(u) \propto u^{2\chi/z}. \qquad (24)$$

To simplify our discussion let us take the limit when $|r_1 - r_2| \to \infty$ for a fixed time interval (the $u \to 0$ limit in the above equations). Then the two-point correlation function becomes

$$G^{(2)}_\infty(r, t; r', t) \propto |r - r'|^{2\chi}. \qquad (25)$$

Assuming now that $\chi = \alpha + i\zeta$ we get

$$G^{(2)}_\infty(r, t; r', t) = \text{Re}\, \tilde{c}\, e^{i\beta} \left|\frac{r - r'}{r_0}\right|^{2(\alpha + i\zeta)}$$

$$= c \cdot |r - r'|^{2\alpha} \cdot \cos\left[\beta + 2\zeta \log\left(\frac{|r - r'|}{r_0}\right)\right]. \qquad (26)$$

In other words, and as advertised, asymptotically the two-point correlation function is a log-periodic function modulated by a power law [5] and [14]. The various constants c, r_0, and β depend on the specific problem that we are considering. However, the two exponents α and ζ are calculable through the renormalization group and depend on *general* features of the system and, perhaps, they are *universal*, although a classification of dynamical critical phenomena in terms of classes of universality to date is not available [12].

6.2 MAXIMAL CORRELATION

Where does maximal two-point correlation occur? This happens for region sizes given by

$$|r_{(n)} - r'| = r_0 \cdot \exp\left\{\frac{1}{2\zeta}\left[\tan^{-1}(\alpha/\zeta) - \beta\right]\right\} \cdot (e^{\frac{\pi}{2\zeta}})^n \qquad (27)$$

or

$$|r_{(n)} - r'| \equiv A \cdot b^n , \qquad (28)$$

with $n = 0, \pm 1, \pm 2, \ldots$. Here A is not fully predictable by the RG, but the "base" b is fully calculable, and both its presence and specific value are a result of the coarse graining.

Equation (27) is our main result. It shows that in the presence of complex fixed points of the renormalization group there are regions of finite size within finite-sized regions, all of which are calculable once the size of any one region is known. The separation between the regions depends only on general properties of the dynamics and noise.

In short,

1. there is a *hierarchy* of regions of average linear sizes $R_{(n)} \equiv |r_{(n)} - r'|$ where correlation on that scale is favored; and
2. the hierarchy can be *classified* according to an *integer* n, with each member of the hierarchy having a unique number.

7   SOME APPLICATIONS

In this section we explicitly illustrate the above using examples occurring in nature. We will consider only two examples from astrophysics, where the equations given in section 4.2 do apply [3, 4], and will mention a few examples where we could conceivably extend the methods described here.

7.1   THE UNIVERSE: FROM MOLECULAR CLOUDS IN THE INTERSTELLAR MEDIUM TO THE HORIZON

Matter in the universe clumps in a variety of structures which are contained within larger structures which, in turn, are contained within even larger structures. This observational property leads us to think of the universe as "hierarchical," in other words, as made up of structures contained within structures (cf. Goldman and Pérez-Mercader [7] and references cited there). For astrophysical objects, this property manifests itself for at least 10 orders of magnitude in longitudinal size and more than 20 orders of magnitude in mass. Such regularity, from molecular clouds in the interstellar medium (ISM) to superclusters of galaxies and beyond, is remarkable (see fig. 2).

It is also known that *within each class* in this hierarchy, the two-point density fluctuation correlation function displays power-law behavior in object separation with an exponent value which, within observational errors, is the same for all the classes and, therefore, perhaps "universal," in spite of the different specific mechanisms that may be at work in the formation of each of the classes. For details see Goldman and Pérez-Mercader [7]. In systems where gravity plays a dominant role, the appropriate order parameter is the local fluctuations in matter

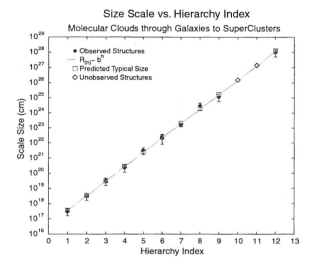

FIGURE 4   Longitudinal size scale vs. hierarchy index n for discrete structures in the universe from molecular clouds through galaxies to superclusters of galaxies and predictions from eq. (28). The value of b is 9.02 ± 0.24.

density, denoted by $\delta\rho(\vec{r}, t)$. Assuming that under coarse graining the gas and dust system develops a complex fixed point (this assumption can be made on the basis of the fact that the equations describing the evolution of the system at large scales can be written in the form (9) which is known to have them), then the two-point correlation function acquires a complex exponent.

Fitting the data to all the known structures gives the results shown in figures 2, 4, and 5. We can see that the results are in excellent agreement with observations.

7.2   PLANETARY SYSTEMS: THE TITIUS-BODE LAW IN THE SOLAR SYSTEM

In its original form, the famous Titius-Bode law of planetary distances was expressed so that the radius of the orbit, $r(n)$, of a given planet in units of 0.1 AU, is given by eq. (2), (4), and (5)

$$r(n) = 4 + 3 \times 2^n \tag{29}$$

where $n = -\infty$ for Mercury, and $0, 1, 2, \ldots$ for the other planets ordered according to the sizes of their orbits around the Sun [10].

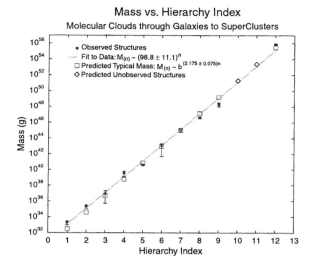

FIGURE 5   Mass vs. hierarchy index n for discrete structures in the universe from molecular clouds through galaxies to superclusters of galaxies and predictions from eq. (28). The value of b is 9.02 ± 0.24.

The "law" is phenomenological and in modern times is restated as follows: the (reduced) distance $d(n)$ to the nth planet from the Sun[7] is given by the power law:

$$d(n) = A \times B^n \times f(n) \tag{30}$$

where $A = 44$ (which is equivalent to a scale of 0.205 if distances are expressed in astronomical units instead of the size of the central object), $B = 1.73$, and $f(n)$ oscillates around 1 with an amplitude of about 0.1.[8] The power-law piece of the Titius-Bode "law," $A \times B^n$, is of the form obtained in eq. (28) from our consideration of the properties of the two-point correlation function. The same phenomenology also holds for the average radii of the orbits of the satellites of Jupiter and Saturn. The values of the best fitted B are different than in the case where the central body is the Sun, but they are of the *same* order of magnitude.

The observed positions of the planets and the predictions of the power law in eq. (30) are shown in figure 6.

We see that the kind of hierarchical behavior obtained from maximal correlation in nonequilibrium phenomena, is very good for describing data at the

---

[7]Here $d(n) \equiv (r(n) - R_0)/R_0$, where $r(n)$ is the radius of the orbit of the nth object and R_0 the radius of the central object.

[8]The oscillatory behavior can be ascribed [10] to "a point-gravitational or tidal evolution starting from *after* the planets were formed" (our italics); this has been interpreted as a result of the chaotic evolution of the solar system through its history.

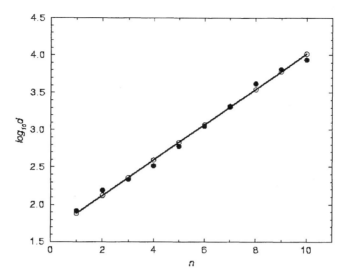

FIGURE 6 The Titius-Bode law. Predicted (open circles) and observed (filled circles) positions of the planets in the solar system. Predicted positions (and the straight line) are computed using the power law with $B = 1.73$, as derived without any object in location $n = 5$ (Asteroid Belt) and $n = 10$ (Pluto). The remaining points represent the other planets.

largest scales in the universe, as well as in the solar system. In fact, one can show that for the only multiplanetary system known to date, υ-Andromedæ, this law also holds and predicts the existence of planets in the system at particular orbits.

7.3 OTHER POSSIBLE APPLICATIONS

Of course, many other hierarchical systems are known: metabolic pathways, reaction networks, proteins, cells and ecologies in biology, fault systems and rivers in geology or corporations and urban areas in socioeconomy, to just mention a few examples. But one particularly interesting and important system where these ideas may find application is in the context of the random networks associated with life.

Recent discoveries made in biology indicate that living systems are organized into a genome, transcriptome, proteome, and metabolome. There is strong evidence that each of these systems is hierarchical by itself, and, in addition, there are modes of interaction among them characterized by very different time scales. This has led to the proposal of a "Life's Pyramid of Complexity" [11], where hierarchical organization is the key to providing the functionality for life to exist and survive. At the same time, hierarchical organization is essential to help explain the observed "bottom-up" universality of life as it uses motifs and pathways assembled into functional modules and large-scale organizational systems such as the cell itself. But hierarchical organization is also the mechanism that seems to operate the "top-down" organism specificity that is observed in the living world.

Is there a way of applying the notions of hierarchical organization emerging during coarse graining, as discussed here, to what is observed in living systems? It would be the manifestation of some coarse-grained dynamics occurring in some particular environment.

This is, of course, an extremely ambitious challenge and a formidable problem, but perhaps some of the notions discussed here could help: the unification of nonlinear stochastic partial differential equations and the renormalization group is a powerful recipe of two separate ideas, each of which has its own trandition of success in science. Together they are more than their simple sum.

8 CONCLUSIONS

We have seen that coarse graining a nonequilibrium many-body system leads to several classes of scaling behavior. The phase transitions that take place in the coarse-graining process can be interpreted as manifestations of emergence. Emergence occurs as the system transits from one fixed point of the RG to another one. In particular, and as is well known, scaling behavior will be reached as one approaches the fixed points. In addition, the type of behavior depends crucially on the properties of the fixed point: for real fixed points one has the usual phenomenology also seen in equilibrium phase transitions.

In out-of-equilibrium systems, the presence of complex fixed points leads to a far richer phenomenology, and this phenomenology naturally describes hierarchical systems. In fact, one sees power-law behavior modulated by log-periodic corrections in the correlation functions. The modulation leads to "gaps" where the correlation is minimized, and "clumps" where the correlation is maximized. Because of this mathematical form of the correlation functions, one has clumps within clumps separated by gaps, which is precisely the structure corresponding to hierarchical behavior. Furthermore, given the properties (such as size) of just one of the members in this hierarchy, one can characterize the full hierarchy.

The above can be shown to describe reasonably well some natural phenomena, which we have illustrated by examining what is known about the universe at scales where gravity plays an important role. In planetary systems where gravitational effects are important but they are in competition with other phenomena like magnetic fields and collective effects such as turbulence, these quantitative arguments also apply, as in the case of the famous Titius-Bode law.

The scenario presented here can be generalized to any extended system where coarse graining can be applied and where complex exponents are present. This includes geophysical, biochemical, biological, and ecological phenomena.

ACKNOWLEDGMENTS

The author thanks Prof. Murray Gell-Mann, Prof. Alvaro Giménez, Dr. Terry Goldman, Dr. David Hochberg, the late Prof. Dennis Sciama, and Dr. Geoffrey West for many discussions on scaling phenomena over the course of the last ten years.

REFERENCES

[1] Barabási, A.-L., and H. E. Stanley. *Fractal Concepts in Surface Growth*. Cambridge: Cambridge University Press, 1995.

[2] Chaisson, E. J. *Cosmic Evolution: The Rise of Complexity in Nature*. Harvard University Press, 2001.

[3] Domínguez, A., D. Hochberg, J. M. Martín-García, J. Pérez-Mercader, and L. Schulman. "Dynamical Scaling of Matter Density Correlations in the Universe." *Astronomy and Astrophysics* **344** (2000): 27.

[4] Domínguez, A., D. Hochberg, J. M. Martín-García, J. Pérez-Mercader, and L. Schulman. "Dynamical Scaling of Matter Density Correlations in the Universe." *Astronomy and Astrophysics* **363** (2000): 373.

[5] Douçot, B., W. Wang, J. Chaussy, B. Pannetier, R. Rammal, A. Vareille, and D. Henry. "First Observation of the Universal Periodic Corrections to Scaling: Magnetoresistance of Normal-Metal Self-Similar Networks." *Phys. Rev. Lett.* **57** (1986): 1235.

[6] Gell–Mann, M., and F. Low. "Quantum Electrodynamics at Small Distances." *Phys. Rev.* **75** (1954): 1024.

[7] Goldman, T., and J. Pérez-Mercader. "Hierarchies in the Large-Scale Structures of the Universe." *Science* (2002): submitted.

[8] Hochberg, D., and J. Pérez-Mercader. "The Renormalization Group and Fractional Brownian Motion." *Phys. Lett. A* **296** (2002): 272.

[9] León, J., J. Pérez-Mercader, and M. F. Sánchez. "Low-Energy Limit of Two-Scale Field Theories." *Phys. Rev. D* **44** (1991): 1167.

[10] Martin-Nieto, M. *The Titius-Bode Law of Planetary Distances: Its History and Theory*. Oxford: Pergamon Press, 1972.

[11] Oltvai, Z. N., and A.-L. Barabási. "Life's Complexity Pyramid." *Science* **298** (2002): 763.

[12] Onuki, A. *Phase Transition Dynamics*. Cambridge: Cambridge University Press, 2002.

[13] Simon, H. *The Sciences of the Artificial*. 3d ed. MIT Press, 1996.

[14] Sornette, D. *Critical Phenomena in Natural Sciences: Chaos, Fractals, Self-Organization and Disorder: Concepts and Tools.* New York: Springer-Verlag, 2000.

[15] Wilson, K. G., and J. Kogut. "The Renormalization Group and the Expansion." *Phys. Rep.* **C-12** (1974): 76.

[16] Zinn-Justin, J. *Quantum Field Theory and Critical Phenomena.* 3d ed. New York: Oxford University Press, 1996.

The Architecture of Complex Systems

Vito Latora
Massimo Marchiori

1 INTRODUCTION

At the present time, the most commonly accepted definition of a complex system is that of a system containing many interdependent constituents which interact nonlinearly.[1] Therefore, when we want to model a complex system, the first issue has to do with the connectivity properties of its network, the architecture of the wirings between the constituents. In fact, we have recently learned that the network structure can be as important as the nonlinear interactions between elements, and an accurate description of the coupling architecture and a characterization of the structural properties of the network can be of fundamental importance also in understanding the dynamics of the system.

---

[1] The definition may seem somewhat fuzzy and generic: this is an indication that the notion of a complex system is still not precisely delineated and differs from author to author. On the other side, there is complete agreement that the "ideal" complex systems are the biological ones, especially those which have to do with people: our bodies, social systems, and our cultures [2].

Nonextensive Entropy—Interdisciplinary Applications
edited by Murray Gell-Mann and Constantino Tsallis, Oxford University Press

In the last few years the research on networks has taken different directions producing rather unexpected and important results. Researchers have: (1) proposed various global variables to describe and characterize the properties of real-world networks and (2) developed different models to simulate the formation and the growth of networks such as the ones found in the real world. The results obtained can be summed up by saying that statistical physics has been able to capture the structure of many diverse systems within a few common frameworks, though these common frameworks are very different from the regular array, or capture the random connectivity, previously used to model the network of a complex system.

Here we present a list of some of the global quantities introduced to characterize a network: the characteristic path length L, the clustering coefficient C, the global efficiency E_{glob}, the local efficiency E_{loc}, the cost Cost, and the degree distribution $P(k)$. We also review two classes of networks proposed: small-world and scale-free networks. We conclude with a possible application of the nonextensive thermodynamics formalism to describe scale-free networks.

2   SMALL-WORLD NETWORKS

Watts and Strogatz [17] have shown that the connection topology of some biological, social, and technological networks is neither completely regular nor completely random. These networks, that are somehow in between regular and random networks, have been named *small worlds* in analogy with the small-world phenomenon empirically observed in social systems more than 30 years ago [11, 12]. In the mathematical formalism developed by Watts and Strogatz, a generic network is represented as an unweighted graph $\mathbf{G}$ with N nodes (vertices) and K edges (links) between nodes. Such a graph is described by the adjacency matrix $\{a_{ij}\}$, whose entry a_{ij} is either 1 if there is an edge joining vertex i to vertex j, or 0 otherwise. The mathematical characterization of the small-world behavior is based on the evaluation of two quantities, the characteristic path length L and the clustering coefficient C.

2.1   THE CHARACTERISTIC PATH LENGTH

The characteristic path length L measures the typical separation between two generic nodes of a graph $\mathbf{G}$. L is defined as:

$$L(\mathbf{G}) = \frac{1}{N(N-1)} \sum_{i \neq j \in \mathbf{G}} d_{ij}$$

where d_{ij} is the shortest path length between i and j, i.e., the minimum number of edges traversed to get from a vertex i to another vertex j. By definition $d_{ij} \geq 1$, and $d_{ij} = 1$ if there exists a direct link between i and j. Notice that if $\mathbf{G}$

is connected, for example, there exists at least one path connecting any couple of vertices with a finite number of steps, then d_{ij} is finite $\forall i \neq j$ and also L is a finite number. For a nonconnected graph, L is an ill-defined quantity, because it can diverge. This problem is avoided by using E_{glob} in place of L, as we will show below.

2.2   THE CLUSTERING COEFFICIENT

The clustering coefficient C is a local quantity of **G** measuring the average cliquishness of a node. For any node i, the subgraph of first neighbors of i, **G$_i$** is considered. If the degree of i, that is, the number of edges incident with i, is equal to k_i, then **G$_i$** is made of k_i nodes and at most $k_i(k_i - 1)/2$ edges. C_i is the fraction of these edges that actually exist, and C is the average value of C_i all over the network (by definition $0 \leq C \leq 1$):

$$C(\mathbf{G}) = \frac{1}{N} \sum_{i \in \mathbf{G}} C_i , \qquad C_i = \frac{\# \text{ of edges in } \mathbf{G_i}}{k_i(k_i - 1)/2} .$$

The mathematical characterization of the small-world behavior proposed by Watts and Strogatz is based on the evaluation of L and C: small-world networks have high C-like regular lattices, and short L-like random graphs. The small-world behavior is ubiquitous in nature and in man-made systems. Neural networks, social systems [6] such as the collaboration graph of movie actors [17] or the collaboration network of scientists [13], technological networks such as the World Wide Web or the electrical power grid of the Western U.S., are only few examples. To give an idea of the numbers obtained we consider the simplest case of the neural networks investigated, that of the nematode *C. elegans*: this network, represented by a graph with $N = 282$ nodes (neurons) and $K = 1974$ edges (connections between neurons), gives $L = 2.65$ and $C = 0.28$ [17]. It is also important to notice that a network such as the electrical power grid of the Western U.S., can be studied by such a formalism only if considered as an unweighted graph, i.e., when no importance whatsoever is given to the physical length of the links.

3   EFFICIENT AND ECONOMIC BEHAVIOR

A more general formalism, valid both for unweighted and weighted graphs (also nonconnected), extends the application of the small-world analysis to any real complex network, in particular to those systems where the Euclidian distance between vertices is important (as in the case of the electrical power grid of the Western U.S.), and that are, therefore, too poorly described only by the topology of connections [8, 9]. Such systems are better described by two matrices, the adjacency matrix $\{a_{ij}\}$ defined as before, and a second matrix $\{\ell_{ij}\}$ containing the weights associated with each link. The latter is named the matrix of physical distances, because the numbers ℓ_{ij} can be imagined as the Euclidean distances between i and j. The mathematical characterization of the network is based on the evaluation of two quantities, the global and the local efficiency (replacing L and C), and a third one quantifying the cost of the network. Small worlds are networks that exchange information very efficiently both on a global and on a local scale [9].

3.1   THE GLOBAL EFFICIENCY

In the case of a weighted network the shortest path length d_{ij} is defined as the smallest sum of the physical distances throughout all the possible paths in the graph from i to j.[2] The efficiency ϵ_{ij} in the communication between vertex i and j is assumed to be inversely proportional to the shortest path length: $\epsilon_{ij} = 1/d_{ij}$. When there is no path in the graph between i and j, $d_{ij} = +\infty$ and consistently $\epsilon_{ij} = 0$. Suppose now that every vertex sends information along the network, through its edges. The global efficiency of $\mathbf{G}$ can be defined as an average of ϵ_{ij}:

$$E_{\text{glob}}(\mathbf{G}) = \frac{\sum_{i \neq j \in \mathbf{G}} \epsilon_{ij}}{N(N-1)} = \frac{1}{N(N-1)} \sum_{i \neq j \in \mathbf{G}} \frac{1}{d_{ij}}.$$

Such a quantity is always a finite number (even when $\mathbf{G}$ is unconnected) and can be normalized to vary in the range $[0, 1]$ if divided by $E_{\text{glob}}(\mathbf{G}^{\text{ideal}}) = 1/(N(N-1)) \sum_{i \neq j \in \mathbf{G}} 1/\ell_{ij}$, the efficiency of the ideal case $\mathbf{G}^{\text{ideal}}$ in which the graph has all the $N(N-1)/2$ possible edges. In such a case the information is propagated in the most efficient way since $d_{ij} = \ell_{ij} \; \forall i, j$.

3.2   THE LOCAL EFFICIENCY

One of the advantages of the efficiency-based formalism is that a single measure, the efficiency E (instead of the two different measures L and C), is sufficient to define the small-world behavior. In fact, the efficiency can be evaluated for any subgraph of $\mathbf{G}$, in particular for $\mathbf{G_i}$ which is the subgraph of the first neighbors of i (made by k_i nodes and at most $k_i(k_i - 1)/2$ edges), and, therefore, it can be used also to characterize the local properties of the graph. The local efficiency of $\mathbf{G}$ is defined as:

$$E_{\text{loc}}(\mathbf{G}) = \frac{1}{N} \sum_{i \in \mathbf{G}} E(\mathbf{G_i}), \qquad E(\mathbf{G_i}) = \frac{1}{k_i(k_i - 1)} \sum_{l \neq m \in \mathbf{G_i}} \frac{1}{d'_{lm}},$$

where the quantities $\{d'_{lm}\}$ are the shortest distances between nodes l and m calculated on the graph $\mathbf{G_i}$. Similar to E_{glob}, E_{loc} can be normalized to vary in the range $[0, 1]$ and plays a role similar to that of C [8]. Small worlds are networks with high E_{glob} and high E_{loc}.

---

[2]$\{d_{ij}\}$ is now calculated by using the information contained both in $\{a_{ij}\}$ and in $\{\ell_{ij}\}$.

3.3   THE COST

An important variable to consider, especially when we deal with weighted networks and when we want to analyze and compare different real systems, is the cost of a network [8]. In fact, we expect both E_{glob} and E_{loc} to be higher (L lower and C higher) as the number of edges in the graph increases. As a counterpart, in any real network there is a price to pay for number and length (weight) of edges. This can be taken into account by defining the cost of the graph $\mathbf{G}$ as the total length of the network's wirings:

$$\mathrm{Cost}(\mathbf{G}) = \frac{\sum_{i \neq j \in \mathbf{G}} a_{ij}\, \ell_{ij}}{\sum_{i \neq j \in \mathbf{G}} \ell_{ij}}. \tag{1}$$

Since the cost of $\mathbf{G}^{ideal}$ is already included in the denominator of the formula above, Cost varies in $[0,1]$ and assumes the maximum value 1 when all the edges are present in the graph. In the case of an unweighted graph, $\mathrm{Cost}(\mathbf{G})$ reduces to the normalized number of edges $2K/N(N-1)$.

With the three variables—E_{glob}, E_{loc}, and Cost—all defined in $[0,1]$, it is possible to study in an unified way unweighted (topological) and weighted networks. And it is possible to define an economic small world as a network having low Cost and high E_{loc} and E_{glob} (i.e., both economic and small world). In figure 1 we report a useful illustrative example obtained by means of a simple model to construct a class of weighted graphs. We start by considering a regular network of $N = 1000$ nodes placed on a circle ($\ell_{i,j}$ is given by the Euclidean distance between i and j) and $K = 1500$ links. A random rewiring procedure is implemented: it consists of going through each of the links in turn and independently, with some probability p, of rewiring it. Rewiring means shifting one end of the edge to a new node chosen randomly with a uniform probability. In this way it is possible to tune $\mathbf{G}$ in a continuous manner from a regular lattice ($p = 0$) into a random graph ($p = 1$), without altering the average number of first neighbors equal to $k = 2K/N$. For $p \sim 0.02 - 0.04$ we observe the small-world behavior: E_{glob} has almost reached its maximum value 0.62 while E_{loc} has not changed much from the maximum value 0.2 (assumed at $p = 0$). Moreover for these values of p the network is also economic; in fact the Cost stays very close to the minimum possible value (assumed, of course, in the regular case $p = 0$).

3.4   SOME EXAMPLES OF APPLICATIONS TO REAL NETWORKS

The neural network of the *C. elegans* has $E_{\mathrm{glob}} = 0.35$, $E_{\mathrm{loc}} = 0.34$, Cost $= 0.18$: the *C. elegans* is an economic small world because it achieves high efficiency both at the global and local level (about 35% of the global and local efficiency of the ideal completely connected case); all of this is at a relatively low cost, with only the 18% of the wirings of the ideal graph. As a second example we consider a technological network, the Boston underground, also known as the Massachusetts

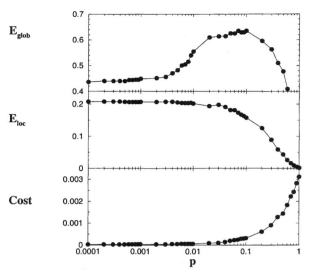

FIGURE 1   The three quantities E_{glob}, E_{loc}, and Cost are reported as functions of the rewiring probability p for the model discussed in the text. The economic small-world behavior shows up for $p \sim 0.02 - 0.04$.

Bay Transportation Authority (MBTA). The MBTA is a weighted network consisting of $N = 124$ stations and $K = 124$ tunnels connecting couples of stations (the matrix $\ell_{i,j}$ is given by the Euclidean distance between stations). For such a system we obtain $E_{\text{glob}} = 0.63$, $E_{\text{loc}} = 0.03$, and Cost $= 0.002$. This means that MBTA achieves the 63% of the efficiency of the ideal subway with a cost of only 0.2%. The price to pay for such low-cost high global efficiency is the lack of local efficiency. In fact, $E_{\text{loc}} = 0.03$ indicates that, unlike a neural network (or a social system), the MBTA is not fault tolerant; i.e., damage in a station will dramatically affect the efficiency in the connection between the previous and the next station. The difference with respect to neural networks comes from different needs and priorities in the construction and evolution mechanisms. When a subway system is built, the priority is given to the achievement of global efficiency at a relatively low cost, and not to fault tolerance. In fact, a temporary problem in a station can be solved in an economic way by other means: for example, walking, or taking a bus from the previous to the next station. Applications to other real networks can be found in Latora and Marchiori [8].

4   SCALE-FREE NETWORKS

4.1   DEGREE DISTRIBUTION

Other important information on a network can be extracted from its degree distribution $P(k)$. The latter is defined as the probability of finding nodes with k links: $P(k) = (N(k))/N$, where $N(k)$ is the number of nodes with k links. Many large networks such as the World Wide Web, the Internet, metabolic, and protein networks, have been named *scale-free* networks because their degree distribution follows a power law for large k [1, 7]. A social system of interest for the spreading of sexually transmitted diseases [10] and the connectivity network of atomic clusters' systems [5] also show a similar behavior. The most interesting fact is that neither regular nor random graphs display long tails in $P(k)$, and the presence of nodes with large k strongly affects the properties of the network [14], for instance its response to external factors [4]. Barabási and Albert [1] have proposed a simple model (the BA model) to reproduce the $P(k)$ found in real networks by modeling the dynamical growth of the network. The model is based on two simple mechanisms, growth and preferential attachment, that are also the main ingredients present in the dynamical evolution of the real-world networks. As an example, the World Wide Web grows in time by the addition of new web pages, and a new web page will more likely include hyperlinks to popular documents with already high degree. Starting by an initial network with a few nodes and adding new nodes with new links preferentially connected to the most important existing nodes, the dynamics of the BA model produce (in the stationary regime) scale-free networks with a power-law degree distribution [1]:

$$P(k) \sim k^{-\gamma}, \quad \gamma = 3.$$

The model predicts the emergence of the scale-free behavior observed in real networks, though the exponents in the power law of real networks can be different from 3 (usually it is in the range between 2 and 3).

4.2   NONEXTENSIVE STATISTICAL MECHANICS

A more careful analysis of the shape of $P(k)$ of many of the real networks considered indicates the presence of a plateau for small k. See, for example, figure 1(a) of Barabási and Albert [1] and figure 2(b) of Liljeros et al. [10]. We have observed that such a plateau for small k and the different slopes of the power law for large k can be perfectly reproduced by using the generalized power-law distribution [16]

$$P(k) \sim [1 + (q - 1)\beta k]^{\frac{1}{1-q}}$$

with two fitting parameters: q related to the slope of the power law for large k, and β [3]. The generalized probability distribution above can be obtained as a stationary solution of a generalized Fokker-Planck equation with a nonlinear diffusion term [15]. Therefore, we believe that it is possible to rephrase the generalized Fokker-Planck equation in terms of a generalized mechanism of network construction, and to implement a model (more general than the BA model) able to reproduce the plateau and the different slopes of $P(k)$.

ACKNOWLEDGMENTS

We thank E. Borges, P. Crucitti, M. E. J. Newman, A. Rapisarda, C. Tsallis, and G. West for their useful comments.

REFERENCES

[1] Barabási, A.-L., and R. Albert. "Emergence of Scaling in Random Networks." *Science* **286** (1999): 509–512.
[2] Baranger, M. "Chaos, Comlexity, and Entropy: A Physics Talk for Non-Physicists." 2001. NECSI Educational Projects, New England Complex Systems Institute. Sept. 2002. ⟨http://www.necsi.org/projects/baranger/cce.html⟩.
[3] Borges, E., V. Latora, A. Rapisarda, and C. Tsallis. "Degree Distributions in Complex Networks." Unpublished manuscript, 2002 (in preparation).
[4] Crucitti, P., V. Latora, M. Marchiori, and A. Rapisarda. "Efficiency of Scale-Free Networks: Error and Attack Tolerance." *Physica A* **320** (2003): 622–635.
[5] Doye, J. P. K. "The Network Topology of a Potential Energy Landscape: A Static Scale-Free Network." *Phys. Rev. Lett.* **88** (2002): 238701-238704.
[6] Girvan, M., and M. E. J. Newman. "Community Structure in Social and Biological Networks." *Proc. Natl. Acad. Sci. USA* **99** (2002): 8271–8276.
[7] Jeong, H., B. Tombor, R. Albert, Z. N. Oltvai, and A.-L. Barabási. "The Large-Scale Organization of Metabolic Networks." *Nature* **407** (2000): 651–654.
[8] Latora, V. and M. Marchiori. "Economic Small-World Behavior in Weighted Networks." Apr. 2002. arXiv e-print archive, Los Alamos National Laboratory. Sept. 2002. ⟨http://xxx.lanl.gov/archive/cond-mat/0204089⟩.
[9] Latora, V., and M. Marchiori. "Efficient Behavior of Small-World Networks." *Phys. Rev. Lett.* **87** (2001): 198701.
[10] Liljeros, F. L., C. R. Edling, N. Amaral, H. E. Stanley, and Y. Aberg. "The Web of Human Sexual Contacts." *Nature* **411** (2001): 907–908.
[11] Milgram, S. "The Small-World Problem." *Psychol. Today* **2** (1967): 60–67.
[12] Newman, M. E. J. "Models of the Small World." *J. Stat. Phys.* **101** (2000): 819–841.
[13] Newman, M. E. J. "Scientific Collaboration Networks." *Phys. Rev.* **E64** (2001): 016131 and 016132.

[14] Pastor-Satorras R., and A. Vespignani. "Epidemic Spreading in Scale-Free Networks." *Phys. Rev. Lett.* **86** (2001): 3200–3203.

[15] Plastino, A. R., and A. Plastino. "Nonextensive Statistical Mechanics and Generalized Fokker-Planck Equation." *Physica* **A 222** (1995): 347–354.

[16] Tsallis C. "Possible Generalization of Boltzmann-Gibbs Statistics." *J. Stat. Phys.* **52** (1988): 479–487.

[17] Watts, D. J., and S. H. Strogatz. "Collective Dynamics of Small-World Networks." *Nature* **393** (1998): 440–442.

Effective Complexity

Murray Gell-Mann
Seth Lloyd

It would take a great many different concepts—or quantities—to capture all of our notions of what is meant by complexity (or its opposite, simplicity). However, the notion that corresponds most closely to what we mean by complexity in ordinary conversation and in most scientific discourse is "effective complexity." In nontechnical language, we can define the effective complexity (EC) of an entity as the length of a highly compressed description of its regularities [6, 7, 8].

For a more technical definition, we need a formal approach both to the notion of minimum description length and to the distinction between regularities and those features that are treated as random or incidental.

We can illustrate with a number of examples how EC corresponds to our intuitive notion of complexity. We may call a novel complex if it has a great many different characters, scenes, subplots, and so on, so that the regularities of the novel require a long description. The United States tax code is complex, since it is very long and each rule in it is a regularity. Neckties may be simple, like those with regimental stripes, or complex, like some of those designed by Jerry Garcia.

From time to time, an author presents a supposedly new measure of complexity (such as the "self-dissimilarity" of Wolpert and Macready [17]) without recognizing that when carefully defined it is just a special case of effective complexity.

Like some other concepts sometimes identified with complexity, the EC of an entity is context-dependent, even subjective to a considerable extent. It depends on the coarse graining (level of detail) at which the entity is described, the language used to describe it, the previous knowledge and understanding that are assumed, and, of course, the nature of the distinction made between regularity and randomness.

Like other proposed "measures of complexity," EC is most useful when comparing two entities, at least one of which has a large value of the quantity in question.

Now, how do we distinguish regular features of an entity from ones treated as random or incidental? There is, as we shall see, a way to make a nearly absolute distinction between the two kinds of features, but that approach is of limited usefulness because it always assigns very low values of EC, attributing almost all information content to the random category rather than the regular one.

In most practical cases, the distinction between regularity and randomness—or between regular and random information content—depends on some judgment of what is important and what is unimportant, even though the judge need not be human or even alive.

Take the case of neckties, as discussed above. We tacitly assumed that effective complexity would refer to the pattern of the tie, while wine stains, coffee stains, and so on, would be relegated to the domain of the random or incidental. But suppose we are dry cleaners. Then the characteristics of the stains might be the relevant regularities, while the pattern is treated as incidental.

Often, regularity and randomness are envisaged as corresponding to signal and noise, respectively, for example in the case of music and static on the radio. But, as is well known, an investigation of sources of radio static by Karl Jansky et al. (at Bell Telephone Laboratories in the 1930s) revealed that one of those sources lies in the direction of the center of our galaxy, thus preparing the way for radio astronomy. Part of what had been treated as random turned into a very important set of regularities.

It is useful to encode the description of the entity into a bit string, even though the choice of coding scheme introduces another element of context dependence. For such strings we can make use of the well-known concept of algorithmic information content (AIC), which is a kind of minimum description length.

The AIC of a bit string (and, hence, of the entity it describes) is the length of the shortest program that will cause a given universal computer U to print out the string and then halt [3, 4, 11]. Of course, the choice of U introduces yet another form of context dependence.

For strings of a particular length, the ones with the highest AIC are those with the fewest regularities. Ideally they have no regularities at all except the length. Such strings are sometimes called "random" strings, although the terminology does not agree precisely with the usual meaning of random (stochastic, especially with equal probabilities for all alternatives). Some authors call AIC "algorithmic complexity," but it is not properly a measure of complexity, since randomness is not what we usually mean when we speak of complexity. Another name for AIC, "algorithmic randomness," is somewhat more apt.

Now we can begin to construct a technical definition of effective complexity, using AIC (or something very like it) as a minimum description length. We split the AIC of the string representing the entity into two terms, one for regularities and the other for features treated as random or incidental. The first term is then the effective complexity, the minimum description length of the regularities of the entity [8].

It is not enough to define EC as the AIC of the regularities of an entity. We must still examine how the regularities are described and distinguished from features treated as random, using the judgment of what is important. One of the best ways to exhibit regularities is the method used in statistical mechanics, say, for a classical sample of a pure gas. The detailed description of the positions and momenta of all the molecules is obviously too much information to gather, store, retrieve, or interpret. Instead, certain regularities are picked out. The entity considered—the real sample of gas—is embedded conceptually in a set of comparable samples, where the others are all imagined rather than real. The members of the set are assigned probabilities, so that we have an ensemble. The entity itself must be a typical member of the ensemble (in other words, not one with abnormally low probability). The set and its probability distribution will then reflect the regularities.

For extensive systems, the statistical-mechanical methods of Boltzmann and Gibbs, when described in modern language, amount to using the principle of maximum ignorance, as emphasized by Jaynes [9]. The ignorance measure or Shannon information I is introduced. (With a multiplicative constant, I is the entropy.) Then the probabilities in the ensemble are varied and I is maximized subject to keeping fixed certain average quantities over the ensemble. For example, if the average energy is kept fixed—and nothing else—the Maxwell-Boltzmann distribution of probabilities results.

We have, of course,

$$I = -\sum_r P_r \log P_r \,, \tag{1}$$

where log means logarithm to the base 2 and the P's are the (coarse-grained) probabilities for the individual members r of the ensemble. The multiplicative constant that yields entropy is $k \ln 2$, where k is Boltzmann's constant.

In this situation, with one real member of the ensemble and the rest imagined, the fine-grained probabilities are all zero for the members of the ensemble other than e, the entity under consideration (or the bit string describing it). Of course, the fine-grained probability of e is unity. The typicality condition previously mentioned is just

$$-\log P_e \lesssim I\,.\tag{2}$$

Here the symbol "$\lesssim$" means "less than or equal" to within a few bits.

We can regard the quantities kept fixed (while I is maximized) as the things judged to be important. In most problems of statistical mechanics, these are, of course, the averages of familiar extensive quantities such as the energy. The choice of quantities controls the regularities expressed by the probability distribution.

In some problems, the quantities being averaged have to do with membership in a set. (For example, in Gibbs's microcanonical ensemble, we deal with the set of states having energies in a narrow interval.) In such a case, we would make use of the membership function, which is one for members of the set and zero otherwise. When the average of that function over all the members of the ensemble is one, every member with nonzero probability is in the set.

In discussing an ensemble E of bit strings used to represent the regularities of an entity, we shall apply a method that incorporates the maximizing of ignorance subject to constraints. We introduce the AIC of the ensemble and call it Y. We then have our technical definition of effective complexity: it is the value of Y for the ensemble that is finally employed. In general, then, Y is a kind of candidate for the role of effective complexity.

Besides $Y = K(E)$, the AIC of the ensemble E (for a given universal computer U), we can also consider $K(r|E)$, the contingent AIC of each member r given the ensemble. The weighted average, with probabilities P_r, of this contingent AIC can be related to I in the following way.

We note that Rüdiger Schack [15] has discussed converting any universal computer U into a corresponding U' that incorporates an efficient recoding scheme (Shannon-Fano coding). Such a scheme associates longer bit strings with less probable members of the ensemble and shorter ones with more probable members. Schack has then shown that if K is defined using U', then the average contingent AIC of the members lies between I and $I + 1$. We shall adopt his procedure and thus have

$$\sum_r P_r K(r|E) \approx I\,,\tag{3}$$

where $\approx$ means equal to within a few bits (here actually one bit).

Let us define the total information Σ as the sum of Y and I. The first term is, of course, the AIC of the ensemble and we have seen that the second is, to within a bit, the average contingent AIC of the members given the ensemble.

To throw some light on the role of the total information, consider the situation of a theoretical scientist trying to construct a theory to account for a large body of data. Suppose the theory can be represented as a probability distribution over a set of bodies of data, one of which consists of the real data and the rest of which are imagined. Then Y corresponds to the complexity of the theory and I measures the extent to which the predictions of the theory are distributed widely over different possible bodies of data. Ideally, the theorist would like both quantities to be small, the first so as to make the theory simple and the second so as to make it focus narrowly on the real data. However, there may be trade-offs. By adding bells and whistles to the theory, along with a number of arbitrary parameters, one may be able to focus on the real data, but at the expense of complicating the theory. Similarly, by allowing appreciable probabilities for very many possible bodies of data, one may be able to get away with a simple theory. (Occasionally, of course, a theorist is fortunate enough to be able to make both Y and I small, as James Clerk Maxwell did in the case of the equations for electromagnetism.) In any case, the first desideratum is to minimize the sum of the two terms, the total information Σ. Then one can deal with the possible trade-offs.

We shall show that to within a few bits the smallest possible value of Σ is $K \equiv K(e)$, the AIC of the string representing the entity itself. Here we make use of the typicality condition (2) that the log of the (coarse-grained) probability for the entity is less than or equal to I to within a few bits. We also make use of certain abstract properties of the AIC:

$$K(A) \lesssim K(A, B) \tag{4}$$

and

$$K(A, B) \lesssim K(B) + K(A|B), \tag{5}$$

where again the symbol $\lesssim$ means "less than or equal to" up to a few bits. A true information measure would, of course, obey the first relation without the caveat "up to a few bits" and would obey the second relation as an equality.

Because of efficient recoding, we have

$$K(e|E) \lesssim -\log P_e. \tag{6}$$

We can now prove that $K = K(e)$ is an approximate lower bound for the total information $\Sigma = K(E) + I$:

$$
\begin{aligned}
K &= K(e) \lesssim K(e, E), & \text{(7a)} \\
K(e, E) &\lesssim K(E) + K(e|E), & \text{(7b)} \\
K(e|E) &\lesssim -\log P_e, & \text{(7c)} \\
-\log P_e &\lesssim I. & \text{(7d)}
\end{aligned}
$$

We see, too, that when the approximate lower bound is achieved, all these approximate inequalities become approximate equalities:

$$
\begin{aligned}
K &\approx K(e, E), & \text{(8a)} \\
K(e, E) &\approx Y + K(e|E), & \text{(8b)} \\
K(e|E) &\approx -\log P_e, & \text{(8c)} \\
-\log P_e &\approx I. & \text{(8d)}
\end{aligned}
$$

The treatment of this in Gell-Mann and Lloyd [8] is slightly flawed. The approximate inequality (7b), although given correctly, was accidentally replaced later on by an approximate equality, so that condition (8b) came out as a truism. Thus (8b) was omitted from the list of new conditions that hold when the total information achieves its approximate lower bound. As a result, we gave only three conditions of approximate equality instead of the four quoted here in (8a)–(8d).

Also, in the discussion at the end of the paragraph preceding eq. (2) of Gell-Mann and Lloyd [8], we wrote $\log K_U(a)$ by mistake in place of $\log K_U(b) + 2\log\log K_U(b)$, but that does not affect any of our results.

Clearly the total information Σ achieves its approximate minimum value K for the singleton distribution, which assigns probability one to the bit string representing our entity and zero probabilities to all other strings. For that distribution, Y is about equal to K and the measure of ignorance I equals zero.

There are many other distributions for which $\Sigma \approx K$. If we plot Y against I, the line along which $Y + I = K$ is a straight line with slope minus one, with the singleton at the top of the line. We are imposing on the ensemble—the one that we actually use to define the effective complexity—the condition that the total information approximately achieve its minimum. In other words, we want to stay on the straight line or within a few bits of it.

All ensembles of which e is a typical member lie, to within a few bits, above and to the right of a boundary. That boundary coincides with our straight line all the way from the top down to a certain point, where we run out of ensembles that have $Y + I \approx K$. Below that point the actual boundary for ensembles in the $Y - I$ plane no longer follows the straight line but veers off to the right.

Now, as we discussed, we maximize the measure of ignorance I subject to staying on that straight line. If we do that and impose no other conditions, we end up at the point where the boundary in the $I - Y$ plane departs from the straight line. As described in the paper of Gell-Mann and Lloyd (who are indebted to Charles H. Bennett for many useful discussions of this manner), that point always corresponds to an effective complexity Y that is very small. If we imposed no other conditions, every entity would come out simple! In certain circumstances, that is all right, but for most problems it is an absurd result. What went wrong? The answer is that, as in statistical mechanics, we must usually impose some more conditions, fixing the values of certain average quantities treated as important by a judge. If we maximize I subject to staying (approximately) on the straight line and to keeping those values fixed, we end up with a meaningful effective complexity, which can be large in appropriate circumstances.

The situation is made easier to discuss if we narrow the universe of possible ensembles in a drastic manner suggested by Kolmogorov, one of the inventors (or discoverers?) of AIC, in work reviewed in the books by Cover and Thomas [4] and by Li and Vitányi [11]. Instead of using arbitrary probability distributions over the space of all bit strings, one restricts the ensembles to those obeying two conditions. The set must contain only strings of the same length as the original bit string and all the nonzero probabilities must be equal. In this simplified situation, every allowable ensemble can be fully characterized as a subset of the set of all bit strings that have the same length as the original one. Here I is just the logarithm of the number of members of the subset. Also, being a typical member of the ensemble simply means belonging to the subset.

Vitányi and Li describe how, for this model problem, Kolmogorov suggested maximizing I subject only to staying on the straight line. In that case, as pointed out above, one is led immediately to the point in the $I - Y$ plane where the boundary departs from the straight line. Kolmogorov called the value of Y at that point the "minimum sufficient statistic." His student L. A. Levin (now a professor at Boston University) kept pointing out to him that this "statistic" was always small and therefore of limited utility, but the great man paid insufficient attention [10].

In the model problem, the boundary curve comes near the I axis at the point where I achieves its maximum, the string length l. At that point the subset is the entire set of strings of the same length as the one describing the entity e. Clearly, that set has a very short description and thus a very small value of Y.

What should be done, whether in this model problem or in the more general case that we discussed earlier, is to utilize the lowest point on the straight line such that the average quantities judged to be important still have their fixed values. Then Y no longer has to be tiny and the measure of ignorance I can be much less than it was for the case of no further constraints.

We have succeeded, then, in splitting K into two terms, the effective complexity and the measure of random information content, and they are equal to the values of Y and I, respectively, for the chosen ensemble. We can think of the separation of K into Y and I in terms of a distinction between a basic program (for printing out the string representing our entity) and data fed into that basic program.

We can also treat as a kind of coarse graining the passage from the original singlet distribution (in which the bit string representing the entity is the only member with nonzero probability) to an ensemble of which that bit string is a typical member. In fact, we have been labeling the probabilities in each ensemble as coarse-grained probabilities P_r. Now it often happens that one ensemble can be regarded as a coarse graining of another, as was discussed in Gell-Mann and Lloyd [8]. We can explore that situation here as it applies to ensembles that lie on or very close to the straight line $Y + I = K$.

We start from the approximate equalities (8a)–(8d) (accurate to within a few bits) that characterize an ensemble on or near the straight line. There the coarse-graining acts on initial "singleton" probabilities that are just one for the original string and zero for all others. We want to generalize the above formulae to the case of an ensemble with any initial fine-grained probability distribution $p \equiv \{p_r\}$, which gets coarse grained to yield another ensemble with probability distribution $P \equiv \{P_r\}$ and approximately the same value of Σ. We propose the following formulae as the appropriate generalizations:

$$K(p) \approx K(p, P), \tag{9a}$$
$$K(p, P) \approx K(P) + K(p|P), \tag{9b}$$
$$K(p|P) \approx -\Sigma_r p_r \log P_r + \Sigma_r p_r \log p_r, \tag{9c}$$
$$-\Sigma_r p_r \log P_r \approx -\Sigma_r P_r \log P_r. \tag{9d}$$

These equations reduce to (8a) through (8d) respectively for the case in which the fine-grained distribution is the "singleton" distribution. Also, it is easy to see that Σ is approximately conserved by these approximate equalities, as a result of our including the last term in eq. (9c).

Equation (9a) tells us that, to within a few bits, the coarse-grained probability distribution P contains only algorithmic information that is in the fine-grained distribution p. Equation (9b) tells us that the ordinary relation between joint and conditional mutual information holds here to within a few bits even though that relation does not always hold for joint and conditional *algorithmic* information.

We can compare this discussion of coarse graining to the treatment in Gell-Mann and Lloyd [8]. There we required three properties of a coarse-graining transformation from p to P: that the transformation actually yield a probability distribution, that if iterated it produce the same set of P's, and that it obey eq. (9d) above. We attained these objectives by maximizing the ignorance associated with the P's while keeping some averages involving the P's equal to the corresponding averages involving the p's (linear constraint conditions).

Here we emphasize that we are generalizing that work to the case where Y is introduced and the sum of Y and I is kept approximately fixed at its minimum value while we maximize I subject to some constraint conditions linear in the probabilities.

Say we start with the singleton ensemble in which only the original string has a nonzero probability and move down the straight line in a succession of coarse grainings until we reach the ensemble for which Y is the effective complexity. The above equations are then applied over and over again for the successive coarse grainings, and they apply also between the original (singleton) probability distribution and the final one.

Alternatively, we can, if we like, regard the transition from P to p as a fine graining, using the same formulae. We can start at the point where the boundary curve departs from the straight line and move up the line in a sequence of fine grainings. In fact, we can utilize the linear constraints successively. We apply first one of them, then that one and another, then those two and a third, and so forth, until all the constraints have been applied to the maximization of I subject to staying on the straight line. Each additional constraint yields a fine graining.

There are at least four issues that we feel require discussion at this point, even though many questions about them remain. Two of these issues relate to certain generalizations of the notion of algorithmic information content.

AIC as it stands is technically uncomputable, as shown long ago by Chaitin [3]. That is not so if we modify the definition by introducing a finite maximum execution time T within which the program must cause the modified universal computer U' to print out the bit string. Such a modification has another, more important advantage. We can vary T and, thus, explore certain situations where apparent complexity is large but effective complexity as defined above (for $T \to \infty$) is small.

Take the example [6] of energy levels of heavy nuclei. Fifty years ago, it seemed that any detailed explanation of the pattern involved would be extremely long and complicated. Today, however, we believe that an accurate calculation of the positions of all the levels is possible, in principle, using a simple theory: QCD, the quantum field theory of quarks and gluons, combined with QED, the quantum field theory of photons and electromagnetic interactions, including those of quarks. Thus, for T very large or infinite, the modified AIC of the levels is small—they are simple. But the computation time required is too long to permit the calculations to be performed using existing hardware and software. Thus, for moderate values of T the levels appear complex.

In such a case, the time around which the modified AIC declines from a large value to a small one (as T increases) is related to "logical depth" as defined by Charles H. Bennett [2]. Roughly, logical depth is the time (or number of steps) necessary for a program to cause U to print out the coded description of an entity and then halt, averaged over programs in such a way as to emphasize short ones.

There are cases where the modified AIC declines, as T increases, in a sequence of steps or plateaus. In that case we can say that certain kinds of regularities are buried more deeply than others.

While it is very instructive to vary T in connnection with generalizing K— the AIC of the bit string describing our entity—we encounter problems if we try to utilize a finite value of T in our whole discussion of breaking up K into effective complexity and random information. Not all the theorems that allow us to treat AIC as an approximate information measure apply to the generalization with variable T.

In addition to logical depth, we can utilize a quantity that is, in a sense, inverse to it, namely Bennett's "crypticity," [2] which is, in rough terms, the time necessary to go from the description of an entity to a short program that yields that description. As an example of a situation where crypticity is important, consider a discussion of pseudorandomness. These days, when random numbers are called for in a calculation, one often uses instead a random-looking sequence of numbers produced by a deterministic process. Such a pseudorandom sequence typically has a great deal of crypticity. A lengthy investigation of the sequence could reveal its deterministic nature and, if it is generated by a short program, could correctly assign to it a very low AIC. Given only a modest time, however, we could fail to identify the sequence as one generated by a simple deterministic process and mistake it for a truly random sequence with a high value of AIC.

The concept of crypticity can also be usefully applied to situations where a bit string of modest AIC appears to exhibit large AIC in the form of effective complexity rather than random information. We might call such a string "pseudocomplex." An example of a pseudocomplex string would be one recording an image, at a certain scale, of the Mandelbrot set. Another would be an apparently complex pattern generated by a simple cellular automaton from a simple initial condition. Note that a pseudorandom string, which has passed many of the usual statistical tests for randomness, is not appreciably compressed by conventional data compression algorithms, such as the one known as LZW [4]. By contrast, a pseudocomplex string typically possesses a large number of obvious statistical regularities and is, therefore, readily compressible to some extent by LZW, but not all the way to the very short program that actually generated the string.

We should mention that a number of authors have considered mutual information as a measure of complexity in the context of dynamical systems [1, 5, 12]. Without modification, that idea presents a conflict with our intuitive notion of complexity. Consider two identical very long bit strings consisting entirely of ones. The mutual information between them is very large, yet each is obviously very simple. Moreover, the statement that they are the same is also very simple. The pair of strings is not at all complex in any usual sense of the word.

Typically, the authors in question have recognized that a more acceptable quantity in a discussion of complexity is mutual algorithmic information, defined for two strings as the sum of their AIC values minus the AIC of the two taken together. If two strings are simple and identical, though very long, their mutual AIC is small.

Of course, identical long strings could be "random," in which case their very large mutual algorithmic information does not correspond to what we usually mean by complexity. EC is still the best measure of complexity.

We can easily generalize the definition of mutual information to the case of any number of strings (or entities described by them). For example, for three strings we have

$$K_{\mathrm{mut}} = K(1) + K(2) + K(3) - K(1,2) - K(2,3) - K(1,3) + K(1,2,3). \quad (10)$$

Under certain conditions we can see a connection between mutual algorithmic information and effective complexity. For example, suppose we are presented not with a single entity but with N entities that are selected at random from among the typical members of a particular ensemble. The mutual algorithmic information content among these entities is then a good estimate of the AIC of the ensemble from which they are selected, and that quantity is, under suitable conditions, equal to the effective complexity candidate Y attributed to each of the entities.

The way the calculation goes is roughly the following. On average the K value for m arguments is approximately $Y + mI$, and the sum in eq. (10) then comes out equal to Y. It is easily shown that such an equality yielding Y holds not just for three entities but for any number N, with the appropriate generalization of eq. (10). The elimination of the I term produces the connection of K_{mut} with the effective complexity candidate.

At last we arrive at the questions relevant to a nontraditional measure of ignorance. Suppose that for some reason we are dealing, in the definition of I, not with the usual measure given in eq. (1), but rather with the generalization discussed in this volume, namely

$$I_q = -\frac{[\Sigma_r (P_r)^q - 1]}{(q-1)}, \tag{11}$$

which reduces to eq. (1) in the limit where q approaches 1. Should we be maximizing this measure of ignorance—while keeping certain average quantities fixed—in order to arrive at a suitable ensemble? (Presumably we average using not the probabilities P_r but their qth powers normalized so as to sum to unity—the so-called Escort probabilities.) Do we, while maximizing I, keep a measure of total information at its minimum value? Is a nonlinear term added to $I + Y$? What happens to the lower bound on $I + Y$? Can we make appropriate changes in the definition of AIC that will preserve or suitably generalize the relations we discuss here? What happens to the approximate equality of I and the average contingent AIC (given the ensemble)? What becomes of the four conditions in eqs. (8a) to (8d)? What happens to the corresponding conditions (9a) to (9d) for the case where we are coarse graining one probability distribution and thus obtaining another one?

As is well known, a kind of entropy based on the generalized information or ignorance of eq. (11) has been suggested [16] as the basis for a full-blown alternative, valid for certain situations, to the "thermostatistics" (thermodynamics and statistical mechanics) of Boltzmann and Gibbs. (The latter is, of course, founded on eq. (1) as the formula for information or ignorance.) Such a basic interpretation of eq. (11) has been criticized by authors such as Luzzi et al. [13] and Nauenberg [14]. We do not address those criticisms here, but should they prove justified—in whole or in part—they need not rule out, at a practical level, the applicability of eq. (11) to a variety of cases, such as systems of particles attracted by $1/r^2$ forces or systems at the so-called "edge of chaos."

ACKNOWLEDGMENTS

This research was supported by the National Science Foundation under the Nanoscale Modeling and Simulation initiative. In addition, the work of Murray Gell-Mann was supported by the C.O.U.Q. Foundation and by Insight Venture Management. The generous help provided by these organizations is gratefully acknowledged.

REFERENCES

[1] Adami, C., C. Ofria, and T. C. Collier. "Evolution of Biological Complexity." *PNAS (USA)* **97** (2000): 4463–4468.

[2] Bennett, C. H. "Dissipation, Information, Computational Complexity and the Definition of Organization." In *Emerging Syntheses in Science*, edited by D. Pines, 215–234. Santa Fe Institute Studies in the Sciences of Complexity, Proc. Vol. I. Redwood City: Addison-Wesley, 1987.

[3] Chaitin, G. J. *Information, Randomness, and Incompleteness.* Singapore: World Scientific, 1987.

[4] Cover, T. M., and J. A. Thomas. *Elements of Information Theory.* New York: Wiley, 1991.

[5] Crutchfield, J. P., and K. Young. "Inferring Statistical Complexity." *Phys. Rev. Lett.* **63** (1989): 105–108.

[6] Gell-Mann, M. *The Quark and the Jaguar.* New York: W. H. Freeman, 1994.

[7] Gell-Mann, M. "What is Complexity?" *Complexity* **1/1** (1995): 16–19.

[8] Gell–Mann, M., and S. Lloyd. "Information Measures, Effective Complexity, and Total Information." *Complexity* **2/1** (1996): 44–52.

[9] Jaynes, E. T. *Papers on Probability, Statistics and Statistical Physics*, edited by R. D. Rosenkrantz. Reidel: Dordrecht, 1982.

[10] Levin, L. A. Personal communication, 2000.

[11] Li, M., and P. M. B. Vitanyi. *An Introduction to Kolmogorov Complexity and Its Applications.* New York: Springer-Verlag, 1993.

[12] Lloyd, S., and H. Pagels. "Complexity as Thermodynamic Depth." *Ann. Phys.* **188** (1988): 186–213.

[13] Luzzi, R., A. R. Vasconcellos, and J. G. Ramos. "On the Question of the So-Called Non-Extensive Thermodynamics." IFGW-UNICAMP Internal Report, Universidade Estadual de Campinas, Campinas, Sao Paulo, Brasil, 2002.

[14] Nauenberg, M. "A Critique of Nonextensive q-Entropy for Thermal Statistics. Dec. 2002. lanl.gov e-Print Archive, Quantum Physics, Cornell University. ⟨http://eprints.lanl.gov/abs/cond-mat/0210561⟩.

[15] Schack, R. "Algorithmic Information and Simplicity in Statistical Physics." *Intl. J. Theor. Phys.* **36** (1997): 209–226.

[16] Tsallis, C. "Possible Generalization of Boltzmann-Gibbs Statistics." *J. Stat. Phys.* **52** (1988): 479–487.

[17] Wolpert, D. H., and W. G. Macready. "Self-Dissimmilarity: An Empirically Observable Measure of Complexity." In *Unifying Themes in Complex Systems: Proceedings of the First NECSI International Conference*, edited by Y. Bar-Yam, 626–643. Cambridge, Perseus, 2002.

Index